The Theory of Plant Breeding

The Theory of Plant Breeding

OLIVER MAYO

Waite Agricultural Research Institute
University of Adelaide

Second Edition

CLARENDON PRESS · OXFORD
1987

Oxford University Press, Walton Street, Oxford OX2 6DP
Oxford New York Toronto
Delhi Bombay Calcutta Madras Karachi
Petaling Jaya Singapore Hong Kong Tokyo
Nairobi Dar es Salaam Cape Town
Melbourne Auckland
and associated companies in
Beirut Berlin Ibadan Nicosia

Oxford is a trade mark of Oxford University Press

Published in the United States
by Oxford University Press, New York

First edition 1980
Second edition 1987

British Library Cataloguing in Publication Data
Mayo, Oliver
The theory of plant breeding.—2nd ed.
1. Plant-breeding
I. Title
631.5'3 SB123
ISBN 0-19-854172-4
ISBN 0-19-854171-6 Pbk

Library of Congress Cataloging in Publication Data
Mayo, Oliver.
The theory of plant breeding.
Bibliography: p.
Includes index.
1. Plant-breeding. 2. Crops—Genetics. I. Title.
SB123.M36 1986 631.5'3 86-12750
ISBN 0-19-854172-4
ISBN 0-19-854171-6 (pbk.)

Set by Cotswold Typesetting Ltd, Cheltenham
Printed in Great Britain by
St Edmundsbury Press
Bury St Edmunds, Suffolk

Preface

This book has three aims. First, I have tried to give an account of plant breeding theory that is accessible both to plant breeders seeking the genetical bases of their work and to geneticists seeking applications of genetical principles in plant breeding. Secondly, I have sought to explain where the theory is inadequate or needs extension. Thirdly, I have tried to show what aspects of plant breeding may be expected to be particularly fruitful in the future.

The book is neither an axiomatic development of quantitative genetics from basic principles to practical techniques, nor a compendium of plant breeding plans and procedures. Writing the latter was not my intention, and as for the former, even now plant breeding is not an exact science, though I have here tried to increase its exactitude. Thinking of Byron rather than Bismarck, perhaps we should consider

science
But an exchange of ignorance for that
Which is another kind of ignorance.

Adelaide O.M.
May 1979–November 1985

Acknowledgements

This book was planned, partly written, and then partly rewritten for the 2nd edition while I was a Visiting Fellow at Wolfson College, Oxford, working in the Genetics Laboratory and the Department of Biomathematics, University of Oxford. I thank the President and Fellows, and Professors W. F. Bodmer, J. H. Edwards, and P. A. Armitage for their generosity and hospitality.

For advice, criticism, and help of various kinds at various times, I wish to thank the following people: Dr P. A. Baghurst, Dr M. G. Bulmer, Dr J. Burley, Professor R. N. Curnow, Professor D. W. Davis, Professor C. J. Driscoll, Dr S. R. Eckert, Dr K. G. Eldridge, Mr T. W. Hancock, Dr A. M. Hopkins, Professor A. Kerr, Professor B. D. H. Latter, Dr C. N. Law, Dr G. M. E. Mayo, Dr M. J. Mayo, Dr A. J. Rathjen, Professor N. W. Simmonds, Dr C. Smith, Dr D. H. B. Sparrow, Dr D. J. Street, Professor C. W. Stuber, Mr D. E. Symon, Mr R. Thompson, Dr G. Thomson, Professor B. S. Weir, and Dr G. N. Wilkinson. I also thank Mrs B. Goldsmith and Mrs H. M. Simpson for patiently typing and retyping the book.

I thank Dr A. Gallais, Professor J. L. Jinks, Professor H. D. Patterson, Professor A. Robertson, Springer-Verlag, and the Editors of *Heredity*, *Canadian Journal of Genetics and Cytology*, *Annales de l'Amélioration des Plantes*, *Genetics*, and *Journal of Agricultural Science* for permission to republish the figures and tables specifically acknowledged in Sections 3.4.2, 4.4.2.2, 4.4.3, 4.6.1, 4.6.2, 7.4.1, and 9.4.2.2.

Contents

1

Introduction

1.1 Purposes of the book

A book on plant breeding is generally justified by an appeal to the King of Brobdingnag, who 'gave it for his opinion, "that whoever could make two ears of corn, or two blades of grass, to grow upon a spot of ground, where only one grew before, would deserve better of mankind, and do more essential service to his country, than the whole race of politicians put together"'. With a continually growing world population, a continually increasing supply of basic foodstuffs is needed, even if the maldistribution of wealth ensures that the most productive regions are the most likely to have problems in disposing of this surplus (Mayer 1984).

This book will not deal with economic problems, nor with detailed methods. Its purpose is three-fold; to give a coherent account of existing genetical theory, without extensive development of proofs, so that the theory (which at the moment is widely dispersed in an enormous number of research papers and books on other topics) is made accessible to geneticists seeking applications and to plant breeders seeking a basis for their work; to show where the theoretical developments are most needed; and to show or at least hint at where they are most to be expected. We might therefore think of the projector of the Academy of Lagado whom Gulliver described as having been 'eight years upon a project for extracting sunbeams out of cucumbers, which were to be put in phials hermetically sealed and let out to warm the air in raw inclement summers. He told me, he did not doubt, that in eight years more, he should be able to supply the Governor's gardens with sunshine, at a reasonable rate; but he complained that his stock was low and entreated me "to give him something as an encouragement to ingenuity, especially since this had been a very dear season for cucumbers"'. This book is intended to be such an encouragement.

After a very brief historical survey, outlining some of the most important directing factors in domestication and past breeding practice, the book is planned to set out the breeder's objectives, to describe the relevant principles of genetical theory, and then to show how the objectives may be met by the theory as it has been developed.

Because the success of plant breeders in recent decades has changed the face of the industry served, attention will be paid to large-scale planning and to genetical conservation, as well as to new methods which do not depend in the same way on genetical theory.

1.2 Historical background

1.2.1 Domestication

Plant domestication is older than recorded history, constituting as it does one of the earliest steps in the development of agriculture. It is interesting to speculate on how and why agriculture arose when it did, but it is outside the scope of this book (see, e.g. Pfeiffer 1976). Table 1.1 shows the time of domestication of eighty major crop plants as listed by Simmonds (1976a), together with the names of some of the most important plants domesticated in each era. The process of domestication involves the recognition that a particular species has something to offer, such as food or fibre, and the development of methods of cultivating the plant, together with methods for processing the crop, where necessary, e.g. extraction of fibre by letting the soft parts of the plant rot away in the case of hemp or jute. Thus, selection, both conscious and unconscious will be a continuous part of these processes. For example, harvesting of any grain crop will be greatly facilitated if the plants do not lodge, i.e. fall over, and the heads do not shatter. However, to some extent, people at the time of divergence from being hunter-gatherers to being farmers would unconsciously facilitate the development of non-shattering, non-lodging grain plants, since when they collected the seeds, they would naturally choose those from erect plants, and the more the head had tended not to shatter, the more seeds would be available from that plant.

Table 1.1

Era of domestication of the major groups of crop plants listed by Simmonds (1976a)

Era	Number of groups	Important examples
Pre-5000 BC	12	Barley; wheat; maize; beans; peas; yams; lentils
Pre-2500 BC	22	Potato; grapevine; millets; rice; rye; sorghum; groundnut; lucerne; cotton; olive; apple
Pre-1 BC	23	Soybeans; flax; citrus; oats; tea; beet†; fig; banana; cocoa
Pre-1000 AD	16	Sugarcane; coffee; tobacco
1000 AD-present	7	Rubber

†Post-1700 for sugar.

Later, the process might become more systematic or purposeful, as is the case now in the development of new pasture grasses. The results of such domestication may be very clearly seen in cultivated maize, which has entirely lost all mechanisms for the dispersal of its seeds, since the kernels are firmly attached to a rigid axis, i.e. the cob, and are not covered by floral bracts, though the ear is enclosed by modified leaf sheaths (Mangelsdorf 1965). In addition, there has been a reduction in branching of inflorescences, a change from many to one or two stalks, the lowering of the position of the female inflorescence on the plant, and a change from relatively hard to relatively soft kernels, all changes leading to an almost complete necessity for the intervention of Man for the continuation of the plant's existence.

In many cases, the ancestors of crop plants are incompletely identified or unavailable for investigation, but where they appear to coexist with the crop plant, as with *Hordeum vulgare* and *Hordeum spontaneum* in the Middle East (Harlan and Zohary 1966), they may yield novel or useful information. Thus, in this case it is clear that the very extensive genetical variation in *Hordeum* has not arisen from different selective practices, since it occurs to a similar extent in wild and domestic species.

From Table 1.1, it may be seen that, however rapidly domestication was actually achieved, certainly a great deal of time has in most cases been available for existing crop plants to be developed very highly, though millenia are not in fact needed. For example, cotton was changed from a perennial crop to an annual crop in less than 600 years (Hutchinson 1965, 1974), while rubber has been domesticated and developed in less than two centuries. Advancement of the wheat–rye hybrid *Triticale* to commercial use in less than 40 years is perhaps not a fair comparison, since both parental species were already highly developed. On the other hand, it is likely that the winged bean *Psophocarpus tetragonolobus*, which is widely grown in south-east Asia and Oceania, will be developed from a domesticated, but highly variable and in some cases semi-wild crop to an advanced crop in a very short time, on account of its great promise as a protein crop for the tropics (National Academy of Sciences 1975). However, a short historical time is a long time for the individual breeder; more than ordinary patience is needed.

1.2.2 Dominant attributes of domestication and early plant breeding

As is well known (see for example Hutchinson 1965, 1974; Simmonds 1976a), a very large proportion of all currently important crop plants were developed in a few basic centres of origin. Since this no doubt reflects the fact that conditions for the development of agriculture were not ideal over the face of the globe, nor was Man evenly spread over the habitable sectors of the globe, the fact illustrated in column 1 of Table 1.2 is perhaps not surprising (cf. Jennings and Cock 1977). Nonetheless,

Table 1.2
Some important attributes of the development of 87 major groups of crop plants (data from Simmonds 1976a)

	Main areas of cultivation distant from centre of origin	Introgressive hybridization involved in development of crop	Polyploidy involved in development of crop	Crop self-incompatible	Crop obligate inbreeder	Breeding for disease resistance	
						Important in past	Currently necessary
Yes	64	51	51	24	12	40	18†
No (or uncertain)	23	36	36	63	75	47	29

† Additional to 40 where disease has already become an important factor.

the important lesson may be drawn that a crop plant may always be expected to have at least as much promise in a region other than its source as it has in that centre of development. The importance of introgressive hybridization should also be noted, and of course it is well known that crossing between wild or weedy races or species and crop species has been significant in the evolution of crop species. The importance of breeding systems is shown by the fact that a very large proportion of major crop plants are self-incompatible, while another substantial group consists of self-fertilizing plants. Development of inbreeding from outbreeding species may well have been encouraged through conscious or unconscious selection, on account of a preference for consistency of type, and also of the effects of inbreeding depression. Finally, from Table 1.2 we may conclude that the development of monoculture inherently must bring increased opportunities for the spread of disease, since breeding for disease resistance has become important in most major crop plants. (It is not of course just monoculture which brings the hazard of disease; in many cases, such as selection for gross enlargement of the fleshy parts of fruits, disease susceptibility may be a natural concomitant of the new plant form.)

Frankel (1947), following Vavilov (1935), listed four phases of scientific plant breeding, as follows: collection and classification, whereby original material would be obtained and systematically described and classified; inter-specific investigation, where the origin, evolution, relationship, and crossing behaviour of major systematic units would be investigated; intra-specific investigation, where the nature, origin, variability, and inheritance of varietal differences within species would be examined; and selection and testing, where the effects of selection would be determined, selection would be carried out, and the selected material would be tested. In a sense, plant domestication has involved all these four phases, but it is now the province and responsibility of plant breeder and agronomist to conduct them efficiently as well as systematically.

1.3 The state of plant breeding

So far, we have said nothing about genetics. Historically, plant development under domestication has not been carried out by geneticists, and indeed it is only in recent times that an adequate theory of response to selection has been developed, while certain areas of successful plant breeding, such as the breeding of hybrids, do not yet have a completely satisfactory theoretical underpinning. Nonetheless, the current situation in plant breeding is that we have a very broad understanding of population and quantitative genetics, and this is applied by groups of specialists very narrowly to the extraordinarily narrow range of species which constitutes the major crop plants used by mankind. Much of the recent success of breeders, for example with cereal grains, has come about through large-scale collection and testing in a range of environments, where the scientific input is statistical rather than genetical. Cytogenetical manipulation, both to incorporate alien material into the species gene pool, and to produce large numbers of homozygous individuals through gametophytic or other haploid culture, are two examples of more sophisticated genetical approaches which are now making significant contributions.

Breeding for disease resistance is an area where the genetical investigation may be of a very high order, but in a sense breeding for disease resistance is a defensive activity, compared with breeding for an improved product, e.g. higher or more balanced protein content in grain, and effort devoted to resistance breeding is effort which cannot be devoted to these more positive aims. Nonetheless, it is obvious that resistance is going to be one of the main objectives of plant breeding for the foreseeable future.

As a crop plant is developed in a particular environment, significant progress in increasing yield becomes harder and harder to achieve, even with the introduction of new material from the outside, and at this stage breeders may return and examine the physiology of their crop, with a view to detecting critical limits to production. Table 1.3 shows a list of suggestions for investigations in cereals with the view of achieving maximum yield potential. The table relates to cereal production in the United Kingdom, so that the aspects to be studied will be the same elsewhere, but the types of experiment may not. For example, under Southern Australian conditions, multiple nitrogen dressings would hardly be considered by more than a tiny proportion of growers. Furthermore, where the soil nitrogen cycle is relatively well understood, as in the United Kingdom, fertilizer loss, say one-third of a dressing of 150 kg ha^{-1} (Jenkinson 1982), may be more important through potential contamination of drinking water than for an increase in the farmer's cost of production.

In vegetable crops grown on a much smaller scale than cereals, investigations of some of the important aspects in Table 1.3 are much more advanced. This is partly because plant size is the major determinant of the size of what is marketed in many crops, especially cabbage, carrot, leek, lettuce, onion, and beet, and the size of the product is a crucial determinant of price (Bleasdale 1982).

As time goes by, it becomes more obvious that basic investigations of the kind just described will become necessary to overcome plateaux in yield for highly developed and important crop plants, and here we are seeing and shall see 'genetic engineering' to regulate plant respiration, to change the nutritive value of seed protein and to cross the sterility

Table 1.3
Suggestions for physiological investigation in cereals (from Wade 1977)

Topic	Aspects to be studied	Type of experiment
1. Optimum 'efficient' plant density	Seed rate, Spatial arrangement, Tillering capability	Plant density, nitrogen and soil type interaction
2. Tiller survival	Nutrition, Moisture supply, Pre-anthesis photosynthesis	
3. Floret survival	Efficient development of productive ears	
4. Maximizing timing of inputs	Nitrate reductase, Tiller development, Ear development	Timing of N Multiple N dressings
5. Partitioning of assimilates	Sink strength, Source capacity, Stem reserves, Control by exogenous growth regulator chemicals	Variety and maximum yield potential at different sites
6. Stress tolerance	Heat stress, Moisture stress, Soil moisture extraction, Root development	

barriers between species in general, as predicted, for example, by Day (1977). While these methods have as yet made only modest contributions to current crop varieties, the pace with which they are being incorporated into breeding programmes is accelerating. Accordingly, the contributions of molecular and cellular biology are assessed in later chapters to show how and where they are altering plant improvement methodology.

2

Objectives

2.1 Introduction

The aim of all plant breeding is to increase yield of products; to be slightly more precise, to achieve the maximum yield of usable products for minimum inputs and at the least hazard to the environment. The emphasis on yield has led to an approach based on the elimination of limits to yield as well as direct selection for the components of yield, from Frankel (1947) onwards. Two qualifications are necessary: product quality and form are not attributes of yield as such, but are subsumed under the heading of utility; and the progress of selection comes under the heading of minimization of inputs, for while we know that maximum long-term gains may come from a modest intensity of selection (Robertson 1970), nonetheless, an initial rapid gain at the expense of later progress is often economically vital. (The implication is that different lines should be set up from a given base, and maintained under high and low intensities of selection.)

In this account, I adopt a rather different approach to show how the genetical determination of a trait may affect the selection methods used and how in turn the methods used may result in certain kinds of variation being preferentially selected. The classification is arbitrary; disease resistance could be treated in Sections 2.2.1, 2.3.1.2, and 2.3.2.2 as well as 2.4.3.

2.2 Single gene traits

Selection for traits determined by few genes of pronounced effects is so much more effective in utilizing genetic variation than selection for traits determined by small effects of many genes that breeders will wherever possible aim to utilize single gene variants in preference to polygenic variation. This may have undesirable effects, both as regards the trait of immediate interest (see the discussion of disease resistance in section 6.2.1) and other traits (e.g. the problems of early generation selection in inbreds, section 7.3.3). However, in general, selection for single gene traits has proved very rewarding and is a prime objective of the breeder.

2.2.1 *Utilization of existing variability*

The most general problem in applied genetics is the size of the initial sample to be screened; the more populations sampled, the more chance of obtaining rare variants having the properties desired, but on the other hand the less is the ability to assess the material adequately. In cases where mode of action of genes is known, or where this is (initially at least) irrelevant, however, very large samples may be screened quite readily. Disease resistance is such a field. It is practicable (e.g. Williams 1977a) to screen many thousand varieties of an inbreeding species (in this case cowpea, *Vigna unguiculata*) for resistance to five or more important diseases, since such resistance is commonly oligogenic (i.e. determined by genes with individually distinguishable effects), and if a particular form of resistance differs in detail from previously delineated systems it can nonetheless be readily evaluated by simple genetical investigations.

Disease resistance may be contrasted with plant height. In the hexaploid bread wheats, great progress has been made by the incorporation of dwarfing genes from the Japanese strain Norin 10. Thus, the fact that there was polygenic variation apparently masking the effects of some major genes determining plant height has not prevented the use of these genes. However, other less obvious effects of these genes could not be investigated. These may be important, and it is only since the recognition that dwarfing is a consequence of gibberellic acid insensitivity that such pleiotropy has become readily recognizable. For example, Gale and Marshall (1975) showed that the *Gai* (gibberellic acid insensitivity) genes in the varieties Minster Dwarf and Tom Thumb operate in the aleurone of germinating grain and prevent the production of α-amylase activity are important, dwarfing genes from these varieties may be more suitable than the widely used Norin 10 dwarfing genes, which do not appear to function in germinating grain. (As dwarfing in Norin 10 appears to be due to only two homoeologous loci (Gale 1977), simple allelic substitutions through back-crossing should allow rapid replacement of dwarfing genes where appropriate).

Many statistical methods exist for the determination of the number of genes affecting a trait, from the early Castle–Wright segregation index (Wright 1952) to genotype assay (Jinks and Towey 1976), and these will be discussed in Chapter 4. Here, we need merely note that they will in general be unsatisfactory by comparison with biochemical genetical methods such as that discussed in the previous paragraph, since the statistical approaches incorporate quite demanding assumptions about the distribution of genotypic effects and other unknowns. The fact that these may be simultaneously estimable is not helpful in assessing the likely worth of specific experiments designed to determine gene number.

Assignment of a very high proportion of observed variation to genotypic causation is often a sufficient indication that major genes should be sought directly.

2.2.2 Mutation breeding

Observable variation in important traits is often very modest. Since the discovery of artificial mutagenesis half a century ago, frequent attempts have been made to utilize it in plant breeding, though not always because existing natural variability has been exhausted; in many cases it has seemed to workers in mutagenesis a logical extension of their other less practical work. So many attempts to utilize artificial mutagenesis in plant breeding have in fact been made that it would be impractical to review the results here (see Chapter 12). What I can do is illustrate those aims which may be well served by use of mutagens.

First, consider mutations affecting specific amino-acids in proteins. For example, as has been widely recognized for many years, most major cereals are deficient in lysine, and in barley, naturally occurring mutants were found with lysine content increased by 10–20 per cent, such as the *lys* gene in the 'Hiproly' strain. Most of these had unsatisfactory agronomic properties, so artificial mutagenesis was used to generate further such variants, partly in the hope that these might have higher lysine content, better disease resistance, higher yield, and so on. One Danish induced mutant, Risø 1508, has 40–50 per cent increased lysine content, and breeding programmes have begun to incorporate it into commercial varieties and to select for modifiers of its yield-depressing and other deleterious effects (e.g. Oram 1976). Others have suggested that alternative crop species provide more promise, e.g. *Amaranthus edulis* already has a very high lysine content (Downton 1973). However, the introduction of a new crop is a much greater problem than the modification of an existing one, and this path has not been followed commercially.

Secondly, there are mutations yielding loss of function. As already noted, many crop plants are self-incompatible, thereby making many techniques such as close inbreeding much slower and more costly. As was shown many years ago (Lewis 1951), most spontaneous self-compatible mutants have a loss of part of the incompatibility reaction. Since loss of function is readily induced by artificial mutagens, self-compatibility is an obvious candidate for achievement thereby; as is indeed the case (de Nettancourt 1969).

Thirdly, there are mutations yielding novelties. In breeding ornamentals, the breeder seeks, perhaps above all, novelty. Treatment with mutagens, by vastly increasing the number of gross aberrations in floral development, allows the production of ever more curious forms.

2.3 Quantitative traits

The general objectives for quantitative traits will of course be as for discretely varying or Mendelian traits, and many of the methods of achieving them will be the same, but one can predict that certain methods, such as mutation breeding, will be more successful when qualitative rather than quantitative differences are sought (cf. Singh and Sharma 1976). The prime quantitative trait, dominating all others in its promotion of methodology and analysis, is yield.

2.3.1 Yield

It has been customary to consider the factors affecting yield under two headings: components of yield and factors limiting yield. I shall do the same, but briefly, since much of the former will be more logically treated in Section 2.4.2, and much of the latter in Sections 2.3.2 and 2.4.3.

2.3.1.1 Components of yield These will differ among crops, but common to all is plant density, since yield is normally measured, in practical terms, as production per unit area. In the case of a cereal, yield will be the product $p\ e\ n\ g$, where p = number of plants per unit area, e = number of ears per plant, n = number of grains per ear, and g = weight of a single grain (Engledow and Wadham 1923, 1924a,b,c; Frankel 1947). Selection could in principle therefore be aimed at increasing one or more of these four factors, without allowing a compensating decrease in the others. As already noted, the process of domestication has frequently involved very substantial changes in one or more of these components.

There is, however, an alternative way of looking at the components of yield, the one above being very much an operational one, ignoring the means whereby e, n, or g is determined. Clearly, these ought to be considered. In cassava, Jennings (1970) has defined the components of yield as number and size of tubers, size and efficiency of canopy, ratio of tops to roots, and duration of the period of dormancy. Some of these variables, such as canopy performance, relate to the physiology of starch deposition more closely than others, while dormancy is largely determined by the length of the dry season in the growing region (Martin 1976). It is important to separate these factors; the last named could better be regarded as a growth-limiting environmental attribute rather than a component of yield. In general, however, this specification of components leads one to consider the appropriateness of plant form, which will be further treated in Section 2.4.2.

In many cases, a breeder may seek to increase a component of yield directly. For example, in developing the new cereal *Triticale*, a wheat-rye hybrid (section 14.4), breeders have sought to increase grain weight directly, partly of course because inadequate grain weight reflected disturbed development as a result of aneuploidy. (Using only filled (non-shrivelled) grain for planting could, in any case, lead to the same result.)

Another example is sugar content in beet and sugarcane. Breeding for sugar content in beet was very successful in the late eighteenth and early nineteenth century, after it was realized that sugar content must be raised for extraction methods to be economic. A more than three-fold increase was achieved by simple mass selection with limited progeny testing. Today, in Australian sugarcane there is substantial scope for selection for sugar content, since this has considerable genotypic variation and is not much affected by competition between plants, whereas yield itself is very markedly affected by such competition (Hogarth 1977).

2.3.1.2 Factors limiting yield Most factors limiting yield are environmental: diseases and pests; drought; flooding; frost; day length; lack of nutrients; length of growing season; number of days with the maximum over a certain temperature; number of nights with the minimum below a certain temperature. The feasibility of environmental modification is not uniform for all these traits. Thus, fertilizer application is generally feasible to overcome lack of nutrients, but water-logging due to rare unseasonable rains may have to be tolerated. Again, if a certain range of winter temperatures is needed for proper fruit-set in certain tree crops it may be impossible to ensure, annually and economically, that such temperatures are achieved. If, on the other hand, crops need protection from frosts while young, but can tolerate it later on, like avocadoes in certain Australian environments, it may be economic to provide such protection early, given the long productive life of such trees.

Breeders will aim to mitigate the effects of limiting factors, but in general these will (apart from disease resistance and earliness) not be central to a breeder's aims, except in the sense discussed in the next section.

2.3.2 Genotype–environment interaction

For the most part, plant breeders have ignored one important aspect of genotype–environment interaction; competition among plants of the same genotype. Selection is usually made on the descendants of plants which perform well as individuals, in the early stages of a selection programme. This will be discussed in Section 2.4.2.

In this section I simply treat two important aims: maximizing response to controllable inputs and minimizing response to uncontrollable inputs.

2.3.2.1 Response to input Fertilizer response is the most clear example: in general plant response to an essential mineral nutrient will be curvilinear, the part of the curve of greatest interest being approximately parabolic with a maximum at some particular level of applied fertilizer. The aim will be to ensure, first, that fertilizer levels do no rise above the maximum (an environmental aim), and, secondly, to change the shape of the curve so that either maximum yield occurs at a lower fertilizer input or the maximum yield is increased or both (genetical aims). Thus, a knowledge of response to fertilizer will be necessary for all genotypes under test. As is discussed in section 3.2, this response will vary with other environmental factors, so that response surfaces will need to be examined.

Most of the varieties of staple cereals which made up the raw material for the 'Green Revolution' were of the type described here. Availability of fertilizer, however, is not always such as to ensure appropriate use of such varieties, and aims have accordingly altered, in cassava breeding at the International Centre for Tropical Agriculture in Colombia (CIAT) for example, to producing varieties which will perform well under existing cultural conditions, though interest in fertilizer response still exists (Martin 1976).

2.3.2.2 Buffering In contrast to the previous section, the breeder's aim in producing well-buffered genotypes will be to ensure that responses to adverse environmental conditions are minimal. Obvious areas of interest concern tolerance of drought, frost, water-logging, and disease. The last-named is considered in Chapters 6 and 13, and in Section 2.4.3, while aspects of drought and water-logging resistance are dealt with in Sections 6.2.2 and 6.3.2.

The theory of genetical homoeostatis, based partly on Waddington's (1957) ideas of canalization, has so far not made much contribution to our understanding of response to environmental stresses in crop plants, but the practical application of heterozygous superiority (heterosis) is one of the triumphs of plant breeding, and is discussed in Chapter 9. Here, we simply note that hybrid production in outbred crop plants provides a means of fulfilling a breeder's aims of high yield and uniformity of field performance and quality.

2.4 Physiological aims

Much of this book is occupied with statistical approaches to the detection and use of quantitative genetical variation; so much so, indeed, that it is worth reiterating here that the breeder's aims should always be to learn the biochemistry and physiology underlying the summaries produced by statistical analyses, for knowledge at the more fundamental level should be easier to apply. It is not usually the case that physiological models (cf. Thornley 1976) can be directly applied in plant breeding, but the attempt will frequently be illuminating.

2.4.1 Biochemical requirements

Height is the archetypal quantitative trait, yet as we have seen an understanding of dwarfing in biochemical terms adds significantly to the breeder's ability to maximize the returns from a shorter plant, despite the remarkable gains already made. Sugar content in sugar cane, despite all the breeding which has gone on, still remains a promising candidate for selection, possibly more promising than yield. Similarly, lowering the content of cyanogenic glycosides in cassava tubers by selective breeding for lower HCN production is clearly simpler than it would be if a bioassay for toxicity were needed. In view of the need for explanation of variability in terms of molecular genetics, it is unfortunate that quantitative genetics is little developed in terms of gene action; work initiated by Seyffert (Seyffert and Forkman 1976) remains a beginning.

Photosynthesis and nitrogen fixation are the two most notable biochemical targets. They are under assault by molecular genetical techniques described in Chapter 11.

2.4.2 Ideotypes

The concept of an ideotype as such, a plant 'with model characteristics known to influence photosynthesis, growth and (in cereals) grain production', was introduced by Donald (1968), but of course plant design for specific purposes is very old. For example, Bailey (1895) recommended that the breeder should 'establish the ideal of the desired variety firmly in the mind before any attempt is made at plant breeding'. However, apart from the fact that the ideotype should incorporate features shown to be agronomically critical, a specific aim in terms of physiology will often reveal defective features of an existing plant type. It could be argued, for example, that long-strawed cereals became obsolete with the invention of mechanical harvesting, and that short-strawed varieties were a belated response to this innovation. (There are many other factors involved; I express the change this way for emphasis.)

Tough-skinned, long-lasting tomatoes ripening in a known time on plants of determinate growth habit are an example of a plant designed for mechanical harvesting and mass processing. That they are tasteless as well is a problem harder to overcome: selection was practised for certain kinds of economically important trait only. Development of current durable North American varieties of nectarine from the original smooth-skinned, fragrant, low-yielding, disease-susceptible mutants has followed the same course. Some of these problems might have been avoided by a less narrow approach; this the ideotype method may provide.

Donald (1968) emphasized that most breeding was concerned with the elimination of defects and the increase of yield. He suggested that this was not enough. For example, early generation selection would generally favour plants competitively successful against other genotypes, whereas in a pure stand minimal competition is desirable. The relationship between plant form and the use of resources might lead to negative relationships between performance at high and low planting density and between competitive ability in mixture and yield in pure stand. Conventional selection methods would not generate such genotypes. (Davies (1977) has suggested that breeding for an ideotype is in a sense breeding to eliminate defects. However, as the relevant defects are noticeable only when the model plant has been developed, this argument obscures the critical role of model specification.)

A consideration of plant growth and competition led Donald to formulate an ideotype for wheat grown under favourable conditions for water and nutrient supply: it should have a short, strong stem; few small, erect leaves; an erect, large ear (i.e. many florets per unit dry matter of tops); awns; and a single culm, with a consequent high ratio of seminal to adventitious roots. Thus, it would be a poor competitor with weeds, would have a low yield per plant, and would require a high seeding rate. As a simple indicator of many of these valuable features, Donald suggested the use of 'harvest index' given by economic yield/biological yield.

The utility of these specifications is readily testable, and many trials have confirmed some features. For example, Syme (1970, 1972) evaluated some semi-dwarf wheats together with some standard Australian varieties and found that the highest yielding types had high harvest index, and that the highest yielding types also had small leaves. However, in his material there was no association, across cultivars, between grain yield and culm length, and the relationship between leaf number and yield (and its components) was not clearly as predicted from the ideotype (Table 2.1).

Hamblin and Donald (1974), using barley, found no relationship between F_3 single plant yield and F_5 field plot yield. In addition, they found a negative relationship between F_5 yield and F_3 plant height or leaf

Table 2.1

Yield and its components and leaf density in four high-yielding wheats (from Syme 1972)

Variety	Yield ($t\ ha^{-1}$)	Ears m^{-2}	Grains ear^{-1}	1000 grain wt (g)	No of leaves m^{-2}
WW15	5.33	540	28.0	31.2	2350
Pitic 62	4.35	429	28.9	32.6	1780
HMR	4.05	512	24.1	31.1	1990
Robin	3.56	531	17.4	34.3	1770

length, especially with added nitrogenous fertilizer. Furthermore, they found that short F_3 plants with short leaves gave low yield, but their descendants in the F_5, still short with short leaves, yielded well. Thus, pure stand yield was clearly inversely related to early generation competitive ability.

Competition may be expected to be more complex in a grass sward. Hayward and Vivero (1984) examined the relationship between dry matter yield in spaced plants, rows, and two sizes of plot, of lines of the pasture grass *Lolium perenne* which had been selected on the yield of spaced plants. Positive correlations were obtained between spaced plant yields, and row and small plot yields, but not with the larger plots which were closer to what would customarily be used to evaluate pasture yields.

Bhatt (1977) and others have examined selection on harvest index. Bhatt, for example, found that wheat could be selected both upwards and downwards for harvest index, 10 per cent differences in each direction being obtained quite readily in the F_2. In the F_3, plants from the low F_2 had low harvest index, but variability over a wide range was seen from the high material in the F_2. Further experiment would be needed to determine whether high harvest index could be readily stabilized, but this seems likely.

The concept does not provide complete guidance for plant design. For example, ear type (two-row or six-row) of barley is determined by a single gene, and in different regions the different phenotypes are preferred, e.g. two-row in Europe, six-row in North America. It is not clear that either type has potentially uniform superiority over the other. Furthermore, if the two types are crossed, development of either type may be disturbed when the two-row or six-row type is recovered by back-crossing (Riggs and Kirby 1978a,b), so that breeders may prefer to work solely with their predominant type, despite potential improvement from crossing with the other.

Overall, this approach to breeding is an intermediate stage on the path towards full utilization of genetical and biochemical knowledge. Implicit in it is the idea that selection should be conducted, where possible, under production conditions. This is a very important idea and can be exploited even where selection is aimed at eliminating faults rather than raising the yield. Thus, Dowker (cf. Freeman and Dowker 1973; Dowker *et al.* 1975) has shown how to select against colour defects in carrots by exploiting genotype–environment interactions. For example, he showed that for purple-topped roots, the greatest differences between genotypes occur at sites with sandy soil, the least at a peat moss site. When selecting to reduce this trait, the breeder should choose a sandy site.

2.4.3 Disease resistance

As we have already seen, mass cultivation of any particular crop plant may be expected to bring disease in its train. Management practices, such as crop or varietal alternation (for annual plants such as the major cereals) or the introduction of predators or parasites of the pathogenic organisms, may alleviate many such problems, but it is the breeder's objective to make the crop resistant to the disease as rapidly and cheaply as possible, without disadvantage to the desirable agronomic properties of existing varieties.

A first step will then often have to be the development of methods of screening existing varieties for resistance to known types of pathogen (e.g. Ellis and Hardman 1975). This has obvious disadvantages: first, screening essentially only detects resistance to identified types, whereas in the field many other types of pathogen may be found; and secondly, the requirement for rapid progress may lead to selection of specific resistance to a given strain or strains, rather than general tolerance to the disease. At this point, we should therefore consider the classification of resistance. As Day (1974) has pointed out, this may be done functionally or genetically. In the former case, resistance may be general or specific (van der Plank's (1963) horizontal or vertical), characterized by the ability to withstand to some extent many races or strains of pathogen (see, for example, Heijbroek 1977) or to withstand almost completely a few races (see section 13.2.2). In addition, disease tolerance, where plants can yield usefully although infected or infested, should be distinguished. Because of the nature of the interaction between host and pathogen (section 6.2.1), specific resistance will initially be more successful, but mutant pathogens will almost invariably arise which overcome this resistance. Thus, a breeder with a long time-horizon may wish to incorporate general resistance or tolerance into his varieties, but the economic exigencies of the real world may prevent this.

An alternative classification of resistance is genetical: resistance may be oligogenic, polygenic, or cytoplasmic. In general, though not universally, specific resistance is oligogenic. Thus, the combination of aims of rapid progress and the isolation of major genes (Section 2.2.1) is such, once again, to ensure that precise detection of specific resistance may actually mean selection against useful general resistance; the breeder should be aware of this and aim to prevent it.

As Ellis (Ellis and Hardman 1975; Crisp *et al.* 1977) and others have emphasized, for certain pests, such as insects, resistance assessment techniques may also fail to discriminate between the biological types of resistance, non-preference, antibiosis, and tolerance. (These are of course not mutually exclusive categories.) It may in some cases be possible to combine more than one type of resistance into a variety, just as it is possible to incorporate several resistance genes into a genotype or a population (section 13.3).

2.5 Time

Plant breeding is a slow business, so that breeders should always be poised to take advantage of methods which allow more rapid production of commercially useful material. For example, as Riley (1974) has emphasized, the early generations of selection in self-fertilized crops are often largely a matter of waiting for the plants to become relatively homozygous and produce sufficient seed for selection among novel combinations of genes to be practicable. Use of haploid culture from F_1 gametes, with production of diploids thereafter, would allow homozygotes carrying some chromosomes from each parent to be examined very rapidly.

To see how this time affects the return from breeding, consider mass selection in an idealized population of outbreeding plants. Then following the analysis of Hill (1971), we can write annual return

$$R = Ma.\frac{ih^2\sqrt{V_P}}{2}$$

where M is current total production, a is the cash value of a unit of extra production, i is the intensity of selection, h^2 the heritability of yield, and V_P the phenotypic variance of yield. This return has to be discounted by the cost of capital (discount rate, d) and have the cost of the breeding programme offset against it, to achieve the profit over the life of the programme given by

$$P = \frac{R}{d^2(1+d)^{y-2}} - \frac{C}{d} - I$$

where I is the initial investment in the programme, C its annual cost, and y is the number of years from the start before a return is achieved.

A breeding programme for a given crop does not in fact proceed by small annual increments, and there are many other limitations which need to be overcome before this kind of model can readily be applied to evaluate actual breeding programmes. Nonetheless, we can see the crucial importance of the variable y: if this is, say 5, and d is high (it is the nominal or money rate of interest, combining rates for real return and inflation; cf. Smith 1978), say 0.15, the total return of the programme is only three-quarters of what it would be if gains started after 3 years, and this might mean the difference between a profitable breeding programme and an unprofitable one. Given the frequent slow response by producers to the availability of new varieties, perhaps a problem for sociologists rather than plant breeders, the importance of time is not always realized. Macindoe and Walkden Brown (1968) give good examples of the phenomenon of slow acceptance of new varieties, for wheat in Australia.

Breeding objectives do not remain constant over time and when they change there will normally be a lag before useful gains are realized. Smith (1985) has shown that in animal breeding there is a very high probability of substantial future economic gain from the development of breeding stocks for objectives different from current ones. In plant breeding, one might recall the change to dwarf and semi-dwarf varieties in cereals, current problems in protein content and type in cereals, and for the future herbicide resistance so that herbicides can be used on growing crops. Costs of varietal development will be small relative to potential returns.

The benefits from n selected stocks are

$$R(1-d)^y\{1-(1-d)^{t-y+1}\}/d - nC$$

where y is as before the lag, or number of years before the benefit is recovered commercially, t is the time in years after which returns may be ignored, C is the cost of developing a line, R is the initial response from one year of selection, and d is the cost of capital. To quantify the benefit in any given case, heroic assumptions are needed, but in general benefits are expected to be substantial. The method in essence reduces uncertainty about the future.

Adoption of such a strategy with regard to objectives would replace the historical state of affairs in which many breeders, working on a small scale in relative isolation, aimed at slightly different objectives, so that general and agronomic diversity in new varieties was substantial.

3

Methods for field experimentation

3.1 Introduction

Many important advances in experimental design have been brought about by the needs of plant breeders, from the work of Eden and Fisher (1927) onwards. Good textbooks on experimental design, such as those by Cochran and Cox (1950), Daniel (1976), Federer (1955), or Montgomery (1976), cover all the standard designs both as to layout and analysis, and it would not be appropriate in this book to deal with them. My aim in this chapter is simply to introduce a number of important topics which relate to the plant breeder's objectives and which may increase his chance of attaining these objectives. Before doing so, however, I note in addition that it had long been of concern (see for example Day 1965) that statistical methodology could only cope with 'ideal' data, while 'real' data would never meet the assumptions inherent in classical least-squares analysis, and transformations of data would frequently have no theoretical justification (cf. Box and Cox 1964). More recently, it has been realized that with judicious combination of the theory of robust procedures (i.e. those which are relatively unaffected by departures from normality: see Box 1953; Hogg 1977a,b; Stigler 1977) and the very powerful computer-based statistical packages such as GENSTAT (Alvey *et al.* 1977), experimental design and analysis should never be limiting factors in what can be achieved in field experimentation. However, improvement is still possible, as discussed in section 3.4.1.

We therefore consider the measurement of plant response to environmental variables, the interaction between plants, the control of environmental variability in field experiments and the problem of adaptation to a range of environments. These form only a very small fraction of the possible topics which might be considered under the heading experimental methods, and indeed many other aspects will be considered in subsequent chapters, but they all vitally affect the plant breeder's performance.

3.2 The measurement of plant response

Plant response to environmental factors may be investigated by purposive variation of known controllable physical and biological factors, or by indirect assessment through comparison among different genotypes in a range of environments not specified by individual physical or biological factors. Genetical aspects of the latter approach are discussed extensively in Section 6.3.3 and experimental considerations in Section 3.4.2; here we are concerned with deliberate variation of particular factors.

For most purposes, factorial designs may be used, or fractional factorials if many factors are to be investigated, since the number of experimental units increases multiplicatively with the number of factors (Daniel 1976; Federer 1955). Problems usually arise with the detection of interactions between two or more factors, and it is here that standard methods most need supplementation, though it could be argued that this is not critical since such high-order interactions often have low repeatability and are hard to interpret.

The response surface technique is aimed at relating an average response to the values of quantitative variables that affect response (Box 1954; Box and Youle 1955; Box and Draper 1975). It was originally developed in the context of optimization; in industrial process control, for example, the aim will be to vary input factors to maximize yield of a particular product at minimum cost. This yield, y, may well vary in a quadratic fashion with a single maximum in response to increasing inputs of one factor, x_1, all others being held constant. Another likely form of relationship is a combination of a linear increase in response and an exponential decline, as was observed for the dependance of grain yield on applied nitrogen in the long term Broadbalk wheat experiment at Rothamsted:

$$y = \alpha + \beta(0.99^{x_1}) + \gamma x_1$$

where y is yield in t ha^{-1} and x_1 is nitrogen applied in kg ha^{-1} (Dyke *et al.* 1982). If a second factor x_2 is varied in the same fashion, a quadratic or a linear plus exponential response may also be observed. When both are varied together and the response is quadratic for each, the joint relationship may have the form

$$y = \mu + \beta x_1 + \beta_{11}x_1^2 + \beta_2 x_2 + \beta_{22}x_2^2 + \beta_{12}x_1x_2.$$

Industrial processes may be strictly controllable, so that a 3×3 factorial will allow estimation of all parameters, though perhaps not with satisfactory precision (cf. for example Daniel 1976; Montgomery 1976). In the case of plant response, say to inputs of any two of nitrogen, phosphorus, and potassium (a frequent type of designed experiment from the work of Eden and Fisher (1929) onwards) this problem of optimization would be magnified by the fact that experimental errors are much greater than in industrial experimentation.

This is not the only problem, although a response surface methodology is clearly of potential benefit in plant breeding, as noted by Knight (1973) for investigations of heterosis where he suggested that differential patterns of environmental response might yield hybrid vigour, on the basis of earlier work on patterns of response (Knight 1971). Another problem arises, for example, in that the method is essentially sequential, i.e. it is a hunting process, whereas plant breeding must move season by season, in discrete, well-defined and well-designed steps. Thus, in one season responses to say nitrogen, phosphorus, and potassium may suggest an optimal range for a set of varieties, but in the next season moisture or temperature may elicit a qualitatively different set of responses. In a fractional factorial in a greenhouse, failure to obtain a response to change in level of a critical nutrient (cf., for example yield response to nitrogen in *Cannabis sativa*; Coffman and Gentner (1977)) might suggest that a different nutrient range be tried, and this could be done, but in a field trial this would require an additional season's work.

3.3 Interaction between plants

Under normal commercial growing conditions, crop plants of almost every kind will interact with each other, affecting yield and other factors, according to their distance apart. It has long been known that the assessment of variability in many species, based as it has been on spaced plants, may not correspond to that found at normal commercial planting rates. For example, Hinson and Hanson (1962) found that at wide spacing the differential response of genotypes to the spacing resulted in an upward bias to the heritability of seed yield in soyabean. The data in Table 3.1 show how heritability estimates are affected by the interaction between the genotype and spacing in four clones of lucerne having different growth habit. While the differences are not in most cases significant, the results are perhaps indicative of a problem.

Elsewhere, we shall be concerned with genotype–environment interaction as it affects response to selection, and as it affects heritability in perennial crops; here we are concerned simply with problems of investigating plant response to spacing.

The relationship of the yield of crops such as cereals to population density is generally clear (Donald 1963; Donald and Hamblin 1976). Biological yield, i.e. total dry weight produced, increases with plant density to a maximum value limited by certain environmental factors which may not in any particular case be known, and at higher densities remains relatively constant, unless such factors as lodging become

Table 3.1

Heritability estimates, obtained by the method of Kehr and Gardner (1960), for two traits in lucerne *Medicago sativa* L. (from Rammah and Böjtös 1976)

Clone	Spacing†	Green yield	Dry yield
C-37	Wide	0.28	0.36
(erect)	Narrow	0.46	0.37
C-244	Wide	0.27	0.24
(semi-erect)	Narrow	0.51	0.52
C-636	Wide	0.60	0.61
(semi-prostrate)	Narrow	0.58	0.62
C-1474	Wide	0.82	0.84
(prostrate)	Narrow	0.44	0.51

†Wide = 30 cm plant-to-plant and row-to-row;
Narrow = 10 cm plant-to-plant and row-to-row.

important. Grain yield, on the other hand, increases to an intermediate maximum, but declines with further increase in plant density. Maximum grain yield occurs at about the point that biological yield ceases to increase.

Standard agronomic practice in any particular area will generally determine seeding or planting rates, but as shown by the work of Donald (1978) and others, it may be advantageous to vary the plant spacing. In the case of planted-out horticultural crops and tree crops optimal spacing is of even greater importance. Thus, experimental designs have been developed from a number of differing points of view to investigate plant spacing and to demonstrate its effects. Nelder (1962) has published systematic designs for spacing experiments in considerable detail. While the use of randomization in agricultural experimentation is advantageous in general, when one comes to investigate spacing, a design like a randomized block may require that half the plants be guard plants, thereby making very inefficient use of spacing (see Gomez and Gomez 1976). Bleasdale (1960, 1967) pointed out that if crops were planted in rows radiating from a point, with the distance between plants along a radius approximately equal to the distance between radii at that point, then a large range of plant densities could be grown in a small area. Such an arrangement would allow approximately rectangular arrangement of plants in any small area, as would be the case both in commercial practice and in conventional randomized designs, and would allow both demonstration plantings to illustrate to growers the effects of different spacings within a small compass and also the estimation of the relationship between biological yield or product yield and plant density through the simple equation

$$y^{-\psi} = \alpha + \beta\rho + \varepsilon.$$

Here, y is plant weight or product yield, ρ is plant density, ε is a residual or error component, and ψ, α, β would be constants for any particular variety and environment. Nelder (1962) developed both fan designs, in which the plants are planted along radii radiating out from a point, and which are suitable for transplanted crops or large-seeded crops, and parallel row designs, with increasing spacing along rows, which are suitable for drilled crops. Carmer (1977) in presenting a number of treatment designs for estimation of optimal plant density for maize grain yield, has suggested that standard experimental designs are appropriate in the case of maize, using the following model, related clearly to that of Bleasdale and Nelder, for the relationship between yield and plant density

$$y' = \rho'\gamma\eta^{\rho'} + \varepsilon.$$

Here, γ and η are constants to be estimated. In this case, the optimum density can be seen to be

$$\rho'_{\text{opt}} = -1/\ln\eta$$

and the maximum yield

$$y'_{\max} = \rho'_{\text{opt}}\gamma\eta^{\rho'_{\text{opt}}}.$$

For any particular very widely grown crop, such methods as Carmer's can be developed, but for preliminary investigations, the approach of Bleasdale and Nelder has many attractions. Gillis and Ratkowsky (1978) have pointed out that the related model of Holliday (1960),

$$y^* = 1/(\alpha^* + \beta^*\rho + \gamma^*\rho^2),$$

has more desirable statistical properties than that of Bleasdale and Nelder, so it may be preferred in some cases. Many specific designs exist for particular interactions between experimental units, e.g. that of Dyke and Shelley (1976) which allows the independent estimation of the effects of treatment to neighbouring units on each side, which may be important in assessing disease resistance or fungicide efficacy. Here also, however, Bleasdale's and Nelder's, or Holliday's methods might have advantages in initial investigations.

3.4 Variability in yield trials

In his first paper on crop variation, Fisher (1921a) was able, by the analysis of yields of plots of wheat over 70 years, to draw attention to slow changes in yield not ascribable to the direct effects of annual fluctuations in the weather, but distinct from the slow but progressive deterioration of plots inappropriately fertilized. Weed infestation was one factor in such changes. This showed very clearly how to realize the aim of any general study of yield variation, i.e. identification of the sources of variation, rather than the study of the influence of any particular set of factors, as discussed in Section 3.2 (cf. Jackson 1967). However, in setting up yield trials, the experimenter has two main considerations: to control variability as much as possible, and thereby to maximize the probability of detecting worthwhile differences in yield in the varieties under test.

3.4.1 Local control

The two main techniques suggested for assessment of and allowance for variation in yield trials are the use of contiguous control plots, relating very much back to the work of Yates (1936), and the use of moving means of contiguous varietal plots, again relating back to early work (Papadakis 1937a).

A number of studies have attempted to investigate first whether control plots are useful in controlling variability, and secondly whether moving means are better. Briggs and Shebeski (1968) showed that the correlation between yield of control plots drops very rapidly with distance, as might be expected from the work of Smith (1938). Baker and McKenzie (1967) also found this. Indeed, in some cases control plots could be shown to increase the residual mean square.

Knott (1972a) and Townley-Smith and Hurd (1973) attempted to investigate the relative merits of plot weight as a proportion of the mean of the nearest n controls and plot weight as the mean of the nearest p neighbours including the plot itself. In both series of experiments, it could be shown that the method of using a moving mean lowered the residual mean square substantially relative to that achieved by using control plots, and it was found that the optimum number of plots varied quite widely. This possibly reflected soil heterogeneity and other such environmental factors. The remarkable heterogeneity of Southern Australian fields has been dramatically demonstrated by Lamacraft (1974) (see also Knight 1978).

There is thus substantial evidence that moving means are in some sense more efficient than control plots, the latter needing 10–40 per cent more plots in any particular design. On the other hand, the moving mean system requires first complete randomization within replicates, making within-family comparison between different lines difficult, and secondly complete harvesting of all plots, which may not be necessary with control plots. It is also biased.

In addition, in neither of the series of experiments mentioned above was covariance of plot mean on controls or on nearest neighbours used.

Generally, plot weights were expressed as percentages of the appropriate control or neighbour mean. This not only has inherent statistical risks (through the possible generation of Cauchy distributions which have undefined variance and other undesirable properties), but is also inefficient by comparison with the use of the appropriate control or neighbour mean as a covariate. (Residuals may also be used (cf. Atkinson 1969).) Knight (1978) has drawn attention to the importance of this approach.

More recently, the problem of bias inherent in existing moving mean techniques has been extensively explored (Bartlett 1978; Wilkinson and Mayo 1982; Wilkinson *et al.* 1983; Wilkinson 1984; Williams 1986). It has been shown that bias can largely be allowed for and that modest improvement in precision over the best current incomplete block techniques should be possible (Patterson and Hunter 1983). The new methods, termed neighbour methods, require assumptions about the form of the variation of soil fertility over a field, for example that it follows a smooth trend, with uncorrelated variation about this trend. Where these assumptions hold, it is possible to prevent the incorporation of treatment (variety) effects into the covariate, the mean of nearest neighbour values, which is used to adjust each plot value. These methods may have their main use in improving the precision of experiments in which fertility patterns are not satisfactorily removed by the original experimental design.

It is well known (see, for example Fisher 1925) that use of a covariate such as the previous year's yield for perennial crops may be very successful in lowering the residual mean square. In general, designs for annual crops are not replicated exactly from year to year, for important and obvious agronomic and logistic reasons (such as the build-up of pathogens in the soil or the difficulty of planting in the identical rows, plots, etc.), so that other less obviously appropriate covariates must be used. This is an area requiring much more investigation, with the new neighbour methods the starting point.

3.4.2 Overall design

From very extensive investigations of the relationship between crop yield and environmental factors (see, for example Williams 1973; Boyd *et al.* 1976), some general but not specifically helpful conclusions may be drawn. First, the scale on which prediction is possible is generally too large to assist with prediction for experimental sites. Secondly, the more variable the yield, the more can be explained by meteorological factors. (This is largely because of the influence of marginal environments.) Since it is not possible to predict performance in specific environments on the basis of climate in the short term, it is therefore clear that breeding objectives as well as selection strategies and site testing should all be initially aimed at the detection and proving of varieties broadly adaptable to the middle range of environments experienced in the growing regions of importance.

For crops grown very widely, such as cereals, this will mean that multi-site and multi-season experiments are essential for yield testing; this is already the case. From the large-scale variety trials which have already been conducted, it is possible then to make specific predictions about the results of such trials for particular crops in particular sets of environments, thereby overcoming some of the problems outlined above. Patterson *et al.* (1977) provide such an analysis for four cereal types in Britain.

Given a set of varieties, the variance of varietal contrasts may be split into components† as follows:

Variety V_{V}
varieties × sites V_{VS}
varieties × years V_{VY}
varieties × sites × years V_{VSY}
sites V_{S}
years V_{Y}
sites × years V_{SY}
residual (plot error) V_{Res}

The computations for such an analysis of variance are described by Comstock and Moll (1963). From this partition, if a new variety is tested at *m* sites in each of *n* years with *q* control varieties, and *r* plots per trial, then the variance of an estimated mean difference between a new variety and *q* control varieties is given by $(1 + q)V/q$ where

$$V = \frac{V_{\mathrm{VS}}}{m} + \frac{V_{\mathrm{VY}}}{n} + \frac{V_{\mathrm{VSY}}}{mn} + \frac{V_{\mathrm{Res}}}{mnr}.$$

It has been found in practice that V_{VY} is very important for all the cereal types listed in Table 3.2 and also for rice (Antonovics and Wu 1978). For spring wheat in Australia, there is evidence that spring rainfall is the main environmental factor in V_{SY}, V_{S}, and V_{Y}, all of which are of major importance (Seif and Pederson 1978). In sugar variety trials in Barbados and Jamaica, Kennedy (1978) has shown that V_{S}, V_{C}, and V_{SC} are all of importance, whereas V_{VS}, V_{VC}, and V_{VSC} are much less important. Here, C refers to 'crops', that is to say, whether the crop harvested is plant cane, the first crop from the vegetatively propagated plants, or ratoon, that is subsequent crops harvested from the canes which have regrown subsequent to the first harvest. Overall, the value of *V* will provide

†Components of variance are usually written σ^2_{V}, σ^2_{VS}, etc., but to preserve a uniform notation when discussing the partition of phenotypic variance, I shall use V_{V}, V_{VS}, etc., throughout.

Table 3.2
Critical percentage differences d (0.025) in three trial systems (from Patterson *et al.* 1977)

	Differences measured between one test variety and means of two controls		
	a	b	c
Spring oats	3.9	4.6	4.4
Spring wheat	6.1	6.5	7.3
Spring barley	5.6	5.9	6.8
Winter wheat	4.7	5.0	5.7

assistance in interpreting the differences between the test variety mean and the mean for q control varieties. If d is the percentage difference between a new variety and the control varieties, then its standard error is given by

$$s = \frac{100\sqrt{\{(1+q)V/q\}}}{\mu}$$

where μ is the mean yield for the crop. For a given level of significance α, a test criterion for accepting a given variety as significantly better than the control is given by the product of s and the standard normal deviate for α. Thus, from the observed values of V, q, and μ, it will be possible to state what percentage difference between a new variety and the controls will be detectable at any given level of significance. (The power of these tests should be adequate.)

Patterson *et al.* (1977) have used this argument to develop critical percentage differences for several trial systems for four crops. They compared the existing trial system (a) with 15 centres, 3 years, and three replicates at each site year combination, with two other systems, one (b) with only 10 centres but otherwise the same, and a third (c) with 15 sites but only 2 years. System (b) needs 10 trials per year each with 96 plots, and system (c) needs 15 per year each with 66 plots. Then Table 3.2 shows the percentage differences needed to be significant at the 5 per cent two-tailed level under United Kingdom conditions. It can be seen that the percentage differences do not increase dramatically with the economies inherent in systems (b) and (c). System (b) would be the cheapest, but on the other hand, system (c) would allow a gain of a year. An alternative approach, also presented by Patterson *et al.* (1977), is to estimate the probability of accepting a new variety under test. Table 3.3 provides a set of acceptance probabilities for the three variety trial systems under discussion.

Table 3.3

Acceptance probabilities in trial systems (a), (b), and (c). (From Patterson *et al.* 1977.) (Reproduced by permission of the Editor, *Journal of agricultural Science*, Cambridge University Press)

		Variety mean yield (as % of control mean) (critical difference = 2.5% of mean of two controls)						
Crop	System	95	97.5	100	102.5	105	107.5	110
Spring oats	a	0.000	0.006	0.106	0.5	0.894	0.994	1.000
	b	0.001	0.017	0.144	0.5	0.856	0.983	0.999
	c	0.000	0.013	0.133	0.5	0.867	0.987	1.000
Spring wheat	a	0.008	0.053	0.210	0.5	0.790	0.947	0.992
	b	0.012	0.065	0.224	0.5	0.776	0.935	0.988
	c	0.021	0.088	0.250	0.5	0.750	0.912	0.979
Spring barley	a	0.004	0.040	0.190	0.5	0.810	0.960	0.996
	b	0.006	0.048	0.202	0.5	0.798	0.952	0.994
	c	0.016	0.077	0.237	0.5	0.763	0.923	0.984
Winter wheat	a	0.001	0.109	0.149	0.5	0.851	0.981	0.999
	b	0.002	0.025	0.163	0.5	0.837	0.975	0.998
	c	0.005	0.042	0.194	0.5	0.806	0.958	0.995

These tables are not applicable to other crops or other sets of environments, and indeed may not be appropriate to future varieties of the four crops discussed, but they indicate the kind of difference which is detectable in a trial of the magnitude described. Ferguson (1962) has presented similar though less detailed results for a range of horticultural crops. Gomez and Gomez (1976) present detailed accounts of many relevant methods with particular applications to rice.

The requirements of distinctiveness, uniformity, and stability for a genotype to obtain Plant Variety Right protection have necessitated the development of agreed standard procedures. The methods used in the United Kingdom are described by Weatherup (1980).

4
Basic quantitative genetics

4.1 Introduction

The approach to quantitative genetics in this chapter is that usually found, namely the partitioning of the phenotypic value of a metric trait into meaningful components, and the same for the variance of a metric trait. In this, I shall use, as far as possible, the notation of Falconer (1982) for its simplicity and clarity, though extensions are of course necessary for many important cases. Bulmer (1980) gives an appropriate account of quantitative genetics for those seeking more detail than the present text can encompass. Becker (1984) gives details of many of the relevant procedures.

It should be emphasized once more that the statistical approach to the analysis of metric traits is not the only one. In Section 4.6, an alternative approach applicable to certain species is discussed, and it is worth mentioning the attempt of Seyffert (1966, and many subsequent papers) to test quantitative models on the basis of known pathways of intermediary metabolism, though this has as yet produced results of limited interest for plant breeding as such (Section 4.7).

4.2 Phenotypic value

The basis of quantitative genetics, from the pioneering work of Fisher (1918) onwards, has been the partitioning of the phenotypic value of some metric trait for an individual into components attributable to the influence of genes and to the influence of the environment. This harks back still further, to Galton's (1874) apposition of nature and nurture as the joint determinants of any trait. In the very simplest case, where genotype and environment do not interact, the phenotypic value P is made up of a genotypic component, G, plus an environmental deviation, E, i.e.

$$P = G + E.$$

The assumption that G and E are independent, though generally wrong

(Fisher and Mackenzie 1923), allows an initially simple analysis, which will be complicated later.

The genotypic value will be determined in most cases by more than one locus, but we begin by considering the effect of just one locus. We can write, considering a population in Hardy–Weinberg equilibrium,

Genotype	A_1A_1	A_1A_2	A_2A_2	Mean (M)
Genotypic frequencies	p^2	$2pq$	q^2	
Metric value	a	d	$-a$	$a(p-q)+2dpq$

In this simple formulation, the two homozygotes take equal and opposite values, by scaling, while d measures the degree of dominance for the trait. Thus, $d = 0$ if there is no dominance; $d = \pm a$ if there is complete dominance; $d < a$ represents incomplete dominance; and $d > a$ over-dominance. Hence, d/a may be used as a scaled measure of the degree of dominance.

If loci do not interact, this description may readily be extended to many loci, so that for example the mean is given by

$$M = \sum_i a_i(p_i - q_i) + 2\sum_i d_i p_i q_i .$$

However, it will at once be observed that this expression, in terms of genotypic value, does not allow analysis of the effect of an individual allele at a locus, and since at every generation new combinations of genes potentially arise, it is important to assess the effect of replacing one gene by another. We need to examine, then, both the average effect of a gene and the average effect of a gene substitution (Fisher 1941). The average effect of a particular allele at a locus is the mean deviation from the population mean of individuals receiving this allele from one parent, together with a random allele at the homologous site, i.e. the mean value of genotypes thus produced less the population mean. The process is as shown:

Gametes	Genotypes generated			Average effect
	A_1A_1	A_1A_2	A_2A_2	
	a	d	$-a$	
A_1	p	q		$\alpha_1 = pa + qd - M = q\{a + d(q-p)\}$
A_2		p	q	$\alpha_2 = -qa + pd - M = -p\{a + d(q-p)\}$

The average effect of a gene substitution is then

$$\alpha = \alpha_1 - \alpha_2 = a + d(q-p).$$

Clearly, α, α_1, and α_2 are all relative values, and cannot be directly measured; what can be measured is an individual plant's breeding value, the sum of the average effects of its genes, estimated by the mean value of its progeny under random mating. In our simple example,

Genotype	A_1A_1	A_1A_2	A_2A_2
Breeding value	$2\alpha_1$	$\alpha_1 + \alpha_2$	$2\alpha_2$
	$= 2q\alpha$	$= (q - p)\alpha$	$= -2p\alpha$

the average influence of dominance evidently depends upon gene frequency.

We can summarize the single locus results as follows:

Genotypes	A_1A_1	A_1A_2	A_2A_2
Frequencies	p^2	$2pq$	q^2
Metric values	a	d	$-a$
Genotypic value	$2q(a - pd)$	$a(q - p) + d(1 - 2pq)$	$-2p(a + qd)$
	$= 2q(\alpha - pd)$	$= (q - p)\alpha + 2pqd$	$= -2p(\alpha + pd)$
Breeding value	$2q\alpha$	$(q - p)\alpha$	$-2p\alpha$
Dominance deviation	$-2q^2d$	$2pqd$	$-2p^2d$

We see here how the tendency of dominance (taking $d > 0$ conventionally) is to increase the disparity between A_1A_2 and A_2A_2 and reduce that between A_1A_1 and A_1A_2 (the meaning of dominance for a qualitative trait).

In a particular experiment, we may cross two true-breeding parental lines, P_1 and P_2, yielding an F_1, self the F_1 plants to form an F_2, and so on, or back-cross the F_1 to $P_1(B_1)$ or $P_2(B_2)$. Now the two parents will differ in a certain number of loci, say k, and the F_1 will be heterozygous at all these k loci, regardless of which parent carried which alleles. The F_1 will therefore differ from the mean of the two parents by $\sum_{i=1}^{k} d_i$. Now the parental values will differ by $\sum_{i=1}^{r} a_i$ where $r \leqq k$, since in general not all the genes increasing the magnitude of the trait will be in one parent. Thus, the ratio

$$\frac{\sum^{k} d_i}{\sum^{r} a_i} = 2\{\overline{F}_1 - (\overline{P}_1 + \overline{P}_2)/2\}/(\overline{P}_1 - \overline{P}_2),$$

which is an extension of the single locus case above and so is often regarded as a measure of average degree of dominance, should rather be regarded as indicating the relative potence of the two parental gene sets

(Mather and Jinks 1971). Thus, we see that even where gene frequencies are known and equal the effect of dominance may not necessarily be partitioned out independently of other factors.

Reverting to the previous case, consider a second locus, B, with effects thus:

Genotypes	B_1B_1	B_1B_1	B_2B_2
Metric values	a_B	d_B	$-a_B$

If $G = G_A + G_B = A_A + A_B + D_A + D_B$, A and D being the additive and dominance contributions to the metric values of a trait, then we can say that the two loci are completely additive in their effect on the trait. Any departure, say I_{AB}, such that $G = G_A + G_B + I_{AB}$, $I_{AB} \neq 0$, is called the epistatic deviation. For all loci,

$$G = A + D + I,$$

where A is the sum of breeding values over all loci, D that of dominance deviations, and I of all interactions or epistatic deviations.

The model rapidly becomes very complex, especially as one attempts to partition I into various second and higher order components. A treatment in terms of variances makes some of the problems simpler. However, before undertaking this, we examine an alternative partition of G. This is in terms of combining ability, an idea first introduced to describe the mean performance of a particular line in a series of crosses of inbred lines of maize. It was then refined by Sprague and Tatum (1942), who contrasted this general combining ability (gca) with specific combining ability (sca), the deviation of the performance of a particular cross from the mean of the gca's of the parental lines (see Gallais (1976a) for a straightforward review of the theory of combining ability). Formally, we can use the definitions of Griffing (1956a), in terms of genotypic values. Consider the two-way array of gametes yielding the genotypic values shown in the body of the following scheme:

	$A_1A_2 \ldots A_m$		Frequencies
A_1	$g_{11}\, g_{12}$	g_{1m}	p_1
A_2			
$\vdots$			
A_m	g_{m1}	g_{mm}	p_m
Frequencies	p_1	p_m	1

The gca effect associated with the ith gamete is then the weighted mean value of genotypes in the ith row. For example,

$$\text{gca}_i = \sum_j p_i g_{ij}$$

$$\text{gca}_j = \sum_i p_i g_{ij}$$

$$\text{sca}_{ij} = g_{ij} - \text{gca}_i - \text{gca}_j$$

$$g_{ij} = \text{gca}_i + \text{gca}_j + \text{sca}_{ij}$$

Thus, we define general combining abilities in terms of gametes. Hence, breeding value will in general be the sum of the gca for the two gametes contributing to the zygote. Dominance deviation will be equal to sca in this simplest case (Griffing 1956a). Departures from such simplicity occur as we come to consider various types of non-additivity.

As Griffing (1956a) showed for two loci, A and B, with alleles A_1, A_2 and B_1, B_2, the genotypic value of the genotype $A_iA_jB_kB_l$ is

$$\text{gca}_i + \text{gca}_j + \text{gca}_k + \text{gca}_l + \text{sca}_{ij} + \text{sca}_{kl}$$

provided that

$$\sum p_i(A)p_k(B)g_{ij} = 0$$
$$\sum p_i(A)p_l(B)g_{il} = 0$$
$$\sum p_j(A)p_k(B)g_{jk} = 0$$
$$\sum p_j(A)p_l(B)g_{jl} = 0$$

This restriction, then, is the condition for additivity over loci expressed formally.

4.3 Phenotypic variance

4.3.1 Components of variance

Problems of unequal gene distribution between parents, of the signs of gene effects, and (more positively) considerations such as Fisher's (1930) fundamental theorem of natural selection (relating advance in fitness under natural selection to additive genetic variance in fitness) make an approach to quantitative genetics through variance both simpler and more appropriate in some ways, though of course in practice it is location not dispersion of a distribution which selection must be primarily aimed to alter.

From the genotypic and breeding values and dominance deviations calculated in the previous section, we can obtain V_G the genotypic variance, V_A, the additive genetic variance, and V_D, the dominance variance, directly. Thus,

$$V_G = p^2a^2 + 2pqd^2 + q^2a^2 - M^2$$
$$= 2pq\alpha^2 + (2pqd)^2.$$

The additive genetic variance is the variance of breeding values, i.e.

$$\begin{aligned} V_A &= p^2[2q\alpha]^2 + 2pq\{(q-p)\alpha\}^2 + q^2(2p\alpha)^2 \\ &= 2pq\{a + a(q-p)\}^2 \\ &= 2pq\alpha^2 . \end{aligned}$$

The variance of dominance deviations is

$$\begin{aligned} V_D &= p^2(2q^2d)^2 + 2pq(2pqd)^2 + q^2(2p^2d)^2 \\ &= (2pqd)^2 \\ V_A + V_D &= 2pq\alpha^2 + (2pqd)^2 \\ &= V_G . \end{aligned}$$

(This is to be expected, since A and D are uncorrelated by definition.)

In the notation of Mather and Jinks (1982), $D = 2V_A$, $H = 4V_D$, and $E_1 + E_2$ is comparable to V_E.

For a tetraploid, manifesting tetrasomic inheritance, we should briefly consider the equivalent values. Take the simplest case, with no dominance, so that $V_G = V_A$, since dominance in tetrasomics is not precisely the same as in disomics. The genotypic values could be as follows:

	$A_1A_1A_1A_1$	$A_1A_1A_1A_2$	$A_1A_1A_2A_2$	$A_1A_2A_2A_2$	$A_2A_2A_2A_2$
Frequency	p^4	$4p^3q$	$6p^2q^2$	$4pq^3$	q^4
Genotypic value	$2a$	a	0	$-a$	$-2a$

$$\begin{aligned} M(T) &= 2a(p^4 + 2p^3q - 2pq^3 - q^4) \\ &= 2a(p-q) \\ &= 2M(D) \\ V_A(T) &= a^2\{4p^4 + 4p^3q + 3pq^3 + 4q^4 - 4(p-q)^4\} \\ &= 4a^2pq \\ &= 2V_A(D). \end{aligned}$$

On the other hand, if the range of genotypes is as before, i.e.

$A_1A_1A_1A_1$	$A_1A_1A_1A_2$	$A_1A_1A_2A_2$	$A_1A_2A_2A_2$	$A_2A_2A_2A_2$
a	$\frac{1}{2}a$	0	$-\frac{1}{2}a$	$-a$

$$\begin{aligned} V_A(T) &= a^2pq \\ &= \tfrac{1}{2}V_A(D). \end{aligned}$$

Thus, the relationship between diploids and tetraploids will depend very much on the overall effects of polyploidy on the scale of the metric. In Table 4.1, results are shown comparing diploid and autotetraploid

Table 4.1
Effect of autotetraploidy on mean and genetic variance of time to ear emergence (days after 1 June) and forage yield (g/plot) in two pasture grasses (from Simonsen 1976, 1977)

		Lolium perenne		*Festuca pratense*	
		Ear emergence	Dry matter yield	Ear emergence	Dry matter yield
Diploid	Mean	13.9	1260	11.6	1938
	V_G	2.767	153 058	2.909	229 389
Tetraploid	Mean	12.6†	1433	10.6	2006
	V_G	2.197	101 040	1.897	824 621

†Significantly different from diploid at 5 per cent level or less.

genetical statistics in two pasture grasses. V_G is shown rather than V_A because V_A could not be estimated for the tetraploids in the experimental design of Simonsen (1976, 1977). Both increases and decreases were obtained: no general theoretical expectation may be applied.

If the genoptypic value G may be partitioned into independent components A (additive), D (dominance), and I (epistasis), then V_G may be similarly partitioned, viz.

$$V_G = V_A + V_D + V_I$$

Furthermore, in the absence of genotype–environment interaction, the phenotypic variance (V_P) is given by

$$V_P = V_G + V_E$$

where V_E is the environmental variance. Other partitions are possible, and the estimation of the components will be considered in more detail in Section 4.4. The variances in combining abilities are, as would be expected, simply related to the additive and dominance variances. As gca is defined in terms of gametic rather than zygotic genotypes, the partition is

$$V_G = 2V_{\text{gca}} + V_{\text{sca}}.$$

In the simplest case, $V_A = 2V_{\text{gca}}$ and $V_D = V_{\text{sca}}$, while if the effects of different loci are non-additive, $2V_{\text{gca}}$ will include portion of the epistatic variance. Griffing (1956a) and other later writers (e.g. Cockerham 1963) have extended these results to take account of many higher order interactions, but it is not clear that such statistical refinements have been very useful in practice. The application of V_A, V_D, etc., to prediction of selection response will be discussed extensively in Chapter 7.

4.3.2 Heritability

In Table 4.1, the genotypic variance was shown for two traits in two pasture species. The absolute magnitude of the genotypic variance is of critical importance in that it determines what effect selection can have, but equally, if the genotypic variance is only a small proportion of the available variability, then environmental changes may be more effective in increasing production (as of course they may be even when observed V_E is small). The heritability of a trait is often used to describe this proportion, though it was originally used by Lush (1937), following Wright (1921a), for more general purposes. Two estimates are commonly used, the narrow heritability, $h_n^2 = V_A/V_P$, and the broad heritability, $h_b^2 = V_G/V_P$. In Table 4.2, broad-sense heritabilities are shown for the traits described in Table 4.1.

Table 4.2
Heritability estimates (h_b^2) for date of ear emergence and forage yield (from Simonsen 1976, 1977)

	Lolium perenne		*Festuca pratense*	
	Ear emergence	Forage yield	Ear emergence	Forage yield
Diploid	0.58	0.64	0.73	0.55
Tetraploid	0.48	0.52	0.48	0.65

4.3.3 Correlated characters

Two traits, such as plant height and total dry-matter yield, may be correlated phenotypically, but we may seek to know whether such a correlation is a manifestation of an underlying genetical correlation or reflects environmental factors (cf. harvest index, discussed in section 2.4.2). What is needed is an estimate of the genetical correlation, r_A, defined as the correlation of breeding values for two traits within individuals in the population. An environmental correlation, r_E, defined as the correlation of environmental (possibly including non-additive genetical) deviations, may immediately be thought of as containing most of the remaining information in the phenotypic correlation, r_P, which is the correlation of phenotypic values:

$$r_P = \frac{\mathrm{Cov}_P}{\sqrt{V_{P_X} V_{P_Y}}}$$

$$= \frac{\mathrm{Cov}_A + \mathrm{Cov}_E}{\sqrt{V_{P_X} V_{P_Y}}}$$

for two traits X and Y. Defining $e^2 = 1 - h^2$, we can then write

$$r_P = h_X h_Y \frac{\mathrm{Cov}_A}{\sqrt{V_{A_X} V_{A_Y}}} + e_X e_Y \frac{\mathrm{Cov}_E}{\sqrt{V_{E_X} V_{E_Y}}}$$

$$= h_X h_Y r_A + e_X e_Y r_E.$$

Now by analogy with the parent–offspring relationships shown in Section 4.4.1,

$$\mathrm{Cov}(X_{\mathrm{Parent}}, Y_{\mathrm{Offspring}}) = \tfrac{1}{2}\mathrm{Cov}_A,$$

while $\mathrm{Cov}_{\mathrm{OP}}(X) = \frac{1}{2}V_A$ and $\mathrm{Cov}_{\mathrm{OP}}(Y) = \frac{1}{2}V_A$, so that

$$r_A = \frac{\mathrm{Cov}(X_{\mathrm{Parent}}, Y_{\mathrm{Offspring}})}{\sqrt{\mathrm{Cov}_{\mathrm{OP}}(X)\mathrm{Cov}_{\mathrm{OP}}(Y)}}$$

This estimate has approximate variance $\dfrac{(1 - r_A^2)}{2} \dfrac{V_{h_X^2} V_{h_Y^2}}{h_X^2 h_Y^2}$ (Robertson 1959).

4.3.4 Path analysis

Correlation, of course, does not imply causation, so that even when the breeding values for two different traits are known to be strongly correlated, the reason will need to be sought among pleiotropy, linkage, sequential steps in a given biochemical pathway and any other relevant possibilities. Biochemical and physiological extensions of the genetical investigations may now appear most promising for many such cases, but the technique of path analysis introduced by Wright (1920, 1921b, 1934a) has also been used widely in attempting to elucidate patterns of association found for traits such as the components of yield.

Path analysis is a form of linear regression carried out on a closed system, i.e. with complete determination (see Kempthorne 1957; Li 1975, for different introductory accounts). It therefore has all the limitations of any linear method in a non-linear world; in this it is like most of the other methods of quantitative genetics. Consider the simple example shown in Fig. 4.1. Here, X_1, X_2, X_3, and Y are variables

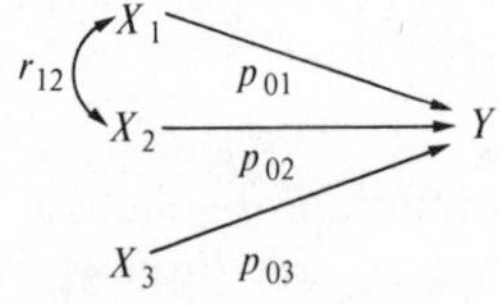

Fig. 4.1. Path diagram for the effects of two correlated variables X_1 and X_2, and a third uncorrelated variable X_3 on a dependent variate Y. See text for explanation of symbols.

standardized to zero mean and unit variance, whence p_{01} and p_{02}, the path coefficients, are the coefficients of simple linear regression of Y on X_1 and X_2. Then,

$$Y = p_{01}X_1 + p_{02}X_2 + X_3$$

by definition, since X_3 is uncorrelated with X_1 and X_2, but X_1 and X_2 are correlated ($r_{12} \neq 0$). For X_1 and X_2, the coefficient of multiple correlation is given by

$$R^2 = p_{01}^2 + p_{02}^2 + 2r_{12}p_{01}p_{02}.$$

Since X_1, X_2, and X_3 completely determine Y, for X_3,

$$R^2 = p_{03}^2 = 1 - (p_{01}^2 + p_{02}^2 + 2r_{12}p_{01}p_{02}).$$

Thus, from a knowledge of part of the system, a knowledge of the whole may be deduced. (Frequently, a variable such as X_3 is introduced as a residual, since in the real world complete knowledge of all components of a system is rare, but this can still allow more precise elucidation of the pattern of interaction of other known factors.)

The actual systems to be studied are far more complex than Fig. 4.1. For example, Sleper *et al.* (1977) considered the relationships among leaf elongation rate, leaf width, leaf area expansion rate, number of tillers per plant, and net CO_2 exchange with total forage yield per plant in tall fescue *Festuca arundinacea* by path coefficients. This allowed the examination of direct effects, with magnitudes given by the path coefficients, and indirect effects through other pathways linked by correlation. It is quite possible for a trait such as tillers per plant to have a substantial direct effect on yield per plant but a negative indirect effect through the reduced yield per tiller, and this is made obvious by the path analysis. However, although path analysis has been widely applied to species ranging from other forage grasses such as crested wheatgrass (*Agropyron cristatum* L. Gaertn.; Dewey and Lu 1959), to cereals (e.g. spring wheat, *Triticum aestivum* L.; Fonseca and Patterson 1968), and legumes (e.g. lucerne, *Medicago sativa* L.; Frakes *et al.* 1961), it is not clear that it has yet been of significant benefit in major breeding programmes, unlike genetical correlations, as used in selection indices (section 7.4.2.4).

4.4 Estimation

Most of the problems of estimation of genetical parameters are problems of experimental design, and so are largely outside the scope of this book, though certain of them are considered where necessary. In general, estimation will be carried out by an investigation of the covariation between values for metric traits in related organisms.

4.4.1 Resemblance between relatives

Species which are readily self-pollinated may most easily have genetical parameters estimated by the dialled cross or related methods (see next section), but this is not as easily achieved for outcrossing species. For these (and also of course for inbreeding species), crosses involving a lower level of inbreeding than self-fertilization must be used, as also must the resemblance between non-inbred related individuals. Indeed, much of the theory depends upon the assumption (rarely met) that tested individuals are a random sample from a larger panmictic population.

We need to use genetical covariance: this is the covariance of the genotypic value of an individual with that of some other of specified relationship. As an example, consider the relationship between a parent of genotypic value X and its offspring of genotypic value Y. Under random mating, treating these values as departures from population values (as before), the offspring mean is (by definition) half the breeding value of the parent. Thus, we seek the covariance of $X = G$ with $Y = \frac{1}{2}A$, i.e. (in the simplest case) of $X = A + D$ with $Y = \frac{1}{2}A$, i.e. with E denoting expectation,

$$\begin{aligned}\text{covariance} &= E\,XY \\ &= E(A + D)\tfrac{1}{2}A \\ &= \tfrac{1}{2}V_A.\end{aligned}$$

This could also be calculated directly from the table:

				Offspring		
				A_1A_1	A_1A_2	A_2A_2
				a	d	$-a$
	A_1A_1	a	p^2	p	q	
Parents	A_1A_2	d	$2pq$	$\frac{1}{2}p$	$\frac{1}{2}$	$\frac{1}{2}q$
	A_2A_2	$-a$	q^2		p	q

$$\begin{aligned}\text{covariance} &= p^3a^2 + p^2qad + p^2qad + pqd^2 \\ &\quad - pq^2a^2 - pq^2ad + q^3a^2 - M^2 \\ &= pq\{a + d(q - p)\}^2.\end{aligned}$$

By similar arguments, covariances for other relationships may readily be derived:

Relationship	Covariance
Offspring and one parent (OP)	$\frac{1}{2}V_A$
Offspring and mid-parent (O$\bar{\text{P}}$)	$\frac{1}{2}V_A$
Half sibs (HS)	$\frac{1}{4}V_A$
Full sibs (FS)	$\frac{1}{2}V_A + \frac{1}{4}V_D + V_{E_C}$

Here, V_{E_c} is that proportion of the environmental variance common to members of a group, as opposed to V_{E_w} which is that which differs among members of a group ($V_E = V_{E_c} + V_{E_w}$). Maternal effects are the most common source of a substantial V_{E_c} component.

In autotetraploids, in the absence of double reduction, the equivalent results are, from Kempthorne (1955, 1957):

Relationship	Covariance
OP	$\frac{1}{2}V_A + \frac{1}{6}V_A$
HS	$\frac{1}{4}V_A + \frac{1}{36}V_D$
FS	$\frac{1}{2}V_A + \frac{2}{9}V_D + \frac{1}{12}V_{Q_1} + \frac{1}{36}V_{Q_2}$

Here, $G = A + D + Q_1 + Q_2$, where Q_1 is the sum of trigenic deviations and Q_2 that for quadrigenic deviations (analogus to dominance deviations), and V_{Q_1} and V_{Q_2} are the corresponding variance components.

Heritability may be estimated in several ways. Considering the relationship between offspring and parent, V_P would be estimated by the parental variance, and so $h^2 = V_A/V_P$ could be estimated by twice the coefficient of simple linear regression of offspring on one parent, $b_{\mathrm{OP}} = \mathrm{Cov}_{\mathrm{OP}}/V_P$. For the regression of offspring on the parental average, $b_{\mathrm{O\bar{P}}} = (\frac{1}{2}V_A)/(\frac{1}{2}V_P)$, so that if maternal effects were absent, this coefficient would be the simplest precise estimator of narrow heritability; such effects are rarely completely absent (Kempthorne 1957), however, so that they should be accounted for.

In the description of covariance derivations given above, we have not considered interactions between loci: these will involve interactions between the additive effects of different loci, *AA* for two loci, *AAA* for three and so on, between dominance deviations, *DD*, *DDD*, etc., and between the additive effects and dominance deviations of two loci, *AD*, etc. Thus, including only two and three locus interactions, we note the following (Kempthorne 1955, 1957):

Relationship	Covariance
OP	$\frac{1}{2}V_A + V_{AA} + \frac{1}{8}V_{AD}$
HS	$\frac{1}{4}V_A + \frac{1}{16}V_{AA} + \frac{1}{64}V_{AD}$
FS	$\frac{1}{2}V_A + \frac{1}{4}V_D + \frac{1}{4}V_{AA} + \frac{1}{8}V_{AD} + \frac{1}{16}V_{DD} + \frac{1}{8}V_{AAA} + \frac{1}{64}V_{DDD}$

Problems will arise with the sampling errors of such estimates since data to measure the various components independently will rarely exist.

For sib analyses, correlation rather than regression methods will be appropriate, essentially using the intraclass correlation coefficient, *t*, of

Fisher (1921b), applied to analyses of variance within and between full or half-sib families. Regression and correlation methods yield the following heritability estimates:

Relationship	Heritability	Variance of estimate
OP	$h^2 = 2b_{OP}$	$4\,V_{b_{OP}} = \dfrac{4}{n-2}\left[\dfrac{V_P(\text{offspring})}{V_P(\text{parent})} - b^2_{OP}\right]$
HS	$h^2 = 4t_{HS}$	$4\,V_{t_{HS}} = \dfrac{8\{1+(m-1)t_{HS}\}^2(1-t_{HS})^2}{m(m-1)(n-1)}$
FS	$h^2 \leqq 2t_{FS}$	

Here, n is the number of families measured, and m the number of individuals tested per family.

4.4.2 *Diallel analyses*

4.4.2.1 Combining ability In a sense, diallel crosses are special cases of analyses using resemblances between relatives, but they are so widespread in plant breeding and so fully developed that they may more properly be treated independently. (Methods for the analysis of unsystematic sets of crosses by the standard least-squares approach have also been developed (e.g. Gilbert 1967), but do not have the same genetical interest, since they usually obtain only general combining abilities.)

A diallel cross consists of all possible crosses between a number of varieties. If there are p varieties, there will be p^2 combinations, consisting of p selfings and $p(p-1)$ crosses, since reciprocal crosses, alternating pollen and ovum parents, may be differentiated by maternal or paternal effects. As p increases, p^2 may become impossibly large. For this reason, many methods have been developed for the examination of partial diallel crosses.

Gilbert (1958) has listed the assumptions needed for diallel analyses, following Hayman's (1954) method. These are: diploid segregation; no reciprocal differences; independent action of non-allelic genes; no multiple allelism; genes independently distributed at random between the parents. As will be seen, reciprocal differences and polyploidy may readily be allowed for, but the other assumptions are more problematic. The analysis of Griffing (1956a), following Yates (1947), being made in terms of combining ability, has less demanding genetical assumptions, and is probably to be preferred.

We follow Griffing in assuming that the experimental material is a random sample from a specified panmictic population. Griffing's analysis ignored the p possible selfings, since these can introduce bias, but as noted by Gilbert (1958), if the particular parents are of interest in themselves, it may be more important to include the selfings. Nonetheless, we examine simply Griffing's analysis for the full and the half diallel cross (ignoring selfings).

Having regard to the model of combining abilities already introduced, we consider $p(p-1)$ F_1s, and write the linear model

$$x_{ij} = \mu + \text{gca}_i + \text{gca}_j + \text{sca}_{ij} + r_{ij} + \text{error}$$

for observations in the cross $i \times j$, where

$i = 1, 2, \ldots p$
μ = population mean
gca_i = general combining ability of the ith line or variety
sca_{ij} = specific combining ability of the cross $i \times j\,(\text{sca}_{ij} = \text{sca}_{ji})$
r_{ij} = reciprocal effect of the cross $i \times j\,(r_{ij} = -r_{ji})$

It is assumed that the error effects are uncorrelated and that they are normally distributed with zero mean and variance V_E.

The analysis of variance is then as shown in Table 4.3.

As Griffing (1956b) showed, the effects may be estimated thus:

$$\hat{\mu} = \frac{X..}{\{p(p-1)\}} \qquad V_{\hat{\mu}} = \frac{V_E}{\{p(p-1)\}}$$

$$\hat{\text{gca}}_i = \frac{\{p(X_{i.} + X_{.i}) - 2X..\}}{2p(p-2)} \qquad V_{\hat{\text{gca}}_i} = \frac{V_E(p-1)}{\{2p(p-2)\}}$$

$$\hat{\text{sca}}_{ij} = \frac{x_{ij} + x_{ji}}{2} - \frac{X_{i.} + X_{.i} + X_{j.} + X_{.j}}{2(p-2)} + \frac{X..}{(p-1)(p-2)} \qquad V_{\hat{\text{sca}}_{ij}} = \frac{V_E(p-3)}{\{2(p-1)\}}$$

$$\hat{r}_{ij} = \frac{x_{ij} - x_{ji}}{2} \qquad V_{\hat{r}_{ij}} = \frac{V_E}{2}$$

Griffing's (1956a) second and simpler analysis, which follows that of Sprague and Tatum (1942), treats only one set of $p(p-1)$ F_1s, using the model

$$x_{ij} = \mu + \text{gca}_i + \text{gca}_j + \text{sca}_{ij} + \text{error}$$

Table 4.3
Analysis of variance

Source	Degrees of freedom	Sums of squares	Expected mean squares
General combining ability	$p-1$	$\frac{1}{2(p-2)}\sum_{i}(X_{i.}+X_{.i})^2-\frac{2}{p(p-2)}X^2_{..}$	$V_E+2V_{sca}+2(p-2)V_{gca}$
Specific combining ability	$\frac{p(p-3)}{2}$	$\sum_{i}\sum_{j,\, i<i}\frac{1}{2}(x_{ij}+x_{ji})^2-\frac{1}{2(p-2)}\sum_{i}(X_{i.}+X_{.i})^2+\frac{1}{(p-1)(p-2)}X^2_{..}$	V_E+2V_{sca}
Reciprocal differences	$\frac{p(p-1)}{2}$	$\sum_{i}\sum_{j,\, i<j}(x_{ij}-x_{ji})^2$	V_E+2V_r
Error from replication	n	..	V_E

Here, $X_{..}=\sum_{i}\sum_{j,\, i\neq j}x_{ij}$ and $X_{i.}=\sum_{j}x_{ij}$

with the definitions as before. This yields the analysis of variance:

Source	Degrees of freedom	Sums of squares	Expected mean squares
General combining ability	$p-1$	$\frac{1}{p-2}\sum_i X_{i.}^2 - \frac{4}{p(p-2)}X_{..}^2$	$V_E + V_{\text{sca}} + (p-2)V_{\text{gca}}$
Specific combining ability	$\frac{p(p-3)}{2}$	$\sum_{\substack{i\ j \\ i<j}}\sum x_{ij}^2 - \frac{1}{p-2}\sum_i X_{i.}^2 + \frac{2}{(p-1)(p-2)}X_{..}^2$	$V_E + V_{\text{sca}}$
Error from replication	n		V_E

In this analysis, $X_{..} = \sum_{\substack{i\ i \\ i<i}}\sum x_{ij}.$

In this later analysis, Griffing considered the different procedures necessary for the so-called 'fixed', 'mixed', and 'random' models; here we ignore this complication, following the view of Yates (1967) and Nelder (1977) that it is in general unnecessary. (In effect, all models where all variance components are estimated are essentially treated as 'random'.) As this view is not universally shared, the reader is referred to the discussion in Nelder (1977) for an airing of the various views. The problem of the reference population from which the parents are chosen, if this is not a very large population from which the parents have been taken at random, is however a very difficult one (Kuehl *et al.* 1968).

The estimation procedures of Griffing (1956b) apply only to complete diallels or complete partial diallels (those described above and immediately below). Their imprecision has been criticized by Feyt (1976), among others. More efficient procedures are discussed by Dhillon and Singh (1978), and Gordon (1979).

From the diallel analysis, it is possible to estimate some of the components of the genetic variance, using the relationships given by Kempthorne (1955), and Levings and Dudley (1963). These are as follows, with offspring–parent covariances included for comparison:

		Coefficients of components of variance						
		V_A	V_D	V_{Q_1}	V_{Q_2}	V_{AA}	V_{AD}	V_{DD}
V_{gca}	Diploid	$\frac{1}{4}$	0	—	—	$\frac{1}{16}$	0	0
	Tetraploid	$\frac{1}{4}$	$\frac{1}{36}$	0	0	$\frac{1}{16}$	$\frac{1}{144}$	$\frac{1}{1296}$
V_{sca}	Diploid	0	$\frac{1}{4}$	0	0	$\frac{1}{8}$	$\frac{1}{8}$	$\frac{1}{16}$
	Tetraploid	0	$\frac{1}{6}$	$\frac{1}{12}$	$\frac{1}{36}$	$\frac{1}{8}$	$\frac{8}{72}$	$\frac{31}{648}$
Cov_{OP}	Diploid	$\frac{1}{2}$	0	—	—	$\frac{1}{4}$	0	0
	Tetraploid	$\frac{1}{2}$	$\frac{1}{6}$	0	0	$\frac{1}{4}$	$\frac{1}{12}$	$\frac{1}{36}$

It can be seen that V_A and V_{AA} may be simply estimated for diploids, but that independent estimates may not be derived for most other components from a combination of diallel and parent–offspring analyses, though if interactions between non-allelic genes are relatively unimportant approximate estimates may be obtained.

The broad sense heritability estimates shown in Table 4.2 were obtained by Simonsen (1976, 1977) as

$$h^2 = (V_{gca} + V_{sca})/\{V_{gca} + V_{sca} + \tfrac{1}{4}(V_m + V_r + V_E)\},$$

where all the components have the meanings given in Griffing's analysis above, except that, in addition, V_m is the variance of maternal effects, obtainable independently in Simonsen's analyses (Gardner 1963). Further, narrow heritability was obtained as

$$h_n^2 = V_{gca}/\{V_{gca} + V_{sca} + \tfrac{1}{4}(V_m + V_r + V_E)\}.$$

For comparison, narrow sense heritabilities for the same traits as in Table 4.2 are shown in Table 4.4, though the imprecision of the various estimates makes definite conclusions difficult.

Table 4.4

Heritability estimates h_n^2 for ear emergence and forage yields (from Simonsen 1976, 1977)

	Lolium perenne				*Festuca pratense*			
	Ear emergence		Forage yield		Ear emergence		Forage yield	
	Diallel	Parent–Offspring	Diallel	Parent–Offspring	Diallel	Parent–Offspring	Diallel	Parent–Offspring
Diploid	0.58	0.44 ± 0.14	0.40	0.46 ± 0.18	0.64	0.78 ± 0.26	0.31	0.14 ± 0.16
Tetraploid	0.32	0.40 ± 0.08	0.39	0.56 ± 0.26	0.48	0.76 ± 0.12	0.29	0.38 ± 0.18

4.4.2.2 Complications The analysis set out above depends, as noted, on the six assumptions set out at the start of Section 4.4.2.1. Some of the assumptions may be violated with impunity; others are essential, especially that of random distribution of genes among parents (Baker 1978); and some may be tested using features of the diallel itself. Hayman (1954), following Mather (1949), showed with a graphical analysis how to approach the problem of dominance. This was extended from diploids to autotetraploids by Dessureaux (1959); the argument below relates only to diploids.

If the assumptions are met, it can be shown that for an $n \times n$ diallel, the variance of all the offspring of the ith parent, i.e. of the ith complete array (row or column), V_i, is simply related to the covariance between these offspring and their non-recurrent parents, W_i. In fact,

$$W_i = V_i + \tfrac{1}{2}V_A - \sum 2pqd^2$$

$$= V_i + \tfrac{1}{2}V_A - \frac{1}{2pq}V_D$$

if gene frequencies at all loci are the same. A plot of W_i against V_i (W_r against V_r, in Hayman's terminology) will therefore display the relationship between V_A and a function of V_D. If dominance is uniform, linearity should be seen. Furthermore, the parabola $W_i^2 = V_P(\text{parent})\,V_i$, delimits the area in which the results may occur, as in Fig. 4.2. (V_P(parent) is the

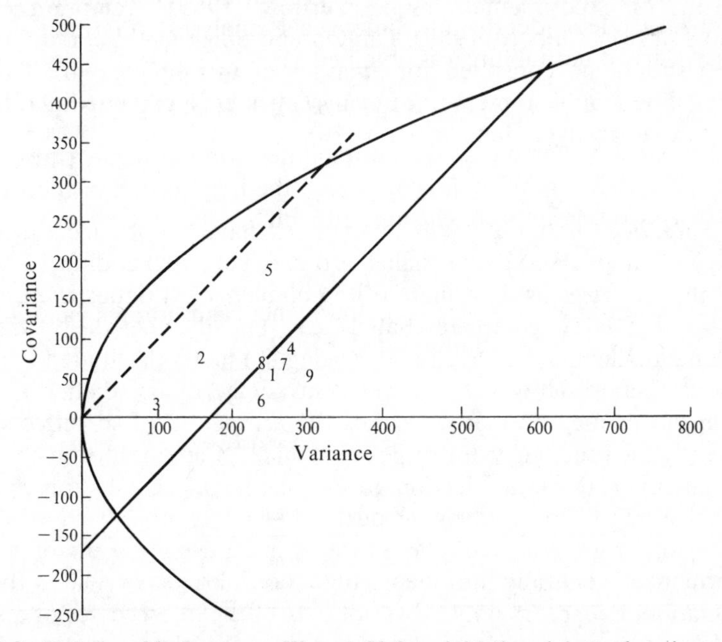

Fig. 4.2. Relationship between W_i and V_i for yield (in g/plot) of a nine parent diallel cross in barley. (Reproduced by permission of the Genetics Society of Canada from Johnson and Aksel 1959.)

parental phenotypic variance.) The position of the pairs of points V_i, W_i relative to the regression line of unit slope through the mean $\overline{V}_i$, $\overline{W}_i$ indicates the nature of dominance. The line through the origin indicates complete dominance: the greater the intercept with the abcissa, the more the tendency to overdominance, as shown in Fig. 4.2, taken from Johnson and Aksel's (1959) analysis of a diallel in barley.

Now, as Gilbert (1958) has noted, W_i and V_i do not fulfil the basic assumptions of independence and normality required for regression analysis, so that departure from simple linearity, used by Hayman to justify ranking the dominance of parents by the sum $W_i + V_i$, may not provide this justification. Gilbert suggested that Bartlett's (1937) test for homogeneity of variance might be applied to V_i as such a test. It is not clear that this advice has always been followed (cf. for example Singh and Sharma 1976), and in any case Bartlett's test is sensitive to departures from normality. Furthermore, the variances of V_i and W_i must increase with V_i (Mayo 1983).

Diallel crosses as set out above were originally intended for analyses of crosses between homozygous lines, but as shown by Griffing (1956b) and Oakes (1967), may be applied to crosses between heterozygous clones, as were used in the diallel crosses in potato shown in Table 5.2.

Most other problems arising with diallel crosses are essentially problems of experimental design; Curnow (1963), Kempthorne and Curnow (1961), Gilbert (1967), Dhillon and Singh (1978), and Gordon (1979) should be consulted for analyses of incomplete partial diallel crosses. The graphical method of analysis has been extended and is discussed exhaustively by Jana (1975, 1976).

4.4.2.3 Related methods As many writers have emphasized (see for example Gilbert 1958), the diallel cross is very demanding of space. Since the size rises as the square of the number of varieties tried, it will be rare that large complete diallels can be advantageously used. A fractional diallel may be much better using the methods alluded to in the previous section. However, the main advantage of the diallel in some circumstances may be that it permits the estimation of sca effects, and frequently the breeding value or gca is of much greater importance, being more amenable to rapid selection. In such cases, the use of a top-cross or poly-cross may be more advantageous.

In a top-cross, selected plants are crossed with testers of known performance, generally in open pollination. Progeny yield only gca information, not sca. A poly-cross consists of the intercrosses of a set of selected plants, again generally by open pollination, planting clones together at random under isolation. Both of these techniques are inherently easier to practise with outcrossing species, and much of the early work on comparison of the methods was carried out using lucerne, which is self-incompatible. Tysdal and Crandall (1948) were able to show a very high correlation between the gca estimates for the two methods, suggesting that this meant that the choice of method would depend on considerations other than genetical ones.

Specific combining ability is of most interest in hybrid breeding (Chapter 9), where the performance of particular crosses is critical. It has long been recommended, therefore, that the simpler crosses be used where sca information is not required (see, e.g., Rojas and Sprague 1952; Hayes 1952). When time is of paramount importance, breeders may begin selection among the progeny generations of the poly- or top-cross, since the estimation generations may represent seasons lost to selection, given fixed resources for breeding work.

In order to obtain more information about combining ability than is provided by these simple crosses but without the labour of a full diallel, Comstock and Robinson (1952) introduced three designs, now called the North Carolina designs I, II, and III.

In design I, two parents are crossed, $P_1 \times P_2$, the resultant F_1 is selfed, and a set of F_2 progeny is crossed to the seed parent, a set of progenies being raised from each F_2-seed parent cross.

In design II, possible only with multi-flowered plants, the same F_2 is obtained as in I, but these plants are then crossed with each other in all possible ways. This design has been least used of the three.

In design III, a set of F_2 progeny is obtained as before, then a set of $F_2 \times P_1$ and $F_2 \times P_2$ crosses is made, yielding a set of pairs of crosses with the same F_2 individual involved in each.

For genetical analysis, the necessary assumptions are essentially those for the diallel mentioned in section 4.4.2.1. From design III, the variance of the mean of the means of the pairs of crosses is an estimate of $V_A + V_E/2$ and the variance of the difference of the means of the pairs of crosses is an estiamte of $V_D + V_E/2$. Then the average level of dominance is given by $\surd(2V_D/V_A)$, since all the assumptions are met, one finds $\surd(2V_D/V_A) = \surd(\sum d_i^2/a_i^2)$. If this is greater than unity, there is over-dominance, if it equals exactly one, there is complete dominance, and if it is zero there is no dominance.

4.4.3 *The number of genes affecting a trait*

If the heritability of a metric trait is very high, this is frequently an indication that a few loci are responsible for much of this observed variability, on account of the improbability of the existence of substantial numbers of loci which affect a trait, but yet are unlinked, do not manifest dominance and do not interact. Even where the distribution of a trait is very close to Gaussian, it may be determined by only a few unlinked loci with additive effects (see Falconer 1982; Thoday and Thompson 1976), and tests of the number of genes (involving higher moments of distributions; see below) are very imprecise. Accordingly, a very high heritability may be used as an indication that specific gene loci should be sought and localized, where possible, though as we have already noted, selection only on major loci affecting a trait will not always be appropriate. Some methods of gene localization are described in Section 4.6.

On occasion, it may be important to know how many genes affect a trait. More properly, we may attempt to estimate the number of effective factors influencing a trait. Effective factors may be defined in terms of the cross between two homozygous lines, P_1 and P_2: at the loci which differ between the lines, some alleles tending to increase the trait of interest will be homozygous in one line, some in the other. Now suppose that all loci are unlinked and have equal effects on the trait of interest, those shown in Section 4.2. Suppose initially that of the k loci at which P_1 and P_2 differ, P_1 contains only loci fixed for the allele increasing the trait, while P_2 contains only loci fixed for the allele decreasing it. Then the difference between P_1 and P_2 will be

$$P_1 - P_2 = 2ka.$$

If we estimate the additive genetic variance of the trait from the F_2, it will be, in the special case considered above,

$$\sum_{i=1}^{k} 2pqa^2\{ = V_P(F_2) - V_E\}$$

and $p = q = \frac{1}{2}$, so that

$$V_A = \frac{ka^2}{2}.$$

Hence,

$$\hat{k} = \frac{(P_1 - P_2)^2}{8V_A}$$

The estimator k must be the minimum number of genes affecting a trait, as we have assumed equality of effects as well as independence of all gene loci. Mather and Jinks (1971) have shown that if one strain has x loci fixed for the '+' allele and $k - x$ fixed for the '−' allele, and conversely for the other strain, then setting $r = (k - 2x)/k$, the estimate is reduced by a factor r^2. Furthermore, if the gene effects are not all equal, and their variance is $V_{\Delta a}$, then the estimate will be reduced independently by a factor $(1 + V_{\Delta a})^{-1}$.

The variance of $\dfrac{(P_1 - P_2)^2}{8V_A}$ will be, approximately,

$$\left(\frac{\mathrm{P}_1 - \mathrm{P}_2}{4V_A}\right)^2 V_{P_1} + \left(\frac{\mathrm{P}_1 - \mathrm{P}_2}{4V_A}\right)^2 V_{P_2} + \left[\frac{(\mathrm{P}_1 - \mathrm{P}_2)^2}{8V_A^2}\right]^2 V_{V_A}$$

$$- \frac{(\mathrm{P}_1 - \mathrm{P}_2)^3}{32V_A^3}\mathrm{Cov}_{\mathrm{P}_1, V_A} + \frac{(\mathrm{P}_1 - \mathrm{P}_2)^3}{32V_A^3}\mathrm{Cov}_{\mathrm{P}_2, V_A}$$

This is likely to be large relative to $\frac{(\mathrm{P}_1 - \mathrm{P}_2)^2}{8V_A}$, so that estimates of k obtained thereby will be most imprecise. An extensive discussion of applications of this index is to be found in Wright (1968).

Many of the problems of this Castle–Wright segregation index (Wright 1934b) have long been recognized (Mather 1949). It is, however, still sometimes advocated as a useful device; see Mayo and Hopkins (1985) for an extended discussion. Some space has been devoted here to its problems so that later developments may be critically compared with it.

Panse (1940), following Fisher *et al.* (1932), introduced another method using F_2 and F_3 second-order statistics to estimate k. The mean genotypic variance (V_G) within F_3 progenies can be shown to be half of the genotypic variance of the F_2 (Fisher *et al.* 1932). Panse showed that the variance of the mean variance within F_3 equalled one-quarter of the sum of squares of the variances of individual factors (or genes, in the limiting case of no linkage or interaction). In Panse's simple example, for two loci with alleles A_1, A_2 and B_1, B_2, the following definitions may be made:

F_3 progenies heterozygous for	Frequency of the progenies	Variance within progeny
A_1 and B_1	$\frac{1}{4}$	$V_G(A) + V_G(B)$
A_1 only	$\frac{1}{4}$	$V_G(A)$
B_1	$\frac{1}{4}$	$V_G(B)$
Neither	$\frac{1}{4}$	0
Mean		$\frac{1}{2}(V_G(A) + V_G(B))$
Variance of this mean variance		$\frac{1}{4}((V_G(A))^2 + (V_G(B))^2)$.

Linkage serves to increase this variance by $\frac{1}{2}x^2 V_G(A) V_G(B)$ where x is the difference between the frequencies of coupling and repulsion gametes.

Ignoring linkage, if the F_2 variance is $\{V_G(A) + V_G(B) + \ldots\}$, and the mean variance within F_3 progenies, say V_3, is $\frac{1}{2}\{V_G(A) + V_G(B) + \ldots\}$, and $V_{V_3} = \frac{1}{4}\{V_G^2(A) + V_G^2(B) + \ldots\}$, and if the segregation in the F_2 is due to k factors with equal variance V_{ef}, then

$$V_3 = \frac{kV_{\mathrm{ef}}}{2}$$

and

$$V_{V_3} = \frac{k(V_{\text{ef}})^2}{4}$$

An estimate of k is then given by V_3^2/V_{V_3}. This estimate will have a very large standard error, like the previous one, though in principle it does not suffer from the same disadvantages as outlined for the Castle–Wright index, apart from sensitivity to unequal gene effects.

The most recent method, called genotype assay by Jinks and Towey (1976; see also Towey and Jinks 1977), is designed to overcome these various problems. The principle is once again simple. It depends upon the detection of the minimum number of factors still segregating after n generations of selfing. Each individual in the nth generation is selfed, then two random progeny are chosen from each and selfed. Thus, progeny means and variances can only differ if their parent was heterozygous at one or more loci.

The probability of individuals in the nth generation being heterozygous at one or more of k loci is

$$Pr[\text{Het}] = 1 - \left[\frac{(2^{n-1} - 1)}{2^{n-1}}\right]^k.$$

Now not all randomly chosen pairs of progeny will have different genotypes, so that only a proportion of heterozygotes will be detected. The more loci at which an individual in the nth generation is heterozygous, the more likely is the detection of heterozygosity. The probability that r loci are heterozygous in the nth generation is

$$Pr[r\,\text{Het}] = \frac{k!}{r!\,(k-r)!}\,\frac{(2^{n-1} - 1)^{k-r}}{(2^{n-1})^k}.$$

The probability that the random pair will differ depends upon r, and is simply $1 - (3/8)^r$. Accordingly, the maximum detectable frequency of heterozygotes is

$$Pr(\text{Max}) = 1 - \{1 - 5/(2^{n+2})\}^k.$$

Now stabilizing selection will tend to maximize the number of identical phenotypes if all additive effects are equal, and so genetical differences will not be manifested. Dominance will also influence the proportion of detectable heterozygotes. The combined effect of these influences was shown by Jinks and Towey to yield a minimum proportion of detectable heterozygotes

$$Pr(\text{Min}) = 1 - (1 - 1/2^{n-1})^k \sum_{r=0}^{k} \frac{\binom{k}{r} \sum_{s=0}^{r} 9^s \left[\binom{r}{s}\right]^2}{(2^{n-3} - 16)^r}.$$

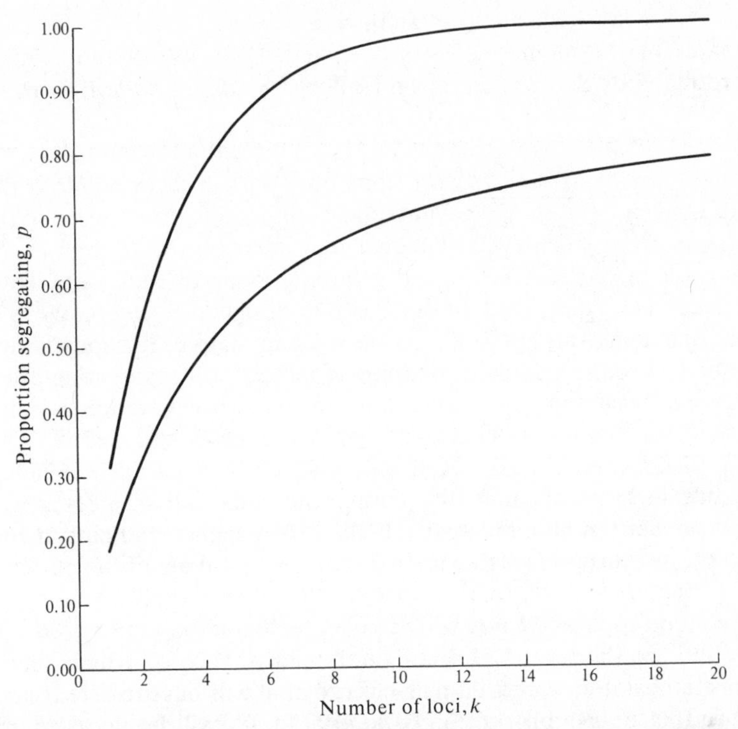

Fig. 4.3. Relationship between number of loci originally segregating (k), and the proportion of these (p) detected in the F_2. Upper curve is maximum detectable, lower curve is minimum detectable. (Reproduced by permission of the Editor, *Heredity, London*, from Jinks and Towey 1976.)

By plotting Pr(Max) and Pr(Min) against k for appropriate generations, we can use the proportion segretating to obtain a range for k (Fig. 4.3), by testing m F_n individuals through two families of l individuals. The precision of the estimate of k will depend on m, while detection of heterozygotes will depend largely on l.

Table 4.5 shows estimates of k by the three methods from two crosses between particular lines of *Nicotiana rustica* (see Hill and Perkins 1969 for details). Genotype assay gives a higher estimate in all cases than the other methods. Clearly, it is also highly imprecise. Estimates show in most cases an increase from F_3 to F_5; this is to be expected because of the extra cycles of gametogenesis available. Jinks and Towey emphasized that because the method detects only genes with effects sufficient to yield a significant difference, given the variability in the material, the range shown will be an underestimate. Extensive simulation for values of m and l from 20 to 50 shows that the method is not very powerful for modest levels of environmental variance which cannot be partitioned out, and is also quite sensitive to unequal gene effects. While genotype

Table 4.5
Estimates of the number of effective factors, k, by different methods. (Reproduced by permission of the Editor, *Heredity, London*; from Jinks and Towey 1976)

Cross	Character	Generation	Castle–Wright	Panse	Jinks & Towey
p3×nil3	Flowering time	F_3	0	3	4–7
		F_5	1		11–20
	Final height	F_3	1	1	2–4
		F_5	3		7–12
nk2×nil1	Flowering time	F_3	0	1	4–9
		F_5	0		9–17
	Final height	F_3	0	1	3–5
		F_5	0		3–5

assay has less defects than the earlier methods and can certainly be refined further (cf. Hill and Avery 1978; Baker 1984; Mulitze and Baker 1985a,b), the question arises whether a precise estimate of the number of effective factors will be practically useful, given the extra work necessary.

A related, apparently not widely used technique is the 'inbred backcross line' method of Wehrhahn and Allard (1965), which has the advantage that the magnitude of effects of the genes detected can be measured (see Section 4.6.1). Park (1977a,b) has developed further related methods which are designed to assess the number of genes affecting a trait over the early generations of selection in order to estimate likely total response, limits to selection depending on the number of loci involved (see Chapter 7).

4.5 Linkage

The importance of linkage in influencing genetical variance has been recognized from the work of Fisher (1918) onwards, and much of the development has been in assessing how linkage affects the maintenance of genetical variation (Section 5.2.1.2) and the response to selection (Section 7.4.2). Here, we shall simply consider first how linkage affects the early generations of a cross between two homozygous lines differing at a pair of linked loci, and secondly, how certain correlations between relatives are influenced by linkage.

Consider two loci of equal effect on a trait, with genotypic values for a metric trait thus:

	A_1A_1	A_1A_2	A_2A_2
B_1B_1	$2a$	$a+d$	0
B_1B_2	$a+d$	$2d$	$-a+d$
B_2B_2	0	$-a+d$	$-2a$

Now suppose that two homozygous lines of equal merit are crossed, thus:

$$\begin{array}{ccc} P_1 & & P_2 \\ \dfrac{A_1B_2}{A_1B_2} & \times & \dfrac{A_2B_1}{A_2B_1} \\ & F_1 & \\ & \dfrac{A_1B_2}{A_2B_1} & \end{array}$$

If the frequency of recombination between A and B is θ, the F_1 wil yield gametes:

Gamete	A_1B_1	A_1B_2	A_2B_1	A_2B_2
Frequency	$\frac{\theta}{2}$	$\frac{1-\theta}{2}$	$\frac{1-\theta}{2}$	$\frac{\theta}{2}$

Unless θ is very small indeed, all genotypes will be represented in the F_2, with mean

$$M = d$$

and genetical variance

$$V_G = 2\theta a^2 + \{1 - 2\theta(1-\theta)\}d^2$$
$$= a^2 + \tfrac{1}{2}d^2 \text{ if } \theta = \tfrac{1}{2}.$$

Thus, the F_2 mean is not affected by linkage, but the variance is, the magnitude of the effect depending upon the degree of dominance at the different loci. (The variance with $\theta = \frac{1}{2}$ also equals the variance with $\theta = (d^2 - 2a^2)/2d^2$, i.e. only when there is overdominance.)

The effects of linkage in such a cross will be maximized when most genes increasing the trait of interest are to be found in one of the parents. While this is generally unlikely *a priori* (Mather 1949), it will be the case when new genetical variation is to be introduced from an inferior into a superior strain (see El-Sayed Osman and Robertson 1968; Cox *et al.* 1984; section 10.6).

Cockerham (1956), using a method of Kempthorne's (1954), has considered the general effects of linkage on covariance between relatives, and reached a number of straightforward conclusions, applicable to diploid panmictic populations in linkage equilibrium. The covariance between relatives where one is a descendant of the other one is unaffected by linkage. By contrast, for relationships where one individual is not ancestral to the other, $\theta < \frac{1}{2}$ will increase the covariance, this increase affecting only the interaction (epistatic) components and not the additive and dominance components. Considering three loci, B_1, B_2, and

Table 4.6

Coefficients of components of genetical variance in the covariance between full sibs (θ = recombination fraction between two loci $\phi = 1 - 2\theta$; from Cockerham 1956)

Source	Coefficient		
	θ unspecified	$\theta = 0$	$\theta = \frac{1}{2}$
Additive (A)	$\frac{1}{2}$	$\frac{1}{2}$	$\frac{1}{2}$
Dominance (D)	$\frac{1}{4}$	$\frac{1}{4}$	$\frac{1}{4}$
$AA\,(A_iA_j)$	$\{1 + (1 + \phi_{ij})\}/8$	$\frac{3}{8}$	$\frac{1}{4}$
$AD\,(A_iD_j)$	$(1 + \phi_{ij})/8$	$\frac{1}{4}$	$\frac{1}{8}$
$DD\,(D_iD_j)$	$(1 + \phi_{ij})^2/16$	$\frac{1}{4}$	$\frac{1}{16}$
$AAA\,(A_iA_jA_k)$	$\{1 + (1 + \phi_{ij})(1 + \phi_{jk})\}/16$	$\frac{5}{16}$	$\frac{1}{8}$
$AAD\,(A_iA_jD_k)$	$\{(1 + \phi_{ij})(1 + \phi_{jk}) + (1 + \phi_{ik})(1 + \phi_{jk})\}/32$	$\frac{1}{4}$	$\frac{1}{16}$
$ADA\,(A_iD_jA_k)$	$(1 + \phi_{ij})(1 + \phi_{jk})/16$	$\frac{1}{4}$	$\frac{1}{16}$
$DAA\,(D_iA_jA_k)$	$\{(1 + \phi_{ij})(1 + \phi_{jk}) + (1 + \phi_{ij})(1 + \phi_{ik})\}/32$	$\frac{1}{4}$	$\frac{1}{16}$
$ADD\,(A_iD_jD_k)$	$(1 + \phi_{ij})(1 + \phi_{jk})^2/32$	$\frac{1}{4}$	$\frac{1}{32}$
$DAD\,(D_iA_jD_k)$	$(1 + \phi_{ij})(1 + \phi_{ik})(1 + \phi_{jk})/32$	$\frac{1}{4}$	$\frac{1}{32}$
$DDA\,(D_iD_jA_k)$	$(1 + \phi_{ij})^2(1 + \phi_{jk})/32$	$\frac{1}{4}$	$\frac{1}{32}$
$DDD\,(D_iD_jD_k)$	$(1 + \phi_{ij})^2(1 + \phi_{jk})^2/64$	$\frac{1}{4}$	$\frac{1}{64}$

B_3 with recombination between them of θ_{12} and θ_{23}, and setting $\phi = (1 - 2\theta)$, Cockerham derived the values shown in Table 4.6 for the components of the covariance between full sibs. The column headed $\theta = \frac{1}{2}$ is an extension of the result given in Section 4.4.1.

Linkage thus tends to increase similarities between progeny of the same parent or parents, as would be expected on account of the increased likelihood of particular chromosome segments being inherited by more than one member of a group of progeny. It will rarely be possible, however, to quantify the bias so engendered, since different problems afflict parent–offspring relationships, such as the problems of environmental control which will be the more necessary should there ben interaction between genotype and environment. Thus, if $h^2_{\text{OP}} < h^2_{\text{HS}}$, it will rarely be clear to what extent linkage has contributed to this inequality. Results such as those in Table 4.4 may not be used to draw inferences about linkage.

4.6 Qualitative methods

Wherever possible the quantitative geneticist must attempt to identify the particular gene loci which are responsible for quantitative variation. Polygenes are not enough; for further advances beyond selection by the methods in Chapter 7, knowledge of gene action at the physiological level will be necessary. The methods described in this section are preliminary attempts to dig deeper than those in Section 4.4.3. Although the methods in the next two sections therefore have the same aim, they are different in essence; in Section 4.6.1 they are purely genetical, in Section 4.6.2 cytogenetic.

4.6.1 Identification of specific gene loci

In the simplest case, correlation may be observed between levels of a metric trait and phenotypes at a known Mendelian locus, either in a crossing programme or in a sample of a population of an outbreeding plant. In the former case, if, say, two inbred lines have genotypes A_1A_1 and A_2A_2 at a particular locus, it may be that the F_2 will show a trimodal or bimodal distribution for the metric trait, and that one mode corresponds to (say) A_1A_1, the others to A_1A_2 and A_2A_2 in the trimodal case or to A_1A_2 and A_2A_2 jointly in the bimodal case. (A demonstrably trimodal distribution would reveal a very large value of *a*.) Such an outcome might reflect effects of the *A* locus or of another locus closely linked to it. In the second case, recombination should occur on selfing A_1A_2 for a sufficient number of generations. This is the basis of the location method of Thoday (1961), following Breese and Mather (1957). He considered the case of two tester loci, *A* and *B*, in the cross

$$\frac{AB}{ab} \times \frac{ab}{ab}$$

If there is one locus, *H*, affecting the trait of interest, and the selected AB strain is high for the trait, the ab tester low, then the gene order will be *HAB*, *AHB*, or *ABH*, each yielding a different pattern of segregation. In the back-cross progeny, only the relative proximity of *H* to *A* and *B* will be detectable, but another generation of crossing whichever of the HA and HB phenotypes are available back to the ab parent will clarify the linkage relationship quite precisely.

More loci than one affecting the metric trait will complicate the picture, but the principle remains the same. It has, however, been shown by McMillan and Robertson (1974) that the method has serious defects. As developed for *Drosophila melanogaster*, which with its short generations and small complement of excellently marked chromosomes is in some ways ideal for genetical manipulations, Thoday's method can both detect non-existent loci and magnify the estimated effects of actual major loci by attributing to them the contributions of undetected closely linked loci.

McMillan and Robertson present the following example to show how this can come about. Consider that the tester loci lie 100 cM apart, and that there are three loci between them, as in Fig. 4.4. In the analysis, scoring crossovers containing *b*, there will be four groups, with the

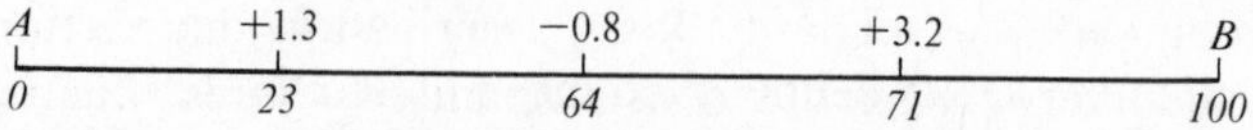

Fig. 4.4. A chromosome containing three genes affecting the metric trait, with effects as shown at distances (shown in italic) from locus *A*. (Redrawn by permission of the Editor, *Heredity, London*, from McMillan and Robertson 1974.)

expected relative frequencies and magnitudes shown in the first two columns of Table 4.7. As the gene at *64* has a negative effect, however, it will be inferred to be at *30*, and this affects all estimates, as shown in columns three and four of the table. In essence, such problems must arise if the tester is superior to the tested strain at some of the loci affecting the trait of interest, and inferior at others. It will generally be impossible to be certain of this, especially in organisms which (like most crop plants) are less well characterized genetically than *D. melanogaster*. McMillan and Robertson presented an investigation of the statistical properties of Thoday's test, but from what has been said already, it should be clear that the method is unlikely to be widely applicable in plant breeding.

Identification of specific gene loci in the manner of Thoday also involves mapping, but is purely genetical, whereas the methods in the next section are also cytogenetical. Where mapping by standard techniques (see for example Mather 1951) is difficult or impractical, a method may be sought which assumes no knowledge on the part of the experimenter of chromosomal (or linkage map) location of any genes. It is the 'inbred back-cross line' technique of Wehrhahn and Allard (1965). This uses a cross of two inbred parental lines to produce an F_1, so in practice will mainly be useful for self-fertilizing species. From the F_1, k successive back-crosses are made to one parent, after which the resulting plants (and their descendants) are selfed for one generation to ensure a

Table 4.7

Cross-overs containing *b* from a back-cross of the chromosome shown in Fig. 4.4 (from McMillan and Robertson 1974) (Reprinted with permission of the Editor, *Heredity, London*)

		Inferred loci	
Expected relative frequencies	Mean values	Position	Effect
23	0	*23*	0.5
41	1.3	*30*	0.8
7	0.5	*71*	2.4
29	3.7	—	—

sufficiently high level of homozygosity. The resultant set, multiplied up to allow field testing, should consist mainly of lines identical to the back-cross parent, together with lines differing at one gene locus, even if ten loci differ between the parents and only three back-crosses are used. From the distributions of the metric of interest the number and magnitude of genes involved will be obtainable. These 'genes' will be somewhere between Mather's effective factors and simple Mendelian genes, but factors isolated by Wehrhahn and Allard's procedure may be reasonably regarded as closer to the latter, though linkage will complicate matters.

In a cross of spring wheat varieties Ramona and Baart 46, Wehrhahn and Allard detected four factors, one of which accounted for 80 per cent of V_A for heading date, while the other three jointly determined about 14 per cent. This result may be compared with the analysis in Section 6.2.2 of heading date as a digenic trait: together they emphasize the value, wherever this is possible, of considering specific gene loci rather than statistical artifacts. It is perhaps only in assessing the limits to selection response that these factors have an application.

4.6.2 Identification of specific chromosomes

The methods of Thoday described above allow chromosomal location, where the markers have themselves already been assigned to chromosomes. Suitable markers will not always be available, especially in polyploids, where conventional linkage analysis may be difficult. In common wheat, *Triticum aestivum* L., an allopolyploid with three homoeologous genomes, A, B, and D (Kihara 1919; Sax 1922; McFadden and Sears 1944; Sears and Rodenhiser 1948; Sears 1952, 1954, 1958; Okamoto 1962; Riley 1965; Chapman and Riley 1966), the tolerance of plants to aneuploidy is such that monosomics and nullisomics can be used to map genes, a plan due to Sears (1953).

The basic method is to establish 21 lines in a variety, each monosomic for one of the 21 chromosomes of the complete complement. Then by crossing these lines to another variety and observing the F_2 segregation, one can locate qualitative genes, in particular, to individual chromosomes. Nullisomy may lead to a specific, identifiable phenotypic change, resulting from loss of a gene on the critical chromosome. Recessives should be manifest in the monosomic F_1, in which the dominant allele has been eliminated with the lost chromosome. Dominants should yield 20/21 3:1 segregations in the F_2s involving the non-critical monosomes. In the critical case, the frequency of F_2 segregants displaying the recessive phenotype is determined by the frequency of nullisomics, usually about 3 per cent of the selfed progeny of a monosomic.

Mapping may be taken further if qualitative traits are under investigation (see Section 4.6.2.1; Driscoll and Sears 1965), but for quantitative traits other methods may be better. One such uses an intervarietal chromosome substitution, where a single pair of chromosomes in one variety is replaced by the homologous pair from another variety (Sears 1953; Law 1966, 1967). Single chromosome heterozygotes are generated, and segregation is observed, as shown in Fig. 4.5.

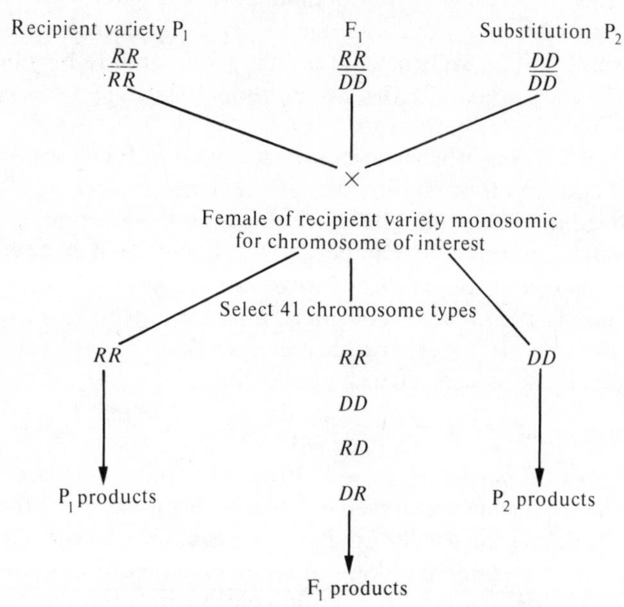

Fig. 4.5. Crosses for the detection of recombination in *Triticum aestivum* by means of inter-varietal chromosome substitutions (see text for interpretation). (Reproduced by permission of the Editor, *Genetics, Princeton*, modified from Law 1966.)

In this figure, *RR* and *DD* are idealized chromosomes in the recipient (P_1) and donor (P_2), respectively, while *RD* and *DR* are recombinant chromosomes. The parental products ($P_1 + P_2$) yield an estimate of the variance of the tested chromosomes in the absence of recombination, while the F_1 products yield an estimate of the variance in the presence of recombination. There are severe statistical difficulties in analysing the bimodal distributions obtained if a factor of large effect lies on the tested chromosome, but where these are surmountable the method may yield information about linked loci on the tested chromosome. If these problems cannot be overcome, then the segregation in the F_1 products still allows assignment of genes of large effect to the tested chromosome.

Table 4.8 shows a number of loci which have been assigned to the chromosomes of common wheat by the techniques described here, or developments of them. Very many more loci have been assigned to

Table 4.8
Factors mapped by cytogenetic techniques in common wheat

Chromosome	Trait	Reference
2B	height	Snape *et al.* (1977)
2B	development response (*Ppd* 2)	Law *et al.* (1983)
2D	development response (*Ppd* 1)	Law *et al.* (1983)
2M†	yellow rust resistance (*Yr*)	Riley *et al.* (1968)
4A	height/gibberellin insensitivity (*Gai/Rht* 1)	McVittie *et al.* (1978)
4D	height/gibberellin insensitivity (*Gai/Rht* 2)	McVittie *et al.* (1978)
5A	time to ear emergence (Vrn_1)	Law *et al.* (1976)
5B	homoeologous synapsis suppressor	Riley and Chapman (1958)
5D	time to ear emergence (Vrn_3)	Law *et al.* (1976)
6B	height	Snape *et al.* (1977)
7B	purple culm (*Pc*)	Law (1966, 1967)
	mildew resistance (*ml*)	
	time to ear emergence (e_1, e_2)	
	leaf rust resistance (*lr*)	
	height (*ht*)	
	grain weight (*gw*)	
	grain number (*gn*)	
	tiller number (*tl*)	

†Actually on chromosome 2 of the M genome, but transferred from *Aegilops comosa* Sibth. to a new breeder's variety of *T. aestivum*, Compair, by Riley *et al.* (1968).

chromosomes; Table 4.8 merely illustrates the range of characters dealt with, and also covers most of the genes whose mapping is described in the text. Snape *et al.* (1977) performed biometrical genetical analyses of the factors affecting height detected in their analysis, but in some senses this obscures the simplicity of the original analysis, which detects genes with effects large enough for them to be likely to be important.

4.6.2.1 Linkage analysis with heteromorphic chromosomes Because of the paucity of good markers in wheat, and the complications of analysis of three homoeologous genomes, classical linkage analyses (Mather 1936) have been adapted to use heteromorphic chromosomes as markers, for centromeric mapping. The back-cross method was introduced by Sears (1962) and the F_2 method by Driscoll (1966), and these have been compared in the manner of Mather (1951) by Wescott *et al.* (1978).

The crosses are set up thus:

Method	Female	Male
Backcross	$\frac{A_1}{A_2}$	× $\frac{A_1}{A_1}$
F_2	$\frac{A_1}{A_2}$	× $\frac{A_1}{A_2}$

For the back-cross, in the absence of dominance, if the recombination fraction is θ, this is estimated by $\hat{\theta}$, the proportion of recombinant types (i.e. telocentrics carrying A_1 and normal chromosomes carrying A_2 in the progeny), with variance $\hat{\theta}(1-\hat{\theta})/n$.

For the F_2, the situation is complicated by the fact that pollen bearing the telocentric effects fertilization in less than the 50 per cent of cases expected. Let the proportion be p, in which case, once more in the absence of dominance, the F_2 will be

	Number of telocentrics		
	0	1	1
A_1A_1	$\frac{1}{2}(1-p)(1-\theta)^2$	$\frac{1}{2}\theta(1-\theta)$	$\frac{1}{2}p\theta^2$
A_1A_2	$(1-p)\theta(1-\theta)$	$\frac{1}{2}(1-2\theta+2\theta^2)$	$p\theta(1-\theta)$
A_2A_2	$\frac{1}{2}(1-p)\theta^2$	$\frac{1}{2}\theta(1-\theta)$	$\frac{1}{2}p(1-\theta)^2$

By the method of maximum likelihood (Bailey 1961), estimates of p and θ are obtained, and the variance of $\hat{\theta}$ is $\hat{\theta}(1-\hat{\theta})(1-2\hat{\theta}+2\hat{\theta}^2)/\{2n(1-3\hat{\theta}+3\hat{\theta}^2)\}$. Thus, the ratio of the two variances is

$$\frac{V_{\hat{\theta}}(B)}{V_{\hat{\theta}}(F_2)}=\frac{2(1-3\hat{\theta}+3\hat{\theta}^2)}{1-2\hat{\theta}+2\hat{\theta}^2},$$

so that the F_2 variance is always less than the backcross variance.

If A_1 or A_2 is dominant, then F_2 estimation of θ may be less efficient than backcross estimation, but of course simultaneous estimation of p is also possible. If A_1 is dominant, the back-cross will be the better method, and crosses should be planned if possible to take account of this fact (Mather 1951).

These techniques are usable essentially only when the plants can tolerate a reasonable level of aneuploidy; in such cases, incorporation of alien genetical material is also more likely to be practicable (Chapter 14). It then becomes desirable to be able to map the alien genes. For example, Driscoll and Anderson (1967) (see also Driscoll and Bielig 1968) located the Transec wheat–rye translocation, which transferred to wheat resistance to wheat leaf rust (*Puccinia recondita* Erikss.) and powdery mildew (*Erysiphe graminis tritici* March.), on chromosome 4A by crosses with Chinese Spring monosomics. Specific location on the β arm required the use of telocentrics for the opposite arms of chromosome 4A (Sears 1954, 1962, 1966), in the manner described above.

These analyses may be greatly disturbed by desynapsis, since some chiasmata terminalize and are lost between diakinesis and metaphase I, and Driscoll (1978) has presented an analysis which takes account of this. However, following the success of Hart and others (see for example, Hart and Langston 1977) in mapping isozyme structural genes in polyploid wheat, more conventional mapping techniques will become more fruitful. This appears likely for hexaploid oats also (Morikawa 1978).

4.6.2.2 Other uses of aneuploidy The aneuploid approach may be used for other purposes, such as the investigation of epistasis in wheat (Snape *et al.* 1975).

The procedure of Snape *et al.* (1975) uses crosses of 21 single chromosome substitution lines of the variety Cappelle–Desprez (D) into the variety Chinese Spring (R). If the 21 substitution lines are termed S_i, $i = 1, \ldots n$, then the following relation holds in the absence of epistasis:

$$\text{mean}(S_i \times D) + \text{mean}(S_i \times R) - 2\,\text{mean}\{S_i \times (R \times D)\} = 0$$

for each *i*. The departure of the observed value from zero is readily tested, so that each chromosome's epistatic contribution may be detected. This laborious analysis is a type of triple test cross (Mather and Jinks 1971), and provides an estimate of the net effect of both the interactions of the substituted chromosomes with others of the complement and the interactions between the other chromosomes of the complement.

To obtain the interaction only of the substituted chromosome with the others, Law (1972) has suggested

$$\tfrac{1}{2}\{\text{mean}(S_i \times R) + \text{mean}(S_i \times D) - \text{mean}(S_i) - \text{mean}(R \times D)\}$$

as a test statistic. Then the difference between the two,

$$\tfrac{1}{2}[\text{mean}(S_i \times R) + \text{mean}(S_i \times D) + \text{mean}(S_i) + \text{mean}(R \times D) - 4\,\text{mean}\{S_i \times (R \times D)\}],$$

measures residual epistatic effects.

The results are very difficult to interpret. For example, for chromosome 6B, interactions among the background chromosomes had a significantly large negative effect on plant height, while the interaction between 6B and the background had a significantly large positive effect, so that the net effect was apparently not different from zero. On the other hand, in most cases the residual epistatic effect was not significantly different from zero, so that the residual cannot be regarded as having the same meaning in all cases.

Significant effects of one kind or another for chromosomes 1B, 1D, 5D, 6B, 6D, 7A, and 7B suggest that these chromosomes be searched for genes affecting plant height. To some extent, this is borne out by the later analysis of Snape *et al.* (1977), where major effects were associated with 5D, 6B, and 7A, as well as with several other chromosomes. It is arguable, however, that most understanding of the genetics of height in wheat has come from the investigation of major genes, especially the dwarfing/gibberellin-insensitivity genes of Norin-10 and Tom Thumb (McVittie *et al.* 1978). The recognition of the relationship between abnormal hormone metabolism and dwarfing and the simple test available for GA-insensitivity allow more rapid selection of appropriate genotypes. There is in some cases a negative association between hormone response and grain yield (Pugsley 1978), but there is no simple relationship. In rice, for example, some dwarfing genes block the production of all GAs, others only certain specific GAs (Suge 1978). Overall, however, selection based on GA-insensitivity can be useful in selection for higher yield (Boyd and Wade 1983).

4.7 Gene products as quantitative traits

In 1942, Fisher could write that, insofar as the genetical analysis was concerned, the number of genes affecting a quantitative trait 'is in fact one of the least influential features of the systems investigated'. However, as will be brought out in Chapter 7, both the limits to selection and the rate of response to selection can depend strongly on the number of genes involved. In general, response will be most rapid and the limit most readily reached the fewer the factors involved. It might therefore be held that where possible selection should be practised upon individual loci (but see the discussion of disease resistance in Section 6.2.1). This will be practicable for qualitative traits, but what are its implications for quantitative traits?

First, given the level of knowledge about the primary gene products, this will usually be impossible, an argument for further biochemical investigation of the determination of quantitative traits.

Secondly, where there is substantial genetical variation in a primary gene product, this may be the result of the action of genes at many other loci. To take an example from another field of applied genetics, the heritability of the proportion of the human variant haemoglobin S {HbS/(HbS + HbA)} in heterozygotes for this trait is 0.88 ± 0.20, as measured by regression of offspring on parent (Nance and Grove 1972). The underlying variability in HbS proportion is substantial, the proportion varying from 27 to 52 per cent, compared with 50 per cent to be expected if each allele is equally active. Thus, as is not surprising, identification of a specific gene locus by observation of variation in its product has not led to a complete elucidation of the genetics of the trait.

Thirdly, it may be of some importance to investigate the results of application of quantitative genetics to individual, clearly defined gene products, in order to test their efficacy and reliability. Extensive tests of quantitative genetical theory applied to model quantitative and other characteristics have been conducted in *Drosophila* (see Falconer 1982; Rendel 1967). In general, the theory is relatively satisfactory (problems regarding prediction of response to selection are discussed in Section 7.4.2), but some of the findings are not relevant to plant breeding, and in addition it is not clear to what extent artifacts may be created by some of the analyses, as in the case of mapping polygenes discussed in section 4.6.1.

In an extensive series of papers (Seyffert 1966, 1971; Jana and Seyffert 1971, 1972; Forkmann and Seyffert 1972, 1977; Seyffert and Forkmann 1976; Forkmann 1977), Seyffert and his colleagues have attempted to treat by quantitative genetical analysis the concentration of anthocyanins by weight in fresh petals of *Matthiola incana.* In a genetical background nearly isogenic for other factors affecting floral pigments, they investigated the effects of four known loci which influence anthocyanin pigment. Their aim was simply to use loci with known qualitative interaction, so that the effects of a small number of diallelic loci would lead to an effectively continuous distribution of anthocyanin pigment phenotypes which would always have known genotypes.

In all cases, known dominance at the level of the single gene could be detected appropriately, as could interaction between the loci, so that overdominance was found to be related to the joint effects of different loci. Double heterozygotes showed higher levels of pigment than the relevant double homozygotes values in nine of twelve cases, and single heterozygotes also were superior in several cases to the appropriate homozygotes. *AD* and *DD* interactions were the main types contributing to the heterosis, particularly the former. Such a precise conclusion is rarely possible with a quantitative trait of unknown determination.

A diallel analysis of four parents each homozygous for a different set of alleles allowed the unambiguous evaluation of some aspects of the (V_i, W_i) graphical method of Hayman (1954) which has been very widely used. It was shown that it is not satisfactory in detecting non-allelic interaction where this is generated by effectively duplicate loci, especially in the presence of strong dominance, a previously undetected problem.

There are still problems with methods, described most recently by Forkmann (1977), even for a trait of such apparent simplicity, but nonetheless this work appears a promising beginning to the careful investigation of some of the assumptions of quantitative genetics as applied to plant breeding. Despite all the work using the methods of quantitative genetics, many of these assumptions remain of doubtful validity. One such is additivity, as noted already, and an illuminating example of Gilbert's (1961) will serve to end this discussion. Consider three models of gene action: additive (A), asymptotic (B), and multiplicative (C), and suppose that eight loci with complete dominance therefore affect the trait of interest thus:

	Number of loci heterozygous or homozygous for the dominant allele								
	0	1	2	3	4	5	6	7	8
A	100	101	102	103	104	105	106	107	108
B	100	102.1	103.8	105.1	106.1	106.9	107.4	107.8	108
C	100	100.97	101.94	102.93	103.92	104.93	105.94	106.97	108

Now if a cross of P_1, having no dominants homozygous, and P_2, having eight dominants homozygous, be performed, the following data and estimates will be obtained:

	A		B		C	
	Mean	Variance	Mean	Variance	Mean	Variance
P_1	100	0	100	0	100	0
P_2	108	0	108	0	108	0
F_1	104	0	108	0	108	0
B_1	106	2	105.8	1.816	103.93	2.000
F_2	106	1.5	107.3	0.451	105.95	1.553
B_2	108	0	108	0	108	0
V_A	1		−0.915		1.105	
V_D	$\frac{1}{2}$		1.365		0.448	
No. effective factors	8		−8.7		7.2	

Case A fulfils the assumptions of Sections 4.3.1 and 4.4.3 completely, and yields correct estimates. Case B is pathological and would be readily detected, while case C, which at the level of the gene product is very different from case A, yields reasonable estimates of V_A, V_D and number of effective factors. Cases A and C would be indistinguishable given a modest level of V_E.

As Gilbert has noted, this simple, not unrealistic example does not inspire confidence in our ability, given methods based on linear models, to detect certain kinds of multiplicative gene action. (Transformation is not satisfactory where neither an additive nor a multiplicative model completely describes the data, given that one seeks to make genetical inferences.)

5

Variation

5.1 The extent of genetical variation

Very few traits in living organisms are invariant. Even those which define the distinction between closely related species may vary. (See Mayr (1963) for an extensive discussion of this problem as it relates to animal speciation. In plants, it is of more concern to taxonomists than to geneticists, since the barriers to hybridization between closely related taxons are generally less severe. Specific categorization is almost impossible in groups such as *Rubus* and *Taraxacum*.) While discrete variation in traits, such as the various floral forms in heterostylic self-incompatibility, may be more obvious on inspection, quantitative traits are essentially variable, and we may in any particular species always expect to find variation in traits of interest to the plant breeder. The problems will be, first, to determine the contributions of different factors to that variation and, secondly, to constrain those factors to optimize the level of the trait of interest: the topic of the bulk of this book. Here, we are concerned merely with the extent of genetical variation, as observed within individual populations. The search for additional variation will be discussed elsewhere (see especially Chapter 11).

5.1.1 Outbreeding organisms

If we consider first Mendelian gene loci, then by far the bulk of available data on variation comes from animal populations. From Table 5.1, it can be seen that very substantial numbers of all structural gene loci must be polymorphic. Evidence about regulatory gene variation in higher eukaryotes is still extremely scanty; what is available is rarely completely satisfactory. For example, the study of the regulation of zein synthesis in maize of Soave *et al.* (1978) was concerned to establish the linkage relationship between *Opaque-2* a regulatory gene, and five structural genes. Because these structural gene loci are located at three or more distinct sites, the authors were unable to draw any inferences about the nature of the regulatory process except the obvious one, that the system they were observing was not the operon model of prokaryotic gene regulation. Indeed, it could not even be concluded that the processes

Table 5.1

Isozyme variation in natural populations of animals (from Powell 1975). H_e is averaged over loci

Class	Number of species	Heterozygosity, H_e, ± standard error
Drosophila	38	0.157 ± 0.007
Other insects	7	0.170 ± 0.027
Other invertebrates	13	0.102 ± 0.021
Total invertebrates	58	0.146 ± 0.009
Fishes	31	0.058 ± 0.006
Amphibians	3	0.105 ± 0.016
Reptiles	9	0.043 ± 0.008
Birds	3	0.031 ± 0.010
Mammals	25	0.039 ± 0.006
Total vertebrates	71	0.050 ± 0.004

observed were regulatory in the same sense as the Jacob–Monod process. Gene regulation is discussed more extensively in Chapter 11.

An equally large gap in our knowledge concerns the relationship between Mendelian gene loci and quantitative traits of agronomic importance. Studies of association have been performed in some cases (e.g. in the slender wild oat, *Avena fatua*, by Hamrick and Allard 1975), but the results are not clear for the most part. This topic will be discussed further in Chapter 9; here we may merely note that where genetical factors are the main determinant of some trait in a given environment, it is to be expected that the Mendelian genes involved will eventually be identified. What we seek in assaying the extent of observed variation for a trait is an idea of how amenable to change by artificial selection that trait will be, how it will respond to changes in the environment, or how easily it may be modified by combinations of genes from different strains.

We do in fact know a great deal about the genetical contributions to many traits of economic importance in almost all important outbreeding crop plants. Table 5.2 illustrates the proportion of phenotypic variance attributable to genetical influences in the cultivated potato, as estimated from diallel analysis (Section 4.4.2). This plant is considered to have a restricted gene pool as compared with many other major crops (Simmonds 1976b). Nonetheless, on the basis of the three trials represented in Table 5.2, several traits of economic importance display substantial potential for selection. Very low heritability for traits such as yield (e.g. $h^2 = 0.003$ for grain yield for winged bean in New Guinea; Khan and Erskine 1978), on the other hand, may still be compatible with

Table 5.2
Heritability, h^2, or seven traits in potato, *Solanum tuberosum* ssp. *tuberosum* (from Tai 1976; Kamiński 1977; Killick 1977)

	h^2		
Trait	Tai	Kamiński	Killick
Total yield	0.26	0.34	0.01
Marketable yield	0.37	—	—
Total number of tubers	0.25	0.28	0.00–0.10‡
Number of marketable tubers	0.31	—	—
Average tuber weight	0.87†	0.29	0.07†
Average weight of marketable tubers	0.88†	—	—
Specific gravity	0.70†	—	0.00–0.05‡

† Significantly greater than zero at the 5 per cent level or lower.
‡Heterogeneous estimates from different replicates.

progress under selection. Simple truncation selection (Section 7.3.2) will be ineffective, but other methods such as family selection (Sections 7.5 and 10.2) may work instead.

5.1.2 Inbreeding organisms

Here, we may not argue by analogy from a large body of genetical information about Mendelian variation in animal populations, but must adopt a different approach. First of all, how much variation is to be found in simple diploids which reproduce largely by self-fertilization? The effect of selfing at a single locus is to halve the frequency of heterozygotes in each generation, no matter how many alleles are present in the population. Since the maximum value for heterozygosity is $(k-1)/k$ for a k-allelic locus with all alleles equally frequent, the level of heterozygosity on complete inbreeding will be generally low, and will be maintained by mutation, a modest level of out-crossing and perhaps by heterozygous advantage in a very few cases.

Table 5.3 illustrates levels of heterozygosity in a variety of species with different breeding systems, mostly outbred. Brown (1978a), summarizing a study of 30 species, noted that outbreeding species tended to show less heterozygosity than expected under panmixia, and inbreeding species more than expected on the basis of structured population inbreeding (Wright 1965; Allard *et al.* 1968). Perhaps the most extensive investigations of genetical variation have been carried out by Allard and his colleagues in inbred species. For example, Hamrick and Allard (1972) presented extensive data on five isozyme loci in *Avena barbata.* They observed polymorphic gene frequences at six of seven localities sampled for all loci, and estimated the frequency of self-fertilization at 0.97 to

Table 5.3
Heterozygosity, H_e, in some natural plant populations

Species	H_e	Attributes	Reference
Phlox drummondii Hook	0.04 } †	Self-incompatible	Levin (1975a)
P. cuspidata Schaele	0.01 }	Self-compatible (facultative inbreeder)	Levin (1975a)
Stephanomesia exigua ssp. *carotifera*	0.09	Outbreeder	Gottlieb (1975)
Lupinus subcarnosus	0.36 } †	Edaphically restricted	Babbel and Selander (1974)
L. texensis	0.10 }	Edaphically restricted but less than *L. subcarnosus*	Babbel and Selander (1974)
Hymenopappus scabiosaeus var. *carymbosus*	0.20	Not edaphically restricted	Babbel and Selander (1974)
H. artemisiaefolius var. *artemisiaefolius*	0.21	Not edaphically restricted but different range from *H. scabiosaeus*	Babbel and Selander (1974)
Conocephalum conicum	0.17‡	Haploid (a liverwort)	Yamazaki (1984)

†Significantly different at 5 per cent level.
‡Equivalent heterozygosity = 1-diploid expected homozygosity.

0.98. As we shall see, this level of inbreeding is such as to eliminate polymorphism unless heterozygotes are at a very great advantage; more precise estimates of the degree of outcrossing are needed before these remarkable findings can be accepted; see Section 8.1. Their implication, however, is clear; for a substantial number of loci in a normally inbred species, heterozygous advantage is even stronger than in the most carefully analysed animal examples.

Genetical variation in quantitative traits of economic importance in crop plants is also very substantial, largely between strains in inbreds. Table 5.4 shows the proportion of measured variability attributable to

Table 5.4
Heritability, h^2, for five traits in oats, *Avena sativa* (from Sampson and Tarumoto 1976)

Trait	h^2†
Days to heading	0.86
Plant height	0.90
Stem diameter	0.79
Thousand-grain weight	0.79
Yield m^{-2}	0.70

†All values significantly greater than zero at 5 per cent level or lower.

genetical influences in oats. Like each of the three sets of results in Table 5.2, these were obtained from one particular experiment only (in this case, an eight-parent partial diallel cross); interaction between genotype and environment, one of the greatest problems and opportunities in plant breeding, will lower heritability, but increase the usable variability in certain circumstances.

5.2 Maintenance

5.2.1 Outbreeding organisms

For about 80 years, geneticists have recognized that in an indefinitely large panmictic population gene and genotype frequencies at any particular locus would remain constant in the absence of mutation, migration, and selection. Genotype frequencies in natural populations are frequently found to agree well with these Hardy–Weinberg equilibrial expectations. For example, Levin (1975a), examining two species of *Phlox* found that the self-incompatible species *P. drummondii* Hook had two of four loci in such agreement, while even the substantially self-fertilized *P. cuspidata* Scheele had one of seven in Hardy–Weinberg equilibrium. However, population size, mutation, and selection influence the gene pool of all populations, even large isolated ones. We shall first consider simple influences on gene frequencies and then go on to consider the interactions between the various factors, leading to a treatment of the maintenance of variability for complex traits.

5.2.1.1 Single locus In a very large population, mutation at a rate u from existing alleles A_1, A_2 ... to new ones A_x, A_y ... will simply lead to an indefinite proliferation of new types, should these be adaptively equivalent, limited only by the number of variants possible. This limit is rarely expected to be a powerful constraint, since even a monomeric gene product will generally consist of a hundred or more amino-acid residues, any one of which may be changed to one of the other 19 possible. However, consideration of the properties of altered polypeptides shows that while a considerable proportion of all possible mutant forms may be functionally equivalent, the others will generally be disadvantageous (Perutz and Lehmann 1968; Harris 1975). Consider any given class of disadvantageous allele, having properties as follows:

	Existing homozygotes	Heterozygotes	Mutant homozygotes
	A_1A_1	A_1A_x	A_xA_x
Viability	1	$1-hs$	$1-s$
Effect on arbitary metric	a	d	$-a$

The equilibrial frequency q for A_x is given approximately by u/hs. Thus, a very small proportion of all alleles detectable will be deleterious. In such a case, the total variance in the arbitrary metric attributable to deleterious alleles would be approximately $2u(a-d)^2/hs$ if measurements were taken before the viability differences were manifest or slightly less if afterwards. Such contributions to the variance will be very modest, relative to those of alleles at higher frequencies, for given allelic effects.

Alleles at higher frequency (polymorphic alleles) will in general have very similar viabilities, if maintained solely by mutation, or will show heterozygous advantage. In the former case, mutation rate u and population size N_e will interact to maintain an effective number of between $\sqrt{(1+8N_e u)}$ and $1+4N_e u$ alleles, depending on the form of mutation to new alleles (Johnson 1974; Ohta 1975). If we consider the simplest case, i.e. two alleles, A_1 and A_x, or two allelic classes, as above but with gene frequencies p and q of the same order [i.e. $p = 0(q)$], it was shown in Section 4.3.1. that, for the arbitrary genotypic effects shown above, the genetical variance for the metric trait is:

$$2pq\{a+d(q-p)\}^2+(2pqd)^2.$$

This is many orders of magnitude greater than in the case where variation is maintained by mutation to deleterious alleles.

In the third case, where variation is maintained by heterotic natural selection, consider this model

	A_1A_1	A_1A_x	A_xA_x
Viability	$1-s_1$	1	$1-s_2$
Metric	a	d	$-a$

The variance here will be as above, if measurement is before viability differences take their effect, or less if made afterward.

Thus, the overall effect of selection at a single locus, in the simplest cases, is to lower the variance in a metric trait affected by that locus. If viability is uncorrelated with the metric, on average, the range of variation will not be so affected.

5.2.1.2 Multiple loci The effects of mutation, selection, and population size on the level of variability determined by many loci have generally been considered in the context of stabilizing selection, since the tendency of natural selection is to minimize variation. Models have so far been relatively simple, since even these tend to be complex to analyse.

We begin by considering the basic additive model, i.e.

Genotype	A_1A_1	A_1A_2	A_2A_2
Metric value	a	0	$-a$

Following Latter (1960), we let fitness for a value of the arbitrary metric x be $1 - e^{-(x^2/2V_f)}$, i.e. an exponential decline occurs in fitness as the metric moves from the optimal value, chosen by scaling to be zero. V_f may be regarded as a measure of the tolerance of the system (organism plus environment) to departures from the optimum, which is the mean. The intensity of natural selection, I, is given by $\frac{1}{2}\ln(1 + V_p/V_F)$ where V_P is the phenotypic variance in the trait.

The change in gene frequency is then

$$\Delta q = \frac{4a^2pq(q-p)}{8V} - u(q-p)$$

where $V = V_f + V_P$ and u is the rate of mutation. At equilibrium, $\Delta q = 0$ and the additive genetic variance in the trait, V_A, which in general is $2a^2pq$, becomes $4uV$. If there are n such loci, the total variance is $4n\bar{u}V$, $\bar{u}$ being the mean mutation rate over all n loci. The proportion of the total variation which is genetical, i.e. the heritability, h^2, is now given by

$$h^2 = 4n\bar{u}(1 + V_f/V_P)$$
$$\cong 2n\bar{u}/I.$$

If $u \cong 10^{-5}$ and $n \cong 100$, then $h^2 \cong 0.5$ requires $I \cong 4 \times 10^{-3}$, i.e. a very low intensity of selection at any one locus.

Now impose artificial directional selection (Chapter 7):

$$\Delta q = \{i2a/(2\surd V_p)\}pq$$

where i is the intensity of selection. From the equilibrial value above, $pq = 2uV/a^2$, so

$$\Delta q = 2iuV/(a\surd V_P).$$

From this, we can see that the change in gene frequency depends as much on mutation rate as on intensity of selection.

Mean fitness at equilibrium was $\surd(V_f/V)$. Selection which changes the population mean by x_0 reduces this to $\surd(V_f/V)e^{-(x^2/2V)}$. In any particular case, then, the conflict between natural and artificial selection will depend critically on the unknown relationship between V_f and V_P. However, in most cases x_0 can be quite substantial before deleterious affects will become important. This assumes that selection begins at the natural equilibrium, whereas if the population mean is already significantly displaced, effects may be more marked. However, the environment will also not be constant, and many traits deleterious in an ancestral species are advantageous in a species where they are desirable to a breeder.

This analysis has neglected the effects of population size and linkage between loci affecting the trait of interest. Latter (1970) has extended it to incorporate population size, for a single locus. Consider a population where the mean value of the metric x is μ (and it is normally distributed about this value with phenotypic variance V_P as before). After selection, the mean value of the metric becomes $\mu(V_f/V)$ and the variance $(V_P V_f/V)$. Using $C = V_P/V$, the coefficient of centripetal selection, since it represents the reduction in both the variance and the deviation from optimum of the mean, Latter (1970) showed, summing over all loci, $\Delta\mu = -\mu Ch^2$, while for a single multiallelic locus,

$$\Delta V_g = -V_g[\tfrac{1}{2}CV_g/V_P]$$

where V_g is the single locus genetic variance for x and h^2 its multilocus heritability. (V_g is assumed small relative to V_P.)

It is now possible to incorporate the effect of population size N, knowing that genetical variance declines by a proportion $1/(2N)$ per generation:

$$\Delta V_g = -V_g[\tfrac{1}{2}CV_g/V_P + 1/(2N)].$$

Over time t, NCg^2 is the main determinant of the rate of decline in V_g from its initial value $g^2 V_P$, since for normally distributed allelic effects,

$$V_g \cong \frac{(g^2 V_P)e^{-t/2N}}{\{1 + NCg^2(1 - e^{-t/2N})\}}.$$

Allowing for mutation,

$$\Delta V_g = -V_g\{\tfrac{1}{2}C(V_g/V_P) + 1/(2N)\} + 2\mu V_u$$

where u is the rate of mutation to new alleles with effects normally distributed about a mean of zero with variance V_u. At equilibrium,

$$\hat{V}_g = \frac{2V_u}{4NC^*}[\surd\{1 + (4NC^*)(4Nu)\} - 1]$$

where

$$C^* = C(V_u/V_P).$$

For small C^*,

$$\hat{V}_g = 4NuV_u.$$

From these results, and developments of them by Bulmer (1972, 1980), it can be argued that mutation pressure alone will not maintain high levels of additive genetic variance.

Unfortunately, it has not yet been possible to incorporate both multiple loci and finite population size into one formal model, but the results so far obtained confirm the critical value of maximizing population size where possible. Lande (1976) and Turelli (1984), however, have considered the effect of genetical linkage on the models discussed above, for the case of infinite population size.

Lande summarized the combined effects of stabilizing selection, mutation, and recombination as follows. Stabilizing selection conceals variability and depletes the expressed component of variance. Mutation increases the expressed component without affecting the hidden component. Recombination changes hidden variability to expressed variability by lowering the correlation between allelic effects at linked loci.

To examine these general verbal conclusions more closely, one needs to consider the covariance of allelic effects, Cov_{ij}, between locus i and locus j of n, with recombination fraction θ_{ij} between loci. The effect of an allele at locus i is x_i. The variance of mutational changes at a rate u_i is V_{u_i}. Then mutation changes Cov_{ij} by an amount $\delta_{ij} u_i V_{u_i}$ per generation, where $\delta_{ij} = 1$ if and only if $i = j$, $\delta_{ij} = 0$ otherwise, i.e. mutation alters variances, but not covariances.

From these assumptions,

$$\Delta\mu = \mu V_g / (V_f + V_E + V_G)$$

analogous to Latter's expressions for $\Delta\mu$ above. ($V_E = V_P - V_G$ is the environmental variance for the trait, and V_G is here the expressed genetical variance taking the effect of linkage into account.)

At equilibrium, the following results hold, approximately,

$$\hat{\mathrm{Cov}}_{ii} = \surd(u_i V_{u_i}) \left[\surd(V_f + V_E + \hat{V}_G) + \sum_{\substack{i=1 \\ j \neq i}}^{n} \frac{\surd(u_i V_{u_i})}{\theta_{ij}} \right]$$

$$\hat{\mathrm{Cov}}_{ij} = \surd(u_i V_{u_i} u_j V_{u_j}) / \theta_{ij}, \qquad i \neq j$$

$$\hat{V}_G = \left[2 \sum_{i=1}^{n} \surd(u_i V_{u_i}) \right] \sqrt{\left[V_f + V_E + \left\{ \sum_{i=1}^{n} \surd(u_i V_{u_i}) \right\}^2 \right]}$$

$$\hat{\rho}_{ij} = \hat{\mathrm{Cov}}_{ij} \Big/ \sqrt{\left\{ \left(\hat{R}_i + \sum_{\substack{j=1 \\ j \neq i}}^{n} |\hat{\mathrm{Cov}}_{ij}| \right) \left(\hat{R}_j + \sum_{\substack{i=1 \\ j \neq i}}^{n} |\hat{\mathrm{Cov}}_{ij}| \right) \right\}}$$

$$2\hat{R}_i = \surd\{u_i V_{u_i} (V_f + V_E + \hat{V}_G)\}$$

where ρ_{ij} is the correlation between the effects of alleles at different loci and $2\hat{R}_i$ is the total expressed variance attributable to locus i.

Lande's results have extremely important implications. The mean effect of alleles at any given locus depends largely on historical or chance effects, although there are overall equilibria for the variance–covariance structure and for the population genotypic and phenotypic means. Hence, drift will be important in determining gene frequencies at individual loci. This contrasts with the results of Ewens and Thomson (1977) for genes with direct effects on fitness, rather than through their contributions to a metric trait stabilized by natural selection. In the direct case, a multilocus equilibrium is to be expected, with each individual gene also at equilibrium. There are also important differences between the results just presented, and Franklin and Lewontin's (1970) result that, for n heterotic loci, the mean fitness of the population depends upon the arrangement of the loci relative to each other, whereas the linkage arrangement is not similarly critical on Lande's model. The form of selection determines the linkage effects.

Lande's number of loci is a critical variable. If there are n_u equally mutable genes, and $V_u = \sum u_i V_{u_i}$, i.e. $V_{u_i} = V_u/2n_u$ in this special case, then

$$\hat{V}_G = \sqrt{\{2n_u V_u(V_f + V_E + n_u V_u/2)\}} + n_u V_u.$$

Thus, the more genes influence the trait, the more expressed variability is maintained (cf. Waddington and Lewontin 1968). Estimation of the effective number of genes determining a trait may then be of some importance (Section 4.4.3). Mutation rates per individual gene are very low: Kahler *et al.* (1984) estimated the rates both for genes affecting morphological traits and for genes affecting isozyme mobility under electrophoresis to be less than 10^{-5}, perhaps less than 10^{-6} per gene per generation. In accord with these limits is the rate per gamete per generation of 0.064 obtained in highly inbred maize by Russell *et al.* (1963). As Lande showed, a minimal estimate of V_u/V_E may then be derived as approximately 10^{-3}. Then, if as few as ten loci affect a trait, the mutational contribution to the variance will be at least one per cent, and probably higher. Turelli (1984) has concluded, in contradistinction, that the amount of variation which can be maintained by mutation pressure is very much smaller, thus agreeing with Latter's and Bulmer's results. Bürger (1986) has provided a more exact argument in support of this conclusion. Turelli has identified the discrepancy as arising largely from the assumption that the single gene equilibrium genetic variance, $\hat{V}_{G_i}$, be much larger than the expected variance of mutation effects at the locus, m_i^2, i.e. $m_i^2 \ll V_{G_i}$ for all i, together with related assumptions. Turelli presents evidence that in most cases $m_i^2 > \hat{V}_{G_i}$ and argues that much more data are needed to resolve the problem completely.

An important problem thus remains unresolved. Alternative explanations based on individual single gene heterozygous advantage (e.g. Gillespie 1984) are not in agreement with extensive data on inbreeding depression (see Chapter 9, and Sved *et al.* 1968). Thus, the maintenance of variability is not well explained as yet.

5.2.2 Inbreeding organisms

Because of non-random seed dispersal about parent plants, and other related effects, even self-incompatible species may be expected to have substantial levels of inbreeding, through sib-crossing in annuals and perennials, and parent–offspring crossing in perennials. Thus, the important contrast between inbreeding and outbreeding species is between those which customarily fertilize themselves, and those which do not. Table 5.3 illustrates the level of heterozygosity in a pair of related species, one inbred, one outbred. Here, we consider the maintenance of heterozygosity on selfing.

It is well known that in a diploid organism the level of heterozygosity at a locus halves every generation on selfing. This can be generalized for polysomic inheritance. In selfing in k-somics, heterozygosity declines as

$$H_e = \frac{2k-3}{2k-2} H'_e$$

where the prime denotes the previous generation. Thus, the approach to homozygosity is substantially slower in polysomic than in disomic organisms. Linkage will also slow the approach to homozygosity. The decline in heterozygosity for a set of loci A^1, A^2, A^3 is as follows:

$$H_{e_1} = \tfrac{1}{2} H'_{e_1}$$
$$H_{e_{12}} = (\tfrac{1}{2})^2 \{1 + (1 - 2\theta_{12})^2\} H'_{e_{12}}$$
$$H_{e_{123}} = (\tfrac{1}{2})^3 \{1 + (1 - 2\theta_{12})^2 + (1 - 2\theta_{13})^2 + (1 - 2\theta_{23})^2\} H'_{e_{123}}$$

where the subscript indicates the locus or loci concerned (cf. Narain 1965, who has extended these results to more loci than three). If recombination is less than 0.5, multiple heterozygosity is greater than in the absence of linkage.

If selfing is incomplete, but occurs to the extent f, with $1-f$ random outcrossing,

$$H_e = \left[1 + \frac{f}{2} - \frac{1}{N}\right] H'_e + \left[\frac{1}{2N} - \frac{f}{2}\right] H''_e \quad \text{(Wright 1969).}$$

Thus, a modest amount of outcrossing will also significantly retard the approach to fixation. However, as noted earlier, $1-f$ is not readily measured with any precision when close to unity (see Brown and Allard 1970). Furthermore, under mixed selfing and random mating, multilocus as well as single locus heterozygosity tends to be concentrated in a few individuals, not dispersed uniformly through the population (Bennett and Binet 1956; Weir and Cockerham 1973) so that even precise knowledge of f may not help in predicting the genotypic composition of a sample from such a population.

The effect of selfing on genetical variance for a metric trait is to increase the variance between lines and to lower that within lines. For example, with the simple model

	A_1A_1	A_1A_2	A_2A_2
Frequency under random mating (R)	p^2	$2pq$	q^2
Frequency on selfing (I)	p	0	q
Metric value	a	0	$-a$

we have the following genetical variances

$$V_G(R) = 2pqa^2$$
$$V_G(I) = 4pqa^2$$

In general, the additive genetical variance, of which this is the simplest case, will be doubled by complete selfing but all the variation will be between lines (Falconer 1982).

Deleterious mutants may be expected to be vary rapidly lost from selfing species. Maintenance of genetical variation within a population will critically depend on the existence and level of heterozygous advantage and on competitive advantage, very probably frequency-dependent.

Hayman and Mather (1953, 1956; also Hayman, 1953; Page and Hayman, 1960), and Haldane (1956) have considered the effects of heterozygous advantage in maintaining variation upon selfing. For a diallelic locus, consider the simple model:

	A_1A_1	A_1A_2	A_2A_2
fertility	$1-s_1$	1	$1-s_2$
viability	$1-t_1$	1	$1-t_2$
productivity = fertility × viability	$1-v_1$	1	$1-v_2$

This equilibrium requires that $v_1, v_2 \neq 1$; $s_1, s_2 > \frac{1}{2}$. If $s_2 > s_1$, A_2 will be lost, and vice versa for A_1. If $v_1 = v_2 < \frac{1}{2}$, a population starting as A_1A_2 entirely (like an F_1) will tend to A_1A_1 and A_2A_2 in the proportion $(1-t_1)$ to $(1-t_2)$, but if $v_1 \neq v_2$, the less fit homozygote will disappear.

If one simply considers selection through viability differences, and supposes that $t_1 = t_2 = t$ and that there is a proportion of selfing as before, then there is an equlibrium level of heterozygosity given by

$$H_e(1-t)(1+f-fH_e) = (1-H_e)(1-f+fH_e).$$

Suppose, for example, that $f = 0.3$ and $t = 0.01$. Then $H_e = 0.42$, i.e. with 30 per cent selfing the equilibrial proportion of heterozygotes is only about 80 per cent of its panmictic value. Such polymorphisms seem likely to be very rare when selfing is almost complete.

These theoretical results, which have not yet been extended satisfactorily to cover the joint effects of mixed selfing and random mating with selection, linkage, mutation, and finite population size, make Allard's findings, mentioned earlier, of very asymmetrical selective forces all the more striking.

In populations of plants which individually are relatively homozygous, but have different genotypes, like stands of mainly inbreeding cereals, it has long been recognized that certain genotypes will be poor competitors, and will disappear, while others will come to dominate (Papadakis 1937b). More recently, it has been shown that individual performance is not absolute, but may be frequency-dependent. In one study on the performance of two F_1 wheat hybrids (Warimek × Halberd and Wariquam × Halberd) grown at varying frequency within stands of the common parental variety, Phung and Rathjen (1976) found a linear decline of about 1 per cent in the hybrid's grain yield per plant for each 1 per cent increase in frequency over the range 4–50 per cent. The decline over the whole range was of the order of 35–40 per cent. Other results have been even more extreme (see, e.g., Khalifa and Qualset 1974).

Many models of frequency-dependent selection have been introduced. It is not clear which is more appropriate to the wheat-breeding results just mentioned, but many of the important aspects of such systems are exemplified by the following simple case. Consider a diallelic locus where fitness depends linearly on gene frequency:

	A_1A_1	A_1A_2	A_2A_2
fitness	$1 - s_1 p$	$1 - \frac{1}{2}(s_1 p + s_2 p)$	$1 - s_2 p$

(Here the frequency of A_1 is $p = 1 - q$.) Each homozygote is at a relative advantage when it is rare, and of course at a disadvantage when frequent. The equilibrium gene frequency is given by $\hat{p} = s_2/(s_1 + s_2)$. This is the same equilibrium as in the case of fixed selective values for the three genotypes in the proportion $1 - s_1$:1:$1 - s_2$. In a random mating population the equilibrium will be stable. In a selfing population, at equilibrium the fitness of the two remaining (homozygous) genotypes will be equal (at $1 - s_1 s_2/(s_1 + s_2)$), but slight departures from the equilibrial frequency through sampling will make them unequal, so that in general the persistence of such variation may be only slightly greater than in the case of fixed selective values considered earlier.

5.3 Implications for breeding methods and results

The most obvious conclusion which may be drawn from this brief sketch of the extent and maintenance of variation is that any sexually reproducing plant population, even a relatively small one, will, unless it has been maintained by strict self-fertilization, possess a significant store of genetical variation for most quantitative traits. Even if there is strict selfing, mutation may still provide a usable source of variation, though for some traits the breeder's net must be cast very wide indeed. The history of plant breeding, of course, makes this conclusion a truism, and there is evidently little theoretical reason to expect the extraordinary degree of difficulty in making progress which is occasionally disclosed in the breeding of established crops. What follows here is essentially a listing of inferences arising from the previous sections, together with a few indications of substantial gaps in our knowledge.

Cross-fertilization may be more important for the maintenance of population mean fitness in inbreds than is generally thought to be the case. As is discussed further in Chapter 8, more work is needed to establish the extent and effects of such crossing, but its hypothetical consequent importance for breeding plans in inbreds is clear.

Selection in polysomics will be slower than in disomics, but there will be more concealed variation on which to work. (Variation in polysomics has not been treated in detail, but selection methods will be discussed where appropriate.)

Mutation can maintain a great deal of variation in a quantitative trait where extreme values of the trait are only modestly affected by stabilizing selection. Even where such selection is strong, linkage allows readily detectable levels of mutation-induced variability to persist. As the environmental variance in a trait increases, so does the amount of variation which can be maintained by mutation (within a single environment), because selection becomes less efficient.

I have concentrated on variation in one environment; substantial genetical differences at specific loci are seen between environments within a single species, but it is also known that single genotypes perform differently in different environments. Much variation can be maintained in this way (see Haldane and Jayakar 1963, and much later work), and it will be the aim of the breeder to use rather than eliminate this variation.

Plant competition is very important. It has generally been ignored in producing new varieties, and this should not be allowed to continue, especially as theory is gradually incorporating the topic (see Sections 6.3.1 and Gallais 1976c).

Breeding systems will alter the way in which variation is expressed; it is not clear in what way they will affect the total available variability, except when this has deleterious manifestations, since this also depends greatly on population size, which is rarely known, except for experimental populations.

Genetical regulation has not been mentioned. It must be important in determining variation, but its role is not clear. Accordingly, the population and quantitative genetical implications of variations in partly elucidated regulatory mechanisms remain unexplored.

It is very evident that there are areas, such as disease resistance, where the store of variation within a species is insufficient to meet the requirements of effective production and there has been much work on transfer of alien gene material for such purposes (see Chapters 13 and 14). Resort will frequently be had to wild relatives (see for example, Hawkes 1977) or populations in centres of diversity of the crop.

6

Interaction between genotype and environment

6.1 Introduction

'Never give children a chance of imagining that anything exists in isolation. Make it plain from the start that all living is relationship,' as Aldous Huxley made one of his characters say in *Island* (1963). Animal breeders can treat their subject in isolation, can adjust the environment up to a point to suit the genotype. In plant breeding, this point is reached much sooner, to say the least. As noted earlier, virtually all major crops are grown more extensively in regions removed from their centres of origin than in those centres, and this applies to most minor crops as well (apart from such oddities as durian and mangosteen (Simmonds 1976a) and a few declining staples and fibre plants). This is not just a truism, nor a series of accidents of history and economics; it also reflects the changing environmental needs of crops cultivated on a grand scale. However, there is limited theoretical ground on which to discuss this topic, and in the present chapter I am concerned to treat those aspects of interaction between genotype and environment which have been placed on a strict genetical basis or which may be analysed by theoretically satisfactory methods. In the former case, the main concern will be with the effects of limited numbers of loci, while in the latter, the approach must be essentially biometrical.

6.2 Single locus interactions

6.2.1 Locus with locus

If we define the genotype which is the subject of investigation then other genotypes become part of the environment for the purposes of the analysis. It is not paradoxical to begin our discussion of genotypic response to the environment by examining locus by locus interactions, for the archetypal case here is Flor's 'gene-for-gene' hypothesis to account for host-pathogen interaction.

In 1942, he began to publish the results of his investigations into the genetics of resistance in flax, *Linum usitatissimum,* and virulence in its rust, *Melampsora lini.* He used two varieties of flax, Ottawa and Bombay, known to have the following response to races 22 and 24 of the rust (S = susceptible, R = resistant):

	Ottawa	Bombay
22	S	R
24	R	S

In one set of experiments (Flor 1946), the two races of rust were crossed and the F_1 and F_2 tested on both Ottawa and Bombay. The results were as follows:

	22	24	F_1		F_2		
Ottawa	S	R	R	R	S	R	S
Bombay	R	S	R	R	R	S	S
Number of cultures				78	27	23	5

In a second set of experiments (Flor 1947), the two flax varieties were similarly crossed and tested on the two races, with these results:

	Ottawa	Bombay	F_1		F_2		
22	S	R	R	R	S	R	S
24	R	S	R	R	R	S	S
Number of plants				109	36	36	12

In each case, the F_2 results are in very good agreement with the segregation ratio, 9:3:3:1, to be expected for two independent genes, yielding dominance for resistance in the host, for avirulence in the pathogen. From these and other similar results, Flor (e.g. 1956) was able to enunciate his 'gene-for-gene' hypothesis, that host and parasite had evolved together, so that for every gene (locus) determining host response to the pathogen, there would be in the parasite a corresponding gene determining pathogenicity. Thus, the observable result of growth or no growth on inoculation of host with pathogen was the clear-cut result of the interaction between specific genes in host and pathogen.

Since then, many host–parasite interactions have been shown or suggested to be primarily determined by gene-for-gene systems (Table 6.1), though the original flax-rust system has long been analysed well past its early illuminating simplicity (Shepherd and Mayo 1972).

The implications of the gene-for-gene hypothesis for resistance breeding were important and far-reaching. First of all, it allowed an orderly classification of specific resistance genes in a number of different crop plants, and the establishment of homologies among different physiological races of their pathogens. Secondly, it became clear that

Table 6.1
Some examples of host–pathogen systems exhibiting gene-for-gene relationships (modified from Day 1974)

Host	Pathogen		References
Linum	*Melampsora lini*	Rusts	Flor (1942)
Triticum	*Puccinia graminis tritici*		Luig and Watson (1961)
Triticum	*P. recondita*		Samborski and Dyck (1968)
Coffea	*Hemileia vastatrix*†		Noronha-Wagner and Bettencourt (1967), Eskes and Carvalho (1983)
Hordeum	*Ustilago hordei*	Smut	Sidhu and Person (1971)
Hordeum	*Erysiphe graminis hordei*	Mildews	Moseman (1959)
Triticum	*E. graminis tritici*		Powers and Sando (1957)
Malus	*Venturia inaequalis*	Other fungi	Boone and Keitt (1957)
Solanum	*Phytophthora infestans*		Black *et al.* (1953)
Lycopersicon	*Cladosporium fulvum* (*Fulvia fulva*)†		Day (1956), Boukema (1981)
Solanum	*Synchytrium endobioticum*‡		Howard (1968)
Solanum	*Globodera rostochiensis*‡	Nematode	Jones and Parrot (1965), Turner *et al.* (1983)
Triticum	*Mayetiola destructor*	Insect	Hatchett and Gallun (1970)
Gossypium	*Xanthomonas campestris* pv. *malvacearum*	Bacteria	Brinkerhoff (1970), Verma and Borkar (1984)
Leguminosae	*Rhizobium*§		Nutman (1969)
Lycopersicon	tobacco mosaic virus‡	Viruses	Pelham (1966)
Lycopersicon	spotted wilt		Day (1951, 1960)
Solanum	potato virus X‡		Howard (1968), Adams *et al.* (1984)

†Together with other genetical variation for resistance in the host.
‡Not yet confirmed.
§Symbiosis rather than parasitism.

evolution of virulence was to be expected and that new strategies of breeding for resistance might be fruitful. These aspects will be discussed further in Chapter 12. Thirdly, after many years during which no simple gene-for-gene system was found in a wild species, the view began to spread that it was an unintentional by-product of breeding for resistance (see Day 1974, for an extended discussion). Thus, choice of the most strongly resistant genotypes, given the problems of selection for such a complex trait, would tend to enhance the effects of given single genes, and possibly (though less certainly) eliminate polygenic variation in disease susceptibility, thereby allowing, in effect coercing even, the pathogen into enhancement of the effects of the most important loci affecting virulence.

Fifty years of breeding for resistance might have taken a different direction had understanding of the genetical interaction between host and pathogen been earlier achieved. As discussed in Section 6.3 below, other approaches to these interactions are now being adopted.

The symbiotic relationship between various plants and bacteria which results in nitrogen fixation (Table 6.1) seems to be controlled by gene-for-gene interactions in certain cases (Nutman 1969). Use of host genotypes which exclude indigenous symbionts may allow a great increase in the efficiency of nitrogen fixation (Devine and Weber 1977).

Plant competition or interaction between plants constitutes a second very important case of interaction between genotypes. In general, few single locus effects have been identified, and it is therefore perhaps unfortunate that the theory, as developed originally by Griffing (1967, 1968a,b) and more recently by Gallais (1976c), has largely been constructed at the single locus level. However, it has important general implications for prediction of selection response, and so is considered in some detail in Chapter 7.

6.2.2 Locus with specific non-genetical agent

Many continuously variable traits are partly determined by major genes. Indeed, it is clear, as was illustrated in section 4.6.2.2, that much past progress has arisen from the use of such genes, and arguable that as much future progress will derive from their use. One such trait is photoperiodic sensitivity. Major genes affecting date of heading have been identified in a number of important cereal crops. A clear example is in the work of Keim *et al.* (1973) on heading time in some winter and spring wheats, following earlier work of Pugsley (1966) on spring wheat.

Keim *et al.* (1973) chose Sonora 64, an early maturing Mexican semi-dwarf spring wheat, as the photoperiod-insensitive parent and crossed it with two photoperiod-sensitive winter wheats, Lancer and Warrior. Initial times to heading in days were:

	Sonora 64	Lancer	Warrior
10-h photoperiod	50.1	172.1	153.6
16-h photoperiod	58.0	72.6	72.2

In the F_1s, all plants had a short time to heading, with no response to photoperiod, but in the F_2 the results with the 10-h photoperiod could be most simply interpreted as

early:late:very late::12:3:1

(Analysis of multimodal frequency distributions, such as the results described here, is not in general a trivial problem. However, in the present context it should simply be noted that segregation in the different crosses allows the interpretation to be tested straightforwardly.) This is the result to be expected from the effects of two unlinked (diallelic) loci displaying both dominance and epistasis. If the loci are called Ppd_1 and Ppd_2, then parental genotypes would be

Sonora	Ppd_1Ppd_1	Ppd_2Ppd_2
Lancer } Warrior }	ppd_1ppd_1	ppd_2ppd_2

Thus, a back-crossing programme including selfing after each back-cross would allow the incorporation of photoperiod sensitivity into an insensitive variety, but the reverse change would require more generations of testing.

Response to extreme soil water levels, whether through drought or flooding, is another area where single gene effects may be important. Schwartz and Freeling (see especially Freeling 1973, 1974; Freeling and Schwartz 1973) have examined aspects of the response of the developing primary root of maize to anaerobiosis produced by flooding. In particular, they have shown that this induces the synthesis of three sets of isozymes of alcohol dehydrogenase, which they have further shown to be the dimeric products of subunits produced by two unlinked loci, Adh_1 and Adh_2. Now one of the sets of isozymes is found in the quiescent embryo and is present in trace amounts in uninduced primary roots. It is the product of Adh_1. It can also be induced by the auxin (or herbicide) 2,4-D under aerobic conditions, whereas Adh_2 is not so affected. Thus, anaerobiosis specifically induces Adh_2. (Of course, other agents may also do so; the specificity is only presumptive.) As it becomes more certain that different isozymes of alcohol dehydrogenase confer tolerance to water-logging (Crawford 1967; Brown *et al.* 1974, 1976; Chew 1984; Gerlach *et al.* 1983), an apparently uncomplicated interaction of this kind should be a useful tool in the investigation of water response.

6.2.3 Locus with non-specific non-genetical agent

In Section 6.2.1, we examined the powerful simplifying hypothesis of Flor (e.g. 1956) for the description of host–pathogen interactions. In recent years, as already noted, the question has arisen whether the search for oligogenic disease resistance has not obscured both quantitative variation in disease response and also hastened the evolution of new virulent races of pathogens. Methods of overcoming this problem will be discussed further in Chapter 12; here we consider a method of analysis which reveals variability in environmental response of different genotypes, though initially without necessarily clarifying what features of the environment engender the differential responses.

The method is that of Finlay and Wilkinson (1963), described in Section 6.3.3 below, and the application is that of Arnold *et al.* (1976), to resistance breeding in cotton. Innes *et al.* (1974) investigated resistance to bacterial blight (blackarm; *Xanthomonas malvacearum*) by a half diallel cross between six inbred varieties, using two different bacterial cultures. (A full diallel analysis was not needed because of the absence of maternal effects on blight resistance.)

There are several important genes, the 'B' genes, which confer resistance to different strains of blight, probably on a gene-for-gene basis, and which constitute more than one gene locus. The diallel analysis, of the type described in Section 4.4.2, indicated not only a substantial degree of dominance, but also a considerable environmental effect, ascribable to the different sites (in the Sudan and in Uganda), but not to any specific factors at those sites.

These results showed the clear importance of testing in more than one environment when assessing disease resistance, and as these workers had shown in an earlier paper (Arnold and Brown 1968; see also Arnold *et al.* 1976), the technique of Section 6.3.3 could be applied to obtain an overall idea of the relationship between environment and host–pathogen interaction. Thus, regression of degree of affection of host in each single host–pathogen combination on the mean of all combinations at each site showed a clear linear relationship in each of three varieties, the slope varying between varieties. (That is, not all varieties were equally responsive to a change in the favourability of environmental conditions to severe infection.) They were also able to show that simple relationships existed between specific environmental factors, such as temperature following inoculation, and disease state at scoring. Thus, a preliminary investigation of overall environmental effects can frequently indicate an approach to a closer analysis, when the different environments are assessed in the light of field experience.

6.3 Whole genotype interactions

6.3.1 Genotype with genotype

Problems of interactions between members of a single species grown together, especially under conditions of intensive monoculture, lie at the centre of current plant breeding practice. As has been frequently noted, selection is generally practised among individuals grown in mixed stands, for crop production in pure stands. Approaches to the solution of these problems are discussed in Chapters 9 and 10; here I shall consider a case where the existence of interaction allows the estimation of certain genetical parameters under conditions where the usual methods (Section 4.4) are inapplicable.

In forest trees, where the inheritance of properties of mature trees is of importance, parent–offspring and even sib analyses may be impossible, for obvious reasons. This can also apply to crops harvested from mature trees, such as coconuts. In 1957, Shrikhande suggested using the interaction between trees to partition out the variance due to all genetical factors. He used an empirical rule obtained by Smith (1938), that the variance between plot means $V(m)$ depended very simply on the individual variance $V(i)$ and the number, m, of plants in the plot:

$$V(m) = V(i)/m^b.$$

Here, b would in effect be a summarizing statistic, accounting for all ways in which the members of the group interacted.

Shrikhande suggested that

$$V(m) = \frac{V_G}{m} + \frac{V_E}{m^b},$$

since the genetical variance of the mean would depend inversely on the sample size, while the environmental variance would alter according to Smith's rule. Some strong assumptions are involved here, in particular that a genetical effect may be independently partitioned out for all trees in a plot, and that it will remain independent of plot effects, for all plot sizes, since Shrikhande (and later Sakai and Hatakeyama 1963) used plots of various sizes to obtain estimates of V_G and V_E.

The method has further been extended by Hühn (1975a), to estimate an interaction variance as well, and we shall consider this model in more detail. We postulate a large stand of plants (trees) of the same age set out as a square grid with no differential damage or thinning through the stand. Furthermore, we assume that a tree's performance is directly influenced only by its four nearest neighbours, and that there are no directional effects on competition. Let

$$y_{ij} = g_{ij} + c_{ij} + e_{ij}$$

where y_{ij} is the phenotypic value (for some trait of interest) of the jth plant (of m) in the ith plot (of p_m); g_{ij} is the genotypic value of the ijth plant; c_{ij} = competitive influence of the four neighbours of plant ij; and e_{ij} = environmental influence of plant ij.

Then it is possible to partition the phenotypic variance of plot means into components depending on V_G, V_E, V_C, the variance due to competition among plants, and r, the average correlation between g_{ij} and c_{ij} for the whole stand, thus:

$$E\{V(m)\} = \frac{V_G}{m} + \frac{(4m - 6\sqrt{m} + 2)}{m^2} V_C + \frac{V_E}{m^b} + \frac{4(\sqrt{m} - 1)r\sqrt{(V_G V_C)}}{m^{3/2}}.$$

This relation can be used as an estimator for V_G, V_C, V_E, r, and b by replacing $E\{V(m)\}$ by the observed values, for different plot sizes $m = 1$, 4, 9, etc., i.e.

$$V(1) = V_G + V_E$$

$$V(4) = \frac{V_G}{4} + \frac{3V_C}{8} + \frac{V_E}{4^b} + \frac{r\sqrt{(V_G V_C)}}{2}, \text{ etc.}$$

Provided that all the assumptions have been met, it should be possible to solve these equations by an iterative least squares procedure or, preferably, an iterative maximum likelihood method such as that of Elston (Kaplan and Elston 1972). However, the sensitivity of the analysis will depend critically on the value of b, which if close to unity will render the method valueless. In addition, partial least squares procedures, such as that of Hühn (1975a,b), which do not yield estimates of errors for the various statistics, will be inappropriate in most cases. This may be seen in the results shown in Table 6.2. It is clear that the method is most unreliable, in that the (unknown) variance of the estimators is very considerable. (While V_G, V_C, and V_E might differ markedly between stands for a given trait, b should not do so, if Smith's empirical rule has general validity.) It may further be held that estimation of r is inappropriate, so that a model not specifying c_{ij} should be fitted, as in the earlier papers cited. Nonetheless, if used with discretion it can provide useful preliminary estimates of genetical parameters, especially for individual stands within well-defined regions. Namkoong and Squillace (1970), in emphasizing some of the problems discussed here, had earlier reached this conclusion.

Table 6.2

'Best' estimates of genetical and other parameters for four traits of Norway spruce in two populations (from Hühn 1975a,b). V_G, V_C, and V_E are shown as proportions of V_P

Trait	Population	r	b	V_G	V_C	V_E
Height	1	−0.2	0.66	0.08	0.16	0.76
	2	−0.6	0.14	0.02	0.62	0.36
Diameter	1	0.8	0.08	0.92	0.02	0.06
	2	0.2	0.02	0.82	0.02	0.16
Crown per cent	1	0.8	0.20	0.90	0.02	0.08
	2	1.0	0.06	0.84	0.04	0.12
Taper	1	−0.8	0.80	0.02	0.10	0.88
	2	0.2	0.54	0.50	0.16	0.34

Hamblin and Rosielle (1978) have introduced the following very simple model for interaction between genotypes:

Genotype	Pure culture genotypic value	Mixed culture genotypic value
A_1A_1	$G_{11} = \mu + a$	$G'_{11} = \mu'' + a'' = G_{11} + \mu' + a'$
A_1A_2	$G_{12} = \mu + d$	$G'_{12} = \mu'' + d'' = G_{12} + \mu' + d'$
A_2A_2	$G_{22} = \mu - a$	$G'_{22} = \mu'' - a'' = G_{22} + \mu' - a'$

For mixed culture, the apparent additive genetic variance is given by:

$$V_A + V_{A'} + 2\text{Cov}_{AA'}$$

where V_A is the additive genetic variance in pure stands, $V_{A'}$ is the additive component of the variance due to competition, and $\text{Cov}_{AA'}$ is the additive covariance between genetical and competitive effects. Also, apparent dominance variance is given by

$$V_D + V_D + 2\text{Cov}_{DD'}$$

with similar definitions to the additive components. Hence,

$$V_G(\text{F}_2, \text{pure}) = V_A + V_D$$

$$V_G(\text{F}_2, \text{mixed}) = V_A + V_D + V_{A'} + V_{D'} + 2\text{Cov}_{AA'} + 2\text{Cov}_{DD'}.$$

This model has the advantage that it allows simple comparison between variation in pure stand and in mixture. As shown in Table 6.3, the underlying genetical variation that will eventually be realized in a relatively pure variety is not well described by the apparent parameter estimates. While this model is developed in the same manner as that in Sections 4.2 and 4.3, it does not allow for frequency-dependent effects, which are likely to be widespread and important, as will be discussed in the next chapter, and thus is unlikely to be generally applicable. Its implications are important nonetheless.

Table 6.3

Components of variance and covariance for seed yield in two barley experiments (from Hamblin and Rosielle 1978; experimental data of Suneson and Ramage 1962)

Experiment	1	2
V_A	92450	30504
$V_{A'}$	8978	64082
$2\text{Cov}_{AA'}$	57620	−88426
Observable 'V_A'	159048	6160
V_D	9801	25122
$V_{D'}$	2401	4957
$2\text{Cov}_{DD'}$	−9702	−20922
Observable 'V_D'	2500	9157
$V_G(\text{F}_2, \text{mixed})$	161548	15317
$V_G(\text{F}_2, \text{pure})$	102251	55626

6.3.1.1 Allelopathy Fuerst and Putnam (1983) have emphasized that interference of one species with another should be subdivided into competition, which leads to a reduction in the resources available to one species, for example a lack of sunlight through shading by a taller or faster growing competitor, and allelopathy, which is active interference with another species through the production of toxins or other chemical agents.

The phenomenon whereby plants of one species or residues of those plants inhibit the germination, growth or development of plants of another species was recognized by Greek writers more than 2000 years ago. Hundreds of such pairwise interactions have been demonstrated (see Rice 1974, 1979 for a review of the topic). However, the name, which strictly refers to two plant species which harm each other, was coined much more recently (Molisch 1937). The current definition, as already implied, is of a one-way effect, rather than a reciprocal relationship. Thus, Young (1983) noted that autotetraploid *Haemarthria altissima* cv. Bigalta had a greater inhibitory effect on *Desmodium intortum* cv. Greenleaf than had the diploid cultivar Greenalta, whereas Greenleaf inhibited neither Bigalta nor Greenalta to a significant extent.

Competition analyses of the kinds discussed in the preceding pages may be used to determine the extent of allelopathy, but given that it is a physiological phenomenon, further analyses will be physiological rather than genetical (Dekker *et al.* 1983). Very extensive biochemical analyses have in fact elucidated the basis of allelopathy in many cases. Indeed, it has been suggested that this biochemical knowledge might be utilized in developing novel synthetic herbicides. In the present context, however, the question arises to what extent allelopathy may be used, for example to inhibit weeds, or avoided, for example in growing successive crops where some of the desired species are allelopathic to others (Newman 1982).

Selection for weed suppression has been attempted, following the observation that some cultivars of certain crops, such as oats, appeared to be markedly inhibitory to weeds, others not. Putnam and Duke (1978), for example, reviewed successful selection for inhibition of weeds by cucumber under well defined experimental conditions, using the techniques of selection to be discussed in later chapters. However, under field conditions the effect was lost, diminished or variable. One problem is that the same cultivar which inhibits certain weeds may stimulate the growth of others. For example, one cultivar of sunflower (*Helianthus annuus*) has been found to inhibit *Abutilon theophrastis* and *Datura stramonium*, and stimulate *Ipomoea purpurea* and *Setaria viridis* (Leather 1983).

Other problems in selection for weed suppression include the facts that the toxins which make the crop or forage plants successful as inhibitors of weed growth may make the product toxic or unpalatable and that weeds will evolve jointly with the crop plants (Newman 1982).

For this latter reason, any breeding effort aimed at blocking the production of toxins by weeds of mixed cultures such as pastures is also unlikely to be successful.

In most cases, avoidance of the deleterious allelopathic effects of crop plants is best achieved at present through agronomic methods, for example by ordering a crop rotation so that a crop whose residues are especially allelopathic is not followed by one which is particularly susceptible. Since plant residues are not necessarily allelopathic, e.g. decomposed couch grass (*Agropyron repens*) rhizomes are allelopathic only if the decomposition has been anaerobic (Newman 1982), management practices may also reduce their deleterious influences.

6.3.2 Genotype with specific non-genetical agent

When experimental design was in its infancy, problems such as 'the manurial response of different potato varieties' stimulated Fisher and others in their consideration of genotype–environment interaction. Differential response to fertilizer represented an interaction between complex but unspecified genotypes and one or more relatively simple environmental agents. In the earliest such analysis, Fisher and Mackenzie (1923) considered the linear relationship

$$y_{ijk} = \mu + g_i + c_j + (gc)_{ij} + \varepsilon_{ijk}$$

where

y_{ijk} = yield of the *i*th potato variety receiving the *j*th fertilizer treatment in the *k*th plot receiving the *ij*th combination
μ = overall mean yield
g_i = effect of *i*th variety
c_i = effect of *j*th fertilizer
$(gc)_{ij}$ = interaction between *i*th variety and *j*th fertilizer
ε_{ijk} = plot effect.

They were able to show that there were significant variety effects and fertilizer responses, but no interaction. Now such a linear form is still frequently used in detecting interaction, and, as we shall discuss in the next section, is frequently rightly criticized because interactions (and indeed main effects) are rarely additive. It is therefore of more than historical interest that Fisher and Mackenzie showed how to fit a product relationship for main effects, and showed that it described the data better than the complex linear model.

Since response to an environmental agent is typically curvilinear with approximate monotonicity below some optimum, other environmental factors being held constant, it may be considered that for the detection of interactions between a single factor and a set of varieties, a simple linear model will be appropriate. However, as soon as additional factors are varied, non-linearity will be likely to manifest itself, and while response surface methods (discussed briefly in Section 3.2) will allow investigation of the observed variability and selection of appropriate genotypes, it will not always be clear what genetical models to employ. To take a very simple example, Dubetz and Bole (1973) tested three spring wheats, Pitic, Manitou, and Kenhi, for yield with and without moisture stress, at five different levels of applied nitrogenous fertilizer. Water stress significantly decreased the yield of all varieties at all fertilizer levels, but whereas without moisture stress all had shown curvilinear increases in fertilizer response, from 0 to 224 kg/ha, under stress only Pitic showed response, the others both performing worse at high nitrogen levels than at intermediate levels. Now the physiology of drought response may be becoming clarified, but nonetheless a genetical model for the results described is lacking. In fact, in one selection programme for drought resistance in Canadian wheats, such interactions were not considered (Hurd *et al.* 1972a,b, 1973), nor were genotypic differences in osmoregulation, though these may also be important (Morgan 1972; Hurd 1964, 1976; Fischer and Wood 1979; Clarke and McCaig 1982). Plant breeding history abounds in examples of successful selection for traits whose physiology is ill-understood, but this does not mean that knowledge of the physiology would not have allowed more rapid success.

6.3.3 Genotype with non-specific non-genetical agent

In a large varietal trial, covering m varieties tested in p environments, it is not possible to specify all differences between environments by known differences in specific physical factors such as soil moisture, nutrient availability, and energy flux. Indeed, as I have already emphasized, it will often not be known which of the many important environmental factors are critical in determining different overall performances among test sites. Multiple regression, using a wide range of independent variables, may allow the determination of the critical factors, especially if some of the techniques for choosing the 'best subset' among these independent variables are employed. Such analyses, however, are fraught with all the perils of non-generality which always attend multiple regression analysis (see, for example, Hocking 1976 and Miller 1984).

To overcome such problems, and obtain a general idea of environmental response, Yates and Cochran (1938) suggested using a biological rather than a physical description of the environment. An assessment could be made of the response of genotypes to the environment by examining the regression of individual varietal performance on varietal means taken over environments. This idea was developed by Finlay and Wilkinson (1963) and is now in widespread use. The description which follows is essentially that of Wright (1971), who has developed the technique further, to allow the assessment of interaction also in terms of regression on genotypic as well as environmental means (see also Wright 1976a,b).

The basic model or data decomposition used in genotype–environment analyses is

$$y_{ijk} = \mu + g_i + e_j + I_{ij} + \varepsilon_{ijk}$$

where y_{ijk} is the kth replicate value at the jth site for the ith variety; e_j is the effect of jth environment; I_{ij} is the interaction between ith variety and jth environment; and the other symbols are as defined earlier. (There may be other complications due to the particular methods of local control and replication used, but they are not relevant to this part of the analysis.)

The model is developed on the basis that the I_{ij} have mean zero, and variance V_I. This is of little assistance in prediction.

The improvement of Yates and Cochran, and Finlay and Wilkinson was to consider how I_{ij} might depend upon e_j; in other words, assuming a linear model, to examine the regression of I_{ij} on e_j, by computing β_i for each of the m genotypes, using

$$y_{ij} = \mu + g_i + (1 + \beta_i)e_j + d_{ij}$$

with y_{ij} the mean over replicates,

$$\beta_i = \left(\sum_j I_{ij}e_j\right) \bigg/ \left(\sum_j e_j^2\right)$$

and d_{ij} a residual. Wright's alternative formulation,

$$y_{ij} = \mu + (1 + \alpha_j)g_i + e_j + d'_{ij},$$

where

$$\alpha_j = \left(\sum_i I_{ij}g_i\right) \bigg/ \left(\sum_i g_i^2\right),$$

allows assessment of the dependence of the interaction upon genotypic values, but in general will not be an appropriate analysis on customary views of causation. Since the α_j and β_i are not independent, Wright (1971) suggested a further modification,

$$y_{ij} = \mu + g_i + e_j + kg_ie_j + d'_{ij},$$

where

$$k = \left(\sum_i\sum_j y_{ij}g_ie_j\right) \bigg/ \left(\sum_i g_i^2 \sum_j e_j^2\right).$$

Now

$$k = \left(\sum_i g_i\beta_i\right) \bigg/ \left(\sum_i g_i^2\right),$$

so that k is useful in the special case where the g_i and β_i are completely correlated. In addition, d''_{ij} is independent of the joint regression of interaction on the g_i and e_j, so that

$$d''_{ij} = \beta'_i e_j + \alpha'_j g_i + \delta_{ij}.$$

The complete analysis of Wright (1971) corresponding to the data decomposition

$$y_{ij} = \mu + g_i + e_j + kg_i e_j + g_i(\alpha_j - ke_j) + (\beta_i - kg_i)e + d'''_{ij},$$

is shown in Table 6.4, and its application to a large experiment of Breese (1969), on forage yield from the pasture grass cocksfoot, *Dactylis*

Table 6.4

Analysis of variance and expectations of mean squares on the assumption of random sampling (from Wright 1971)

Source	Degrees of freedom	Sums of squares	Expected mean squares
Genotypes (g)	$m-1$	$S_1 = rp\sum_i g_i^2$	$V_{\text{Error}} + rV_I + rpV_G$
Environments (e)	$p-1$	$S_2 = rm\sum_i e_j^2$	$V_{\text{Error}} + rV_I + rmV_E$
Interaction (ge)	$(m-1)(p-1)$	$S_3 = r\sum_{ij} y_{ij}^2 - S_1 - S_2 - mp\mu^2$	$V_{\text{Error}} + rV_I$
Joint regression (k)	3	$S_4 = k^2 r\sum_i g_i^2 \sum_j e_j^2$	$V_{\text{Error}} + rV_{\delta_1} + rmpV_K$
Residual	$(m-1)(p-1)-1$	$S_3 - S_4$	$V_{\text{Error}} + rV_{\delta_1}$
Residual regression on genotypes (α')	$p-2$	$S_5 = r\sum_j \alpha_j^2 \sum_i g_i^2 - S_4$	$V_{\text{Error}} + rV_{\delta_2} + rmV_{\alpha'}$
Residual regression on environments (β')	$m-2$	$S_6 = r\sum_i \beta_i^2 \sum_j e_j^2 - S_4$	$V_{\text{Error}} + rV_{\delta_2} + rpV_{\beta'}$
Residual (δ)	$(m-2)(p-2)$	$S_3 - S_4 - S_5 - S_6$	$V_{\text{Error}} + rV_{\delta_2}$
Error	$r(mp-1)$	From replication	V_{Error}

$$\mu = \sum_{ij} y_{ij}/mp, \quad g_i = \sum_j (y_{ij}/p) - \mu, \quad e_j = \sum_i (y_{ij}/m) - \mu$$

$$\beta_i = \left(\sum_j y_{ij}e_j\right) \Big/ \left(\sum_j e_j^2\right) - 1, \quad \alpha_j = \left(\sum_i y_{ij}g_i\right) \Big/ \left(\sum_i g_i^2\right) - 1$$

$$k = \sum_i \beta_i g_i \sum_i g_j^2$$

Table 6.5

Analysis of Breese's (1969) data on yields of cocksfoot at different sites from seven harvests over two seasons (from Wright 1971). Yields were given by Breese as G.M. Prod. (g/plant)

Source	Degrees of freedom	Mean squares/100	Variance components	'Heritabilities'
Genotypes (g)	4	887 ***	$V_G = 1938$	
Environments (e)	13	1494 ***	$V_E = 9473$	$\frac{V_G}{V_P} = 0.13$
Interaction (ge)	52	73 ***	$V_I = 1900$	$\frac{V_G\,V_E}{V_P} = 0.76$
Joint regression (k)	1	714 ***	$V_K = 311$	$\frac{V_G + V_E + V_K}{V_P} = 0.79$
Residual (d)	51	60		
Residual regression on genotypes (α')	12	14	$V_{\alpha'} = 0$	
Residual regression on environments (β')	3	422***	$V_{\beta'} = 898$	$\frac{V_G\,V_E\,V_K\,V_{\beta'}}{V_P} = 0.85$
Residual (δ)	36	48	$V_{\delta_2} = 967$	$k = 0.005$
Error	207	16	$V_{\text{Error}} = 1600$	

*** = significant at the 0.1% probability level.

glomerata, is shown in Table 6.5. The broad sense heritability, $h^2 = V_G/V_P$, is only 0.13, thereby indicating that selection purely on this basis would be slow and unrewarding. However, when information respecting specific independent environmental effects is included, the usable 'heritability' immediately rises to 0.76, statistically significant improvement for prediction of progress under selection; this will be discussed further in Chapter 7. In this particular experiment, both the joint regression and the individual residual regressions were relatively unimportant in increasing 'heritability' which is not unexpected, given the modest value of V_G, V_I, and V_K relative to V_E. However, this will not necessarily always be the case. In the analysis of Table 6.4, the joint regression term, which corresponds to a linear-by-linear interaction, is essentially Tukey's term for removable non-additivity (Tukey, 1949), and a highly significant value indicates that the analysis should be carried out on the logarithmic scale. On this scale, a significant value for this term mainly relates to genotypes showing relatively more variation in poorer environments as against better environments, which probably occurs because whatever factors limit yield in different environments do not affect all genotypes equally.

This analysis is both powerful and intuitively appealing, and it is hardly surprising that it has, as noted by Hill (1976), been applied to most important field crops, including barley, wheat, maize, pasture grasses, peas, and tobacco. (It can also be used for mixtures of species, examining the dependence of yield of species *i* in the presence of species *j* on the mean yield of species *i*, which may be very important for forage plants; Jacquard and Caputa 1970; Harper 1977.) We should therefore consider its limitations.

First, like any regression technique, it should not be over-interpreted; prediction outside the range of environments tested will be inappropriate, and it may not be obvious that such departures have occurred. Secondly, as argued strongly by Knight (1970, 1973), the technique, especially in the form advanced by Finlay and Wilkinson, constrains interactions into a linear form, which as we have seen will frequently be inappropriate. Thirdly, as implied already, the technique will be particularly useful for preliminary screening of large numbers of genotypes (Finlay and Wilkinson used 277), but to pass from such a summary to a specific understanding of individual interactions will not by possible.

Moll *et al.* (1978) have noted that the regression of varietal performance on site mean depends both upon the actual magnitude of responses to environments and on the correlation of the responses of individual varieties between environments, so that a simple interpretation may not in general be possible. However, in their own investigation on yield in maize the latter phenomenon was less important, which is the case of greater promise.

These, and many other problems discussed by Hill (1976), do not detract from the technique's general usefulness and its importance in initiating a fresh examination of genotype–environment interaction over the last 20 years.

7
Response to selection

7.1 Introduction

Plant improvement by selection must take one of two forms: selection among existing populations for desirable traits or desirable levels of metric traits, or selection within a population (which may be descended from an experimental cross) of the most desirable parents from which to establish new lines or varieties. The former process will generally precede the latter, being essentially a screening of the populations available.

7.2 Selection of the best subset of a set of populations

If a trait is a discrete or all-or none trait, or if it can be treated as such by use of a suitable threshold, screening is made relatively simple. For example, plants may be designated as resistant (R) or susceptible (S) to some disease, and a large number of different unknown genotypes screened in an experiment of relatively simple design. Thus, Williams (1977a) was able to screen 5000 lines of cowpeas (*Vigna unguiculata*) for resistance to one minor and four important diseases, as shown in Table 7.1. In a second experiment, Williams (1977b) screened some of the same material for resistance to cowpea yellow mosaic virus, and found 55 completely resistant, 38 partly resistant. Of these 93 lines, 16 were also resistant to anthracnose, rust, cercospora leaf spot, and bacterial pustule (row 1 of Table 7.1). Such a success rate is of course more likely in highly variable material from a number of environments which has not been grown in very large scale monoculture; nonetheless, it is clear that success will in general depend upon the breeder's ability to screen large amounts of material.

7.2.1 Single gene traits

The resistance detected in Table 7.1, is, by and large, likely to be oligogenic, i.e. due to a few major loci, and simple breeding tests of the lines with multiple resistance will establish linkage relationships, after which incorporation of resistance into commercial varieties will be relatively

Table 7.1

Field resistance to five diseases in cowpeas (from Williams 1977a; R = resistant)

Anthracnose (*Colletotrichum lindemuthianum* Sacc. & Magn. Bria. & Car.)	Cercospora leaf spot (*Cercospora cruenta* Sacc.)	Rust (*Uromyces vignae* Bard.)	Bacterial pustule (*Xanthomonas* spp.)	Target spot (*Corynespora casiicola* (Berk. & Cent.)	No. of lines
R	R	R	R	R	28
R	R	R	R		128
R	R	R	R		208
	R	R			278
	R		R		253
		R	R		257
	R	R	R		373
	R				327
		R			337
			R		319

straightforward. The main requirement will be a large enough number of populations for the initial screening to have a high probability of success. If rare traits are sought, screening may demand enormous numbers. For example, in screening for three discrete traits each at a frequency of 10^{-4}, over 400 000 lines would be needed to be 95 per cent confident of obtaining at least one of each type. (This example essentially relates to the problem of screening for dominant mutants at a number of loci.)

If a desired phenotype is recessive, identification of plants having the phenotype allows the immediate establishment of a pure-breeding strain. If the desired phenotype is dominant, on the other hand, progeny testing, either by selfing or by back-crossing to the recessive, will be needed to establish the required strain.

In the case of polyploids, the testing procedure is not as simple, and warrants a brief description of one published method for autotetraploids (Bos 1976). This is included here rather than in later sections because it is a screening rather than a selection procedure.

The procedure for incorporation of, say, A into a line homozygous for a is thus:

$$\begin{array}{ccc} aaaa & \times & AAAA \\ \mathrm{P}_1 & & \mathrm{P}_2 \\ & AAaa & \\ & \mathrm{F}_1 \quad \times & \mathrm{P}_1 \\ AAaa & Aaaa & aaaa \\ (1+2\alpha)/6 & (4-4\alpha)/6 & (1+2\alpha)/6 \end{array}$$

i.e. $A:a::(5-2\alpha)/6:(1+2\alpha)/6$ at the phenotypic level, where α is the frequency of double reduction (see Fisher 1947).

In each back-cross, then, *AAaa* and *Aaaa* plants will be crossed to *aaaa*. After sufficient back-crosses (say four) to ensure that the genome is almost completely that of P_1, selection against *a* will be practised. Simplex (*Aaaa*) and duplex (*AAaa*) plants may be separated by segregation analysis with back-crosses to P_1, while the tested plants themselves are maintained vegetatively. *AAaa* will yield the progeny shown above, while *Aaaa* will yield

AAaa	*Aaaa*	*aaaa*
$\alpha/4$	$(2-2\alpha)/4$	$(2+\alpha)/4$

i.e. $A:a::(2-\alpha)/4:(2+\alpha)/4$ at the phenotypic level. The probable duplex plants are then selfed, yielding (if actually duplex)

AAAA	*AAAa*	*AAaa*	*Aaaa*	*aaaa*
$\left(\frac{1+2\alpha}{6}\right)^2$	$\frac{8(1-\alpha)(1+2\alpha)}{36}$	$\frac{18-24\alpha+24\alpha^2}{36}$	$\frac{8(1-\alpha)(1+2\alpha)}{36}$	$\left(\frac{1+2\alpha}{6}\right)^2$

A large number of progeny [at least 180, to be 99 per cent sure of obtaining at least one quadruplex (*AAAA*) plant] must be grown and vegetatively maintained. These progeny plants are then crossed to the nulliplex P_1, and sufficient progeny raised to discriminate quadruplex and triplex (*AAAa*) from duplex plants. Triplex plants yield

AAaa	*Aaaa*	*aaaa*
$(2+\alpha)/4$	$(2-2\alpha)/4$	$\alpha/4$

Now, if $\alpha \leqq 0.05$, the probability of observing no nulliplex progeny among n is less than 0.95^n, which requires that n be large, e.g. at least 60 to be 95 per cent sure of detecting a triplex. Each potential triplex or quadruplex would require extensive testing, and confidence in the results would depend upon a prior knowledge of the magnitude of α. Bos accordingly considers that back-crossing the non-segregating progenies will be more rewarding, having already detected and eliminated duplexes by the previous generation of backcrossing. His programme would still require many thousands of plants to be grown and an extra generation of back-crossing: perhaps what becomes clearest is that incorporating a dominant into an autotetraploid requires so much effort, relative to the diploid, that other techniques, such as cytogenetic manipulation, may be warranted. Back-crossing in diploids is described in Chapter 10.

7.2.2 Metric traits

For any particular metric trait, the problem in screening before beginning a breeding programme is that of choosing the best t of k available populations. The nature of the k populations is often dictated by chance. For example, in seeking to assess the possibility of combining certain desirable traits in filberts (*Corylus* spp.) by crossing, Thompson (1974) was forced to include crosses which had been made in investigations of incompatibility (see Thompson 1971), since filberts do not begin to yield nuts until several years old. At the opposite extreme, breeders at the various international centres for research into specific crops may have global collections on which to draw, and rely on descriptions in data banks to obtain an initial set of k populations likely to be appropriate for some particular purpose; these descriptions will rarely be specific enough for accurate prediction for given environments.

Once the k populations are chosen, and are under test for the metric trait of interest, the problem is more narrowly defined. Frequently the metric will be distributed in an at least approximately Gaussian fashion, and much work has been carried out by statisticians on the problem of choosing the best t of k normally distributed populations. Wetherill and Ofusu (1974), reviewing much of this work, suggested some general considerations which are relevant to the plant breeding context.

First, anthropomorphic argument must be avoided. There is no question of 'fairness' involved in rejecting the $t+1$th variety of k if t only are to be chosen, even though the tth may not be significantly better than the $t+1$th. This implies that one must accept the exclusion of populations not shown to be worse than those chosen.

Secondly, although the aim is to optimize a particular metric such as yield, other factors like disease susceptibility may at any time cause a previously favoured variety to be rejected. This has the implication that selection for more than one attribute simultaneously may be desirable (see Section 7.4.2.4).

Thirdly, final decisions are never possible. Selected varieties are always being retested, experimentally or commercially, whenever they are grown. Thus, one application of the chosen selection procedure will never be enough. Taking this and the previous desideratum together, one can conclude that setting $t=1$, i.e. picking the top-ranking line, will always be a hazardous procedure, yet it is what the release of a new commercial variety must usually imply.

Fourthly, even when selection is very much based only on the metric of interest, there will usually be some previous information on the properties of the k populations. This may make it easier to assume properties of the distributions of the unknown population parameters, thereby aiding in experimental design and interpretation.

I have treated these criteria for a rational selection programme at some length because they emphasize certain limitations on what can be achieved by the breeder. Details of the various methods will not be presented here, as the literature is voluminous (see Bechofer 1970; Wetherill and Ofusu 1974 for details). Instead, I shall briefly treat one approach, that of Finney (1958, 1966), as it is to some extent particularly aimed at the plant breeder's problem.

Finney was specifically concerned with the overall strategy of variety trials, involving at first very large numbers of small plots, then as the best are detected and the worst discarded, the more promising may be multiplied in number and grown in larger plots, gaining greater precision. While he has treated an annual crop, the approach is so general that it could be adapted to be applied to others. Essentially, he deals with selection on a single criterion, usually yield, but this could in principle be an index combining several traits (Section 7.4.2.4). As emphasized by Curnow (1961), this analysis relates to varietal yield which is reproducible, i.e. it is strictly appropriate only for selection among pure lines of a self-pollinated crop, among hybrids of such lines or among clones of asexually propagated crops.

A cohort of k varieties is obtained. Over n years these are to be reduced to t, so that $t = pk$, where p is a small fraction chosen by the breeder. The process thus over n years reduces k to t by successive selections. In each year, a proportion p_i is retained so that

$$p = \prod_{i=0}^{n} p_i.$$

The total experimental area A will be assumed fixed, and each year a portion A_i is used, i.e. $\sum_{i=0}^{n} A_i = A$. The error variance is assumed to be V_e, and that of the varietal means V_i, with

$$V_i = \frac{p_0 p_1 \ldots p_{i-1} \gamma V_e}{A_i}$$

where γ is a constant of proportionality. It can be shown that the maximum gain possible in one cycle of selection following an initial culling depends critically on γp. If p_1 is optimal at a greater value than p, then an initial random discard may help. For example, if $\gamma = 5$ and $p = 0.1$, then the optimum is $p_1 = 0.178$, so $p_0 = 0.56$ will improve the process. The improvement is modest, however, compared with $p_0 = 1$, $p_1 = 0.1$, a more natural procedure. The smaller the value of p, nonetheless, the greater this gain: for $p = 0.01$ and $\gamma = 5$, the optimal p_0 and p_1 are 0.16 and 0.063, respectively. This yields a 35 per cent increase in the maximal gain in yield.

For two-stage selection, the initial random discard is not helpful for large p, as might be expected, but even for large γ (and hence large gains in precision as the trial proceeds), the improved gains are small. The rule $p_0 = 1, p_1 = p_2 = \sqrt{p}$, and $A_i = \frac{1}{2}A$, Finney showed to be generally useful.

For multi-stage selection, these results could be generalized. For $n > 2$, set $p_0 = 1$. As a first approximation, use $p_i = (p)^{1/n}$, and $A_i = A/n$. In departing from these values, optima will lie in the direction yielding

$$p_1 < p_2 < \ldots < p_n$$
$$A_1 < A_2 < \ldots < A_n$$

With fixed total resources, there will be little benefit in going beyond three or perhaps four stages. The more stages used, the more likely is it that variety × year interactions will become important sources of bias or inaccuracy (Finney 1984).

If selection is not very intense, the distribution resulting from a few stages of truncation selection will not be markedly non-Gaussian (Finney 1956, 1961). Furthermore, initial non-normality will not be a major problem unless skewness is extreme (Curnow 1961).

7.2.2.1 Selection by nearest neighbour methods Just as variability in field trials may be more effectively partitioned by using differences between neighbouring plots in various types of covariance analysis (Section 3.4.1), so may these differences be used to select potentially superior genotypes. In the grid method (Gardner 1961), the field is divided into sections (grids) and the best-yielding plants in each grid are retained for further trials, crosses, etc. In the honeycomb method (Fasoulas 1973), the field layout is regularly hexagonal, and plants are chosen on the basis of their performance relative to their six nearest neighbours. In this layout, it is possible to have a check variety as one of the six neighbours of every tested plant, and to have all genotypes each others' neighbours equally frequently.

Such methods promise to control variability precisely in a highly relevant way, so that their general lack of use has been disappointing (Bos 1983a,b). Two of the reasons are as follows: the methods are inflexible and may be applied more at the screening stage than at the variety trial stage for field crops; and growing plants in the absence of competition will allow different types of genotype–environment interaction to occur than those relevant in a commercially grown pure stand. They may also not allow multi-stage selection procedures such as Finney's described earlier to be carried out precisely, since the proportion retained is a property of the layout, rather than the breeder's choice.

7.3 Selection in inbreeding species

The greatest problem of selection in species which are largely self-fertilized is closely related to that considered in the previous section, since there should be little genetical variation within a variety, and varieties may therefore be chosen for possession of desirable single gene traits or for their rank on the metric scale of interest. Most difficulties will arise with the combination of traits by crossing or with the performance of progeny of crosses to the parental varieties.

7.3.1 Single gene traits

In principle, provided that phenotypes may readily be distinguished, selection for single gene traits should be simple. If the problem consists of selecting among existing varieties, then the discussion in Section 7.2 indicates the appropriate approach. If it consists of choosing appropriate genotypes at two or more loci, crosses will need to be made of the following form:

$$A_1A_1B_1B_1 \times A_2A_2B_2B_2$$

Here, the desired genotype may be taken as $A_1A_1B_2B_2$. Suppose that A and B are linked, with recombination fraction θ. Then the F_2 will consist of

$$\begin{array}{ccccccc} \dfrac{A_1B_1}{A_1B_1} & \dfrac{A_1B_1}{A_1B_2} & \dfrac{A_1B_1}{A_2B_1} & \dfrac{A_1B_1}{A_2B_2} & \dfrac{A_1B_2}{A_1B_2} & \dfrac{A_1B_2}{A_2B_1} & \dfrac{A_1B_2}{A_2B_2} \\ \frac{1}{4}(1-\theta)^2 & \frac{1}{2}\theta(1-\theta) & \frac{1}{2}\theta(1-\theta) & \frac{1}{2}(1-\theta)^2 & \frac{1}{4}\theta^2 & \frac{1}{2}\theta^2 & \frac{1}{2}\theta(1-\theta) \end{array}$$

$$\begin{array}{ccc} \dfrac{A_2B_1}{A_2B_1} & \dfrac{A_2B_1}{A_2B_2} & \dfrac{A_2B_2}{A_2B_2} \\ \frac{1}{4}\theta^2 & \frac{1}{2}\theta(1-\theta) & \frac{1}{4}(1-\theta)^2 \end{array}$$

Of these, $\frac{1}{4}\theta^2$ will have the required genotype. The number, n, of plants grown in the F_2 need then be sufficient to ensure that $\left(1-\dfrac{\theta^2}{4}\right)^n$ is smaller than some required value, say 0.05. If $\theta = 0.01$, this gives $n \cong 120\,000$. It will, therefore, be better with tight linkage to self all plants carrying both A_1 and B_2 and obtain the required double homozygotes in the F_4. Thus, genotypes

$$\frac{A_1B_1}{A_1B_1}, \frac{A_1B_1}{A_2B_1}, \frac{A_2B_1}{A_2B_1}, \frac{A_2B_1}{A_2B_2} \text{ and } \frac{A_2B_2}{A_2B_2}$$

may be discarded, leaving a proportion $\frac{3}{4}\theta^2 + \theta(1-\theta) + \frac{1}{2}(1-\theta)^2$, i.e. approximately $\frac{1}{2}$ for θ small, approaching $\frac{9}{16}$ as θ increases towards $\frac{1}{2}$. Thus, if obtaining the genotype $A_1A_1B_2B_2$ is the main selective requirement, it may be obtained in the F_4 or F_5 with high probability by discarding about a half of all progeny in the F_2, F_3, and F_4 and selfing the rest. Since a simple criterion for reducing the amount of material to be carried forward each generation is often required, selection on qualitative traits is usually practised in early generations. Possible conflicts with other aims are discussed in Section 7.3.3.

7.3.2 Metric traits

A good theoretical account of response to selection in self-fertilized species is that of Pederson (1969a,b). A more precise approach to the problem (using transition matrices) has been taken by Curnow (1978), but it deals with the effects of only a few loci, so that I shall follow Pederson's account here. (The progress of selection under regular inbreeding in normally outbred species has been extensively studied (see for example Cornelius and Dudley 1976), but will be more appropriately dealt with in Chapter 9.)

Consider a population sufficiently large for random effects to be unimportant, with k loci each of individually small effect contributing to the metric trait of interest. Because artificial selection theory, originally developed for panmictic populations, generally allows expression of response in terms of covariance between relatives (see Section 7.4.2), it will be convenient for comparative purposes to do the same for autogamous species.

Pederson (1969a) has shown that the covariance between genotypic values after t and t_2 generations of self-fertilization will be difficult to estimate in general, but is the critical determinant of response to selection: after one generation of selection with intensity i,

$$\bar{X}_1\,(\text{selected}) = \bar{X}_1\,(\text{unselected}) + i\,\frac{\text{Cov}_{G_{0,1}}}{\sqrt{V_{P_0}}}$$

i.e.

$$R = i\,\frac{\text{Cov}_{G_{0,1}}}{\sqrt{V_{P_0}}}$$

and after t_2 generations, $t_2 - t$ under selection, t of selfing without selection,

$$\bar{X}_{t_2}(\text{selected}) = \bar{X}_{t_2}(\text{unselected}) + i \sum_{t=0}^{t_2-1} \text{Cov}_{G_{t,t_2}} / \sqrt{V_{P_t}}.$$

Here, $\bar{X}_t$, denotes the mean and V_{P}, the phenotypic variance in the tth generation, $\text{Cov}_{G_{t,t_2}}$ the covariance between genotypic values in the tth and t_2th generations of selfing without selection. (This latter point is an approximation, depending upon the contributions of individual loci being individually small, as assumed.)

Thus, prediction of $\bar{X}_{t_2}$ (selected) requires a knowledge only obtainable by raising experimentally many generations of selfed plants; in general, it will be more effective to conduct selection experiments without predictions than to carry out the estimation experiments first. However, two special cases of relevance to the breeder may be examined more simply.

First, assume that there is random mating in the base population, and that there are no dominance or epistatic effects. Then, the population mean does not change in the absence of selection, and Pederson (1969a) has shown that

$$\overline{X}_{t_2}(\text{selected}) = \overline{X}_{t_2}(\text{unselected}) + i \sum_{t=0}^{t_2-1} \{2-(\tfrac{1}{2})^t\} V_A / [\{2-(\tfrac{1}{2})^t\} V_A + V_E]^{\frac{1}{2}}.$$

Thus, response here depends upon the additive genetic variance in the base population, V_A, and on the environmental variance, V_E, both of which may readily be estimated.

In a second simplification, Pederson (1969a) has considered the case of only two alleles per locus, as would arise from the crossing of two homozygous lines or varieties. Setting the genotypic frequencies of A_1A_1, A_1A_2, and A_2A_2 equal to P, Q, and R, respectively, he obtained

$$V_{P_{t_2}} = \frac{1}{P+R}\{1-(\tfrac{1}{2})^t Q\}\{V_A + (\tfrac{1}{2})^{t_2} V_D\} + V_E$$

and

$$\mathrm{Cov}_{G_{t,t_2}} = \frac{1}{P+R}\{1-(\tfrac{1}{2})^t Q\}\{V_A + (\tfrac{1}{2})^{t_2} V_D\}.$$

Now, consider selection beginning with the F_2, where $P = \frac{1}{2}Q = R = \frac{1}{4}$:

$$\overline{X}_{t_2}(\text{selected}) = \overline{X}_{t_2}(\text{unselected}) + \sum_{t=0}^{t_2-1} \frac{2\{1-(\frac{1}{2})^t\}\{V_A + (\frac{1}{2})^{t_2} V_D\}}{[2\{1-(\frac{1}{2})^t\}\{V_A + (\frac{1}{2})^{t_2} V_D\} + V_E]^{\frac{1}{2}}}.$$

Thus, dominance effects decline relative to additive effects as selection proceeds.

7.3.3 Early generation selection

Under self-fertilization, progress towards homozygosity is rapid, the frequency of heterozygotes at any single locus halving in each generation. (See Franklin (1977) for a general discussion of the effect of linkage etc.) In any particular cross, plants containing many favourable genes (in heterozygous state) will be far more common in the F_2 and F_3 than in later generations. For this reason, Shebeski (1967) and others have strongly advocated early generation selection for yield.

However, the effect of continued selfing on the genetic variance of a trait is rather different. To take a simple illustrative example, consider a cross between two parents thus:

	P_1		P_2
Genotype	A_1A_1		A_2A_2
Metric value	a	$\times$	$-a$
Variance	V_E		V_E
		F_1	
Genotype		A_1A_2	
Metric value		d	
Variance		V_E	
		F_2	
Genotype	A_1A_1	A_1A_2	A_2A_2
Metric value	a	d	$-a$
Variance	V_E	V_E	V_E

The F_2 variance among genotypes will be

$$V_A + V_D + V_E = \tfrac{1}{2}a^2 + \tfrac{1}{4}d^2 + V_E.$$

In the F_3, this changes to $\frac{3}{4}a^2 + \frac{3}{16}d^2 + V_E$. In the limit it becomes $a^2 + V_E$. Thus, the variance among lines increases, while the genetical variance within lines declines to zero. In the previous section, we saw, in a more general case, how response to selection depends on V_A and V_D; here we wish to note the particular problems of selection in early generations.

These are of two kinds: those which arise from the experimental conditions; and those which arise from the elimination of genetical variation within plants. Depending upon the kind of selection design used (see Chapter 10), plants in the F_2 and other early generations may be grown singly or in small groups, but they will obviously not be grown in large plots. Later generations, however, will be grown in larger plots as the experimenter strives to build up seed stocks and assess field performance. V_E will therefore be different at different stages, thereby confounding the estimation of V_G and the ranking of genotypes, especially if, as noted earlier, performance as single plants and in pure stand is not constant. In addition, heterosis may be important in early generations, as found by Lupton and Whitehouse (1957) in wheat. Competition among plants, which may also be important, is discussed in Sections 5.2.2 and 7.5.

Taking the second point, in the single locus example used here, while $V_P(F_2) > V_P(P_1)$ or $V_P(P_2)$, it is also true that, unless $|d| > a$, the best genotype available will be one of the two homozygotes, but it will in general, as already discussed, be most unlikely that all alleles contributing favourably to the metric of interest are to be found in one parent. In the F_2, therefore, values more extreme than either parent may well be obtained, and $V_P(F_2) > V_P(P_1)$ or $V_P(P_2)$ in general. Take the simple model with two loci, considered already in Section 4.5 thus:

	A_1A_1	A_1A_2	A_2A_2
B_1B_1	$2a$	$a+d$	0
B_1B_2	$a+d$	$2d$	$-a+d$
B_2B_2	0	$-a+d$	$-2a$

Suppose that P_1 has genotype A_1A_1 B_2B_2 and P_2 has genotype A_2A_2 B_1B_1. Then the generation means and genotypic variances will be (assuming $\theta = \frac{1}{2}$):

Generation	Mean	Variance
P_1	0	0
P_2	0	0
F_1	$2d$	0
F_2	d	$a^2 + \frac{1}{2}d^2$
F_3	$\frac{1}{2}d$	.
F_4	.	.
.	.	.
.	.	.

In the F_2, values from $2a$ to $-2a$ should be observed. Thus, selection as early as the F_2 should be worthwhile for metric traits, provided that the problems with heterosis and V_E are surmountable, and that linkage between loci affecting the metric trait is unimportant.

Because of the caveat respecting linkage, Hanson (1959) suggested intercrossing for at least one, preferably several, generations after the F_1, so as to break down linkage combinations from the parental lines. This might be especially advantageous if early generation selection were for other reasons difficult.

Results obtained have been equivocal. We have already been (Section 5.2), in another context, how performance varies in early generations according to how plants are grown, i.e. singly or in a stand. In an earlier trial, where F_3, F_4, and F_5 were grown under more similar conditions than in Hamblin and Donald's work, statistically significant correlation coefficients were obtained of 0.59 between F_3 lines and related F_4 bulk means and 0.56 between F_3 lines and related F_5 bulk means (de Pauw and Shebeski 1973). Knott and Kumar (1975) have suggested, on the basis of an experiment using two crosses in wheat, that F_3 yield testing will not justify the extra work, since F_3, F_5 correlations are rarely as high as de Pauw and Shebeski's, and since, in addition, the top F_6 lines produced by single seed descent will be as good as top F_5 lines produced by F_3 and F_4 yield testing.

More directly relevant to the problem of breakdown of unfavourable linkages is the work of Altman and Busch (1984), who found that three cycles of intercrossing in spring wheat produced no real sign of increased V_G and no gain in mean yield on selection. They concluded that enforced intercrossing could not be justified as a routine procedure.

Several workers (Pederson 1974; Bos 1977; Sneep 1977; Stam 1977) have considered the problem further from a theoretical point of view. Although random mating for a number of generations will increase the ultimate selection response if linkage is moderate and selection weak (cf. Section 7.4.2; see also Stam 1977), the general conclusion is that generations of intercrossing will waste valuable time and that some selection in early generations will be better than no selection. It will generally be selection for qualitative traits, however, and yield selection in the F_2 may be too early.

Other conditions have been summarized by Sneep (1977). Yield tests may be useful in and after the F_3. Both bulk breeding and single seed descent (Section 10.5) will result in loss of variability by drift without unmanageably large trials. Special problems exist for selection in the F_3. In a cereal, for example, it will be necessary to grow large F_3 lines (at least 200 plants per line), to sow three rows per line at normal density, and to select for yield in competition with similar types (e.g. by using only the yield of the middle row).

This is a problematical area where much more work is needed. It is discussed further in Chapter 10.

7.4 Selection in outbreeding species

The response to selection is much simpler to describe in random mating populations than in others, so that our treatment in this section will concentrate on panmictic populations undergoing selection, in order to illustrate complexities such as selection for more than one trait, in more than one environment, or under competition among plants.

7.4.1 Single gene traits

The dynamics of gene frequency change under natural selection were first extensively investigated by Haldane (see Haldane 1927, 1932). In many cases, his and other results may be used to describe change in response to artificial selection. Consider a diallelic locus in a panmictic population large enough for random fluctuations in gene frequency to be negligible relative to systematic changes. Suppose that the proportions of the three genotypes remaining after selection are as follows:

	A_1A_1	A_1A_2	A_2A_2
Frequencies	p^2	$2pq$	q^2
Proportion remaining after selection	α	β	γ

(Here, $0 \leqq \alpha \leqq 1$, $0 \leqq \beta \leqq 1$, $0 \leqq \gamma \leqq 1$). The change in the frequency of A_2 will be

$$\Delta q = q - \frac{pq\beta + q^2\gamma}{p^2\alpha + 2pq\beta + q^2\gamma}.$$

Progress under selection will depend upon the values taken by α, β, and γ:

Type of selection	Dominance	α	β	γ	Δq
Against A_2	Absent	1	$1 - s_2/2$	$1 - s_2$	$-\frac{1}{2}s_2q(1-q)/(1-s_2q)$
Against A_2 (recessive)	Present	1	1	$1 - s_2$	$-s_2^2q(1-q)/(1-s_2^2q)$
Against A_1 (dominant)	Present	$1 - s_1$	$1 - s_1$	1	$s_1q^2(1-q)/\{1-2(1-q^2)\}$
Against homozygotes	Absent	$1 - s_1$	1	$1 - s_2$	$pq(s_1p - s_2q)/(1 - s_1p^2 - s_2q^2)$

From these expressions, we can easily see how artificial selection can be retarded by natural selection. Suppose, for example, that we seek to eliminate a certaint trait, manifested by a proportion K of genotype A_2A_2 (autosomal recessive). Then selection will proceed (in the absence of progeny testing) at the rate

$$\Delta q = -Kq^2(1-q)/(1-Kq^2)$$

if the popuplation was previously in Hardy–Weinberg equilibrium. If the population was in equilibrium, but with gene frequencies maintained by selection against homozygotes, then selection to maintain gene frequencies $p = s_2/(s_2 + s_1)$ and $q = s_1/(s_2 + s_1)$ would conflict with the removal of K A_2A_2. Table 7.2 shows the decline in q in the presence and absence of heterozygous advantage. The effects on Δq will largely be modest, at least unless extremely disadvantageous phenotypes are sought, but a substantial lowering of mean fitness will be discernible. For plants grown in mass monoculture, the meaning of a decline in fitness will depend upon the actual aspect of fitness (growth rate, competitive ability, disease resistance, etc.) which is summarized by s_2 and s_1.

Table 7.2

Effect of heterozygous advantage on artificial selection against a recessive phenotype. See text for symbols

Initial equilibrial					Δq		
					(a) heterozygous advantage only	(b) selection against A_2A_2 only	(a) and (b)
p	q	s_1	s_2	K			
0.5	0.5	0.1	0.1	0.5	0	− 0.071	− 0.067
				1.0	0	− 0.167	− 0.155
0.1	0.9	0.09	0.01	0.5	0	− 0.068	− 0.068
				1.0	0	− 0.426	− 0.424

7.4.2 Metric traits

If the average value of a trait in the two parents of a particular individual is X and the individual's value is Y, and if Y is normally distributed with variance constant over all values of X, then (treating X and Y as departures from their appropriate means) the relationship between X and Y is described by the simple linear regression equation

$$Y = b_{O\bar{P}}X.$$

Now in the previous chapter it was explained how $b_{O\bar{P}}$ estimates narrow sense h^2, so that we may write

$$Y = h^2X.$$

Suppose that a group of parents is chosen with mean value S. Then S, the selection differential, measured as a departure from the population mean is shown in Fig. 7.1.

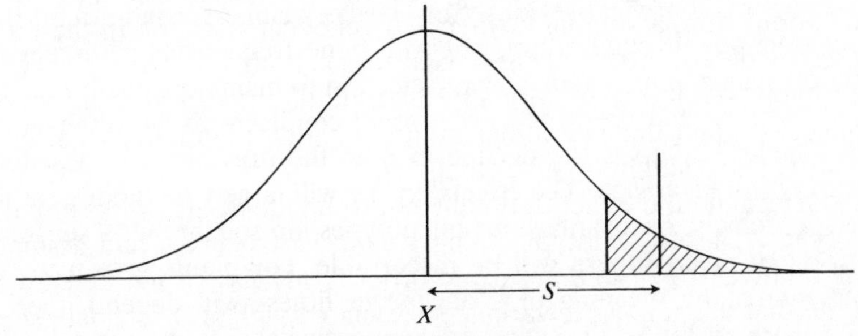

Fig. 7.1. The selection differential, S, measured as a departure from the population mean.

Then the departure of the progeny mean from the population value will be expected to be $h^2 S$, usually written R, the response to selection.

This may be expressed as

$$R = ih^2 \sqrt{V_P}$$
$$= ih\sqrt{V_A},$$

where i is the intensity of selection, measured in units of the standard deviation of the standard normal distribution. This distribution has been tabulated (e.g. Table II1 of Fisher and Yates 1963) or is easily computed, and so given intensities may readily be compared and converted into selection differentials. Furthermore, it may frequently be desired to select a proportion, p_s, of the population, to retain for breeding. This may be converted to a selection intensity i by Smith's (1969) approximation:

$$i = 0.8 + 0.41 \ln \left(\frac{1}{p_s} - 1\right).$$

Simmonds (1977) has tabulated this approximation, and modified it for very low values of p_s.

For the single locus example considered in the previous section and elsewhere, Robertson (1956), following Haldane (1927), has shown that the change in gene frequency in one generation for an intensity of selection i will be

$$\Delta q = \frac{ia}{\sqrt{V_P}} pq.$$

Now it is clear that selection, if effective, will reduce V_A, and hence h^2, so that the expression for response discussed above will only be accurate, if at all, for prediction over one generation. In addition, it is assumed that the factors contributing to V_A are unlinked. To see how the expression may be generalized and extended, we shall examine the work of Griffing (1960) and Robertson (1960, 1970).

Griffing (1960) considered only the case of selection in an infinitely large population, assuming further that genetical variation in the trait of interest is determined by genes, each of small effect, at many loci; that phenotypic variability is normally distributed about a zero mean, with variance V_P; and that the initial random-mating population is at equilibrium at all loci and between all loci.

To determine the effect upon alleles at one locus, A, suppose that the genotype A_iA_j has genotypic value g_{ij} and variance V_{ij}, and assume g_{ij} small relative to $\sqrt{V_P}$ so that second order terms may be ignored. Further assume that the genotypic variance associated with A_iA_j is small, so that $V_{ij} \cong V_P$.

Selection may be considered in general as shown in Fig. 7.2 Here,

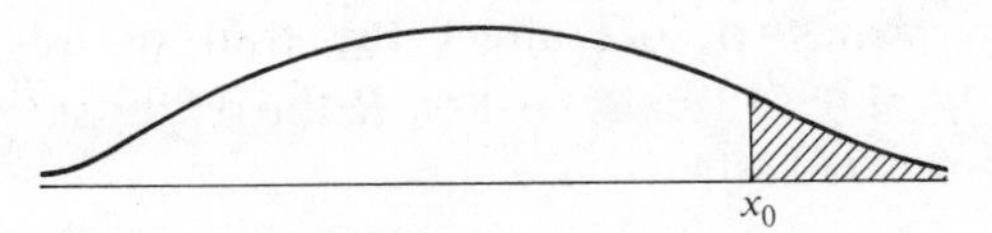

Fig. 7.2. Truncation selection of a continuous trait; the shaded portion indicates the fraction chosen as parents.

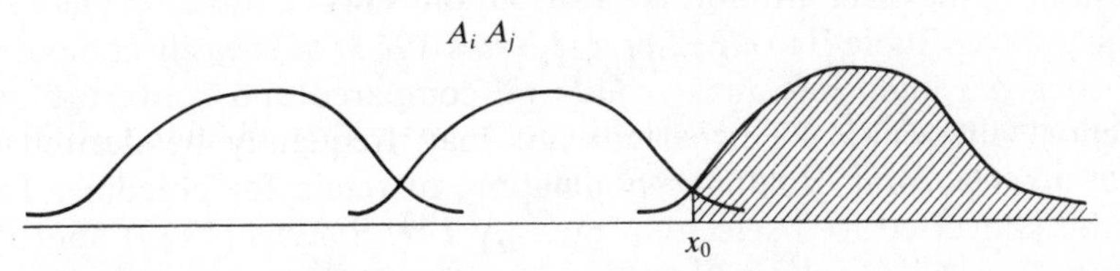

Fig. 7.3. Truncation selection applied to a population made up of several subpopulations, where genotype (A_iA_j) determines the location of a subpopulation.

individuals with metric values $x > x_0$ are chosen and mated at random to produce the next generation. This may be portrayed in terms of subpopulations as shown in Fig. 7.3.

Here one sees directly that as g_{ij} affects the position of the A_iA_j subpopulation so it determines how likely A_iA_j individuals are to be selected. The selective value, w_{ij}, of A_iA_j is defined to be proportional to the probability of an individual of genotype A_iA_j surviving selection,

$$P(x > x_0) = \frac{1}{\sqrt{V_P 2\pi}} \int_{x_0}^{\infty} e^{-\{(x-g_{ij})^2/2V_P\}} dx$$

From this, Griffing showed that for a selection differential S

$$\begin{aligned} w_{ij} &\cong 1 + Sg_{ij}/V_P \\ &= 1 + ig_{ij}/\sqrt{V_P}. \end{aligned}$$

The genotypic mean of selected parents is then $\mu_s = iV_G/V_P$. The progeny mean is obtained as approximately $\mu_1 = ih^2\sqrt{V_P} = R$ as before ($\mu_0 = 0$ by definition). Griffing, however, has been able to show further that in the absence of mutation, and provided natural selection does not affect the trait of interest, the response after n generations is given by

$$\begin{aligned} R_n &\cong nR_1 \\ &= nih^2\sqrt{V_P}. \end{aligned}$$

(This is provided that i is small, since intense selection will reduce V_G and affect response, see Bulmer 1980.) If natural selection favours one homozygote at the locus, then response to artificial selection will be accelerated or impeded according as the allele favoured contributes positively or negatively to the metric which the breeder seeks to increase.

Suppose now that two loci affect the trait of interest, which is determined as the sum of two gametes, say A_iB_k and A_jB_l, since the loci may be linked. Genetic variance is

$$V_G = V_A + V_D + V_{AA} + V_{AD} + V_{DD}$$

since interactions are possible between additive and dominance contributions at the two loci. Starting at complete gametic equilibrium, the effect of one cycle of selection is to yield a response

$$R = i(V_A + \tfrac{1}{2}V_{AA})/\sqrt{V_P}$$

which incorporates an additional factor over that obtained with one locus. In general, Griffing showed that

$$R_n = i\left\{V_A + \frac{(1-\theta)}{2}V_{AA}\right\}\Big/\sqrt{V_P}$$

for recombination fraction θ between loci A and B.

For n cycles of selection, R_n lies between $ni(V_A + \frac{1}{2}V_{AA})/\sqrt{V_P}$ and $i\left[nV_A + \frac{\{2-(\frac{1}{2})^{n-1}\}}{2}V_{AA}\right]\Big/\sqrt{V_P}$. The influence of V_{AA} declines with time, so that the response to selection does not remain linear for many generations.

Generalization for an arbitrary number of loci is less easy. However, Griffing's argument yields the result that between any two random mating generations of selection,

$$\mu_n = \mu_{n-1} + 2i\text{Cov}_{\text{OP}}/\sqrt{V_P}.$$
$$= \mu_{n-1} + 2ih^2\sqrt{V_P}.$$

The additive genetical variance is always the critical determinant of response to selection.

The contribution of fresh mutation to V_G has been discussed in Section 5.2.1.2. Hill (1982) has considered the influence of mutation on response to selection, and has concluded that on the most optimistic assumptions, mutation will rarely be significant within twenty generations. Thus, it may have had an effect on the very long-term response in the experiments to be discussed below, but cannot be considered in planning a breeding programme with the aim of medium-term variety release.

Robertson (1960, 1970) has been concerned to establish both the nature of and the rate of approach to selection limits, i.e. to find out how far and how fast artificial selection may go before usable genetical variation is exhausted. The problems particularly explored by Robertson are the effects of population size and linkage.

His analysis used only the following variables: population size, N; number of diallelic loci, n; equal effect of all loci, a', the difference between the two homozygotes; initial frequency of the desirable allele at a locus, q; i; V_P; d, chromosome length in map units; c, the map distance between adjacent loci; and l the overall length of the genome.

Then using C as a subscript to refer to a particular chromosome, Robertson showed that in the absence of recombination the expected total gain for low values of Nih_C

$$= 2Nih_C^2\sqrt{V_{P_C}}$$

$$= 2N.\text{ change in one generation,}$$

for small $ih_C^2\sqrt{V_P}$.

For high Nih_C, sample size becomes the dominant determinant of response, as shown in Table 7.3. The problem may be very clearly stated, and the table illustrates it: no matter how intense the selection, the best possible outcome is fixation of the best gamete originally available.

Results with free recombination may be contrasted with these results. When Nih_C is small, there is no difference between the expected gain, since in essence only the best gamete is chosen. Thus,

$$L_0 = L_f = 2Nih_C^2/\sqrt{V_{P_C}}$$

where L_0 and L_f are the limits with zero and 0.5 recombination between loci. Hence, $L_0/L_f \cong 1$, in this case. For any particular n and q, there is a value of Nih_C above which the chance of fixing all desirable alleles is close to unity, when

$$L_f = \{2n(1-q)/q\}^{\frac{1}{2}}\sqrt{V_G}$$

giving

$$\frac{L_0}{L_f} \cong 3[q/\{2n(1-q)\}]^{\frac{1}{2}}$$

Table 7.3

The expected superiority, in units of $\sqrt{V_{G_C}}$, of a line homozygous for the best gamete from an initial sample of a given size (from Robertson 1970). Reprinted with permission from Springer–Verlag and Professor A. Robertson

Number of gametes in original sample	Expected superiority, from the normal curve (multi. by $\sqrt{2}$ since the (diploid) line is homozygous for the best gamete)
10	2.18
20	2.64
40	3.05
80	3.44
160	3.76

Thus, the relationship depends upon the number of loci and their initial frequencies. Intermediate linkage values yield, as is to be expected, intermediate results; the two L_0/L_f ratios shown illustrate the extreme range of possibilities.

Robertson was able to show that certain combinations of the eight variables affected the selection limit: Nih_C, Nl, n, and q. However, it was not possible to achieve analytic descriptions of the approach to the limit, though several general conclusions could be drawn.

The optimal intensity of selection is never very great. For $n = 10$, $q = 0.2$, and $l = 27.5$ cM, for example, i should lie between 0.5 and 0.8 for all Nih_C and Nl. For small Nih_C, maximum advance is approximately $2Nih_C$, while for large Nih_C, it is closer to $3Nih_C$. If $Nih_C < 0.05$, linkage is relatively unimportant in retarding the rate of approach to the limit. Overall, $i = 0.5$ is optimal for unlinked loci, while i should be greater than 0.5 if there is linkage.

Population size can be chosen so as to ensure a certain level of response (Nicholas 1980). For selection of intensity i and heritability h^2, the population size required for the chance of success to be α after t generations is approximately $2\langle z_\alpha^2/[t\{ih(R_c/R - 1)\}^2]\rangle$ where achieving R_c/R, a chosen proportion of the predicted response R, is defined as success and z_α is the standard normal deviate corresponding to α. Thus, if $R_c/R = 0.95$, $i = 2$, $h = 0.25$, $\alpha = 0.95$ and $t = 4$, the necessary size is approximately 800.

Plateaux in selection with persistent genetical variation may come about from at least three causes: natural selection which favours an intermediate optimum; natural selection which favours multiple heterozygotes, which are likely to have intermediate phenotypes for a quantitative trait; and differential variability of different genotypes, which will contribute to the estimated V_A and V_G, but not to R. The former two are discussed extensively by Nicholas and Robertson (1980), partly in terms of the model of Section 7.4.1, but also considering the effect of finite population size. They show that a plateau will generally occur well before the phenotype is reached that would have been predicted on the basis of selective neutrality. There is evidence for the third phenomenon, though from animal populations, not plants (Fleischer *et al.* 1983).

These results are only a very general guide, as were those of Finney in Section 7.2.2. We must consider how experimental results relate to them. Many experiments have been carried out with the specific aim of testing quantitative genetical theory. Problems still not completely solved include asymmetry of response to selection aimed at increasing or decreasing a trait and fluctuations in response, with apparent plateaux reached followed by further response. Falconer (1982) has considered these problems extensively, and should be consulted for a detailed review. Here, we shall simply examine examples relating to these important problems.

Table 7.4
Mean gain and predicted gain per generation in oil and protein percentage in four selected lines of maize (from Dudley and Lambert 1969)

			h^2‡	
Population	Actual gain	Predicted gain†	Oil	Protein
Illinois high oil	0.16	0.13	0.3	0.11
Illinois low oil	− 0.07	− 0.02	0.17	0.28
Illinois high protein	0.22	0.26	0.40	0.21
Illinois low protein	− 0.09	− 0.16	0.16	0.23

†Given by $ih\sqrt{V_P}$, where h^2 is broad sense heritability.
‡Broad sense heritability; V_A not presented.

One of the most striking long-term selection experiments on plant material has been the Illinois maize selection experiment, which for over 80 generations has been aimed at increasing and decreasing the oil and protein content of maize kernels. In 1969, Dudley and Lambert reported the results shown in Table 7.4 for 65 generations of the experiment.

It can be seen that average response differs in the increasing and decreasing lines, that actual response differs from expected response, and that genetic variance persists, for h^2 was measured recently in the experiment, not just initially. Over the course of the experiment, response was not consistent, nor precisely linear, yet it has continued for many generations more than should have been necessary to exhaust usable genetic variation. Equally, of course, the experiment has gone on for long enough for both advantageous recombination and mutation to have had significant effect.

Shorter term experiments have generally shown reasonable agreement with theory, though frequently illustrating other problems. For example, Harris *et al.* (1972) reported the results of nine cycles of mass (truncation) selection in maize, starting with the variety Hay's Golden. In 1969, which was a good season, the selected material outperformed Hay's Golden by 5.66 t/ha to 4.20 t/ha, but in 1970, a year of wind, drought, blight, and hail, the equivalent values were 2.98 and 2.77. Statistical procedures must be robust indeed to overcome such fluctuations; generally, the approximations involved and experimental errors will mean that effects of selection may be detected as significant increases or decreases only, with commercial growing needed to assess the true improvement.

This is illustrated further by an experiment of Dudley *et al.* (1963) on selection in two populations of lucerne for a number of traits, including resistance to rust and tolerance of leafhopper yellowing. A selection proportion of $p_s = 0.05$ was used, yielding $i = 2.01$ for a single trait,

Table 7.5
The effect of seven cycles of selection on means and genetic variances on two traits in two populations of lucerne (from Dudley *et al.* 1963)

Selection cycle	Rust resistance (selected trait)						Yield in g/plant (unselected trait)					
	A			B			A			B		
	Mean	V_G	$h^{2a,b}$	Mean	V_G	h^{2a}	Mean	V_G	h^{2b}	Mean	V_G	h^{2b}
2	6.5	3.02	59	5.6	4.79	72	117.8	1657	61	122.1	1743	64
3	6.4	3.28	61	4.5	6.02	80	131.3	1804	62	114.8	1753	64
4	6.5	3.51	62	3.8	4.17	77	122.7	1206	53	121.9	1448	59
5	3.6	4.61	79	2.7	2.38	72	130.1	1455	55	132.1	998	46
6	2.6	2.15	70	2.4	1.94	70	135.0	2215	65	135.6	1640	58
7	2.9	2.03	67	2.5	2.39	74	143.3	2092	63	134.4	1495	56
8	2.8	2.24	71	2.1	1.19	62	152.7	2561	68	136.5	1451	55

a. Broad sense heritability.
b. No significant decline.

though selection was not actually on this basis, so that predicted response cannot be precisely shown. Table 7.5 shows the results of seven cycles of selection on one selected trait, resistance to rust [*Uromyces striatus* Schroet. var. *medicaginis* (Pass.) Arth.], together with the behaviour of an unselected trait, dry matter yield per plant.

It can be seen that both rust resistance and yield increased significantly in both populations, although yield was not a selection criterion. That there was no commensurate decline in the genetic variance for yield suggests that the increase was not obtained through genetic correlation between yield and rust resistance, but rather that the latter constituted a factor limiting yield. (In the event, population B was released as a new variety, Cherokee, with increased yield and tolerance to leafhopper yellowing as well as rust resistance.)

7.4.2.1 Selection against extremes In certain situations, it will be desirable to select intermediate phenotypes of some particular trait. This will arise particularly with selection indices, where allowance can be made for undesirable correlated responses (see the next sections). For example, selection for grain yield in oats may increase both plant height and time to heading (Rosielle *et al.* 1977), while selection for grain yield in maize may increase the moisture content at harvest time (Compton and Bahadur 1977). It is of some interest to know what will happen, in general, if artificial selection is of the type described, when it occurs in nature, as stabilizing.

For a single gene trait, where the heterozygote is intermediate on the metric scale, it has been shown by Curnow (1964) that, in general, 'Selection of individuals in any region near the population mean will favour the least variable genotype irrespective of its position on the scale of mean values.' Now Robertson (1956) had earlier shown that the general tendency will be to fix one or other homozygote, so that from these two results it can be concluded that a homozygote will be fixed unless the heterozygote is less variable than either homozygote.

Multiple loci show the same tendency in a more marked form. As an example, consider a four locus model examined by Mayo (1970). A polymorphism at four equivalent loci A_i, $i = 1, \ldots 4$, each with two possible alleles, was treated by simulation. In the diploid, genotypes at each locus contribute to a score thus:

A_iA_i	$A_iA'_i$	$A'_iA'_i$	$i = 1, \ldots 4$
$-a$	0	a	

The scores are summed over the loci, to give a total value ka, and the fitness is given by $1 - \dfrac{k^2a}{16}$, so fitness is inversely proportional to the deviation from the optimum genotypes, of which there are several kinds, represented by $A_iA'_i$, $i = 1, \ldots 4$, $A_jA'_jA_kA'_kA_lA_lA'_mA'_m$, and $A_jA_jA_kA_kA'_lA'_lA'_mA'_m$. One expects that the only completely stable equilibrium will be with all individuals having the third constitution, i.e. with all loci fixed. In the tetraploid, the situation is the same but expanded: each locus contributes to the score thus:

$(A_i)_4$	$(A_i)_3A'_i$	$(A_i)_2(A'_i)_2$	$A_i(A'_i)_3$	$(A'_i)_4$	$i = 1, \ldots 4$
$-a$	$-\frac{1}{2}a$	0	$\frac{1}{2}a$	a	

The range of possible fitnesses is the same, with smaller intervals between genotypes, and the stable equilibria are again expected to be with all genotypes multiple homozygotes with scores of zero.

Two population sizes, 50 and 100, were used. Techniques of simulation were those of Mayo (1966, 1975). Table 7.6 shows mean times to fixation. Trials were run until all loci were fixed. In no case in more than 30 trials did the populations of size 100 fix with anything less than the optimum fitness, but as the results do not differ much between levels of selection it is clear that random effects are most important when there is selection. The time to fixation in the absence of selection is greater than that with selection in each case, both for diploids and for tetraploids, and for each particular population size, this effect is greater for diploids than for tetraploids. This illustrates the point that multi-locus polymorphisms are inherently less stable than single-locus ones, though of course this is here compounded by the fact that there exist stable, invariant maximum fitness states with all loci fixed. This means that any departure at one

Table 7.6
Mean time in generations, T, to fixation of loci, with population size N and selective disadvantage, s, for the least fit genotype

N	s	Diploid	Tetraploid
50	0	113.0 ± 4.6	205.8 ± 30.5
	0.2	67.0 ± 14.8	162.5 ± 20.5
	0.4	67.2 ± 7.5	169.2 ± 21.4
100	0	211.6 ± 15.8	335.0 ± 45.6
	0.2	158.7 ± 16.1	255.5 ± 28.7
	0.4	161.7 ± 14.9	251.5 ± 27.1

locus from the initial, equilibrial equal gene frequencies for the two alleles will tend to induce a compensating change in the opposite direction at other loci, thus still further disturbing the equilibrium.

In Chapter 5, the maintenance of variability under stabilizing natural selection is treated, and the importance of linkage emphasized. Because artificial selection is generally so much more intense, it may be concluded that artificial stabilizing selection will tend to decrease variability substantially, without necessarily producing balanced, optimal genotypes as in the example of Mayo (1971), even when these potentially exist in the initial population.

7.4.2.2 Correlated response to selection In Table 7.5, response to selection for rust resistance in lucerne was shown, together with simultaneous changes in yield. In this particular case, it seemed unlikely that the improved yield resulted from genetical correlation between disease resistance and yield, but in many cases a correlated response to selection will be the result of such genetical correlation. Such correlation may also be used for indirect selection by means of an index.

If we consider the change in the breeding value, A_X, of a trait Y, when selection is on trait X, then the regression of A_Y on A_X will give the response of Y as

$$CR_Y = b_{A_Y A_X} R_X.$$

Now

$$\begin{aligned} b_{A_Y A_X} &= \mathrm{Cov}_A / V_{A_X} \\ &= r_A \surd(V_{A_Y} / V_{A_X}) \end{aligned}$$

so that

$$\begin{aligned} CR_Y &= \{r_A \surd(V_{A_Y}/V_{A_X})\}(ih_X \surd V_{A_X}) \\ &= ih_x r_A \surd V_{A_Y} \end{aligned}$$

and comparing indirect with direct selection,

$$CR_Y/R_Y = \frac{ih_X r_A V_{A_Y}}{ih_Y V_{A_Y}} = \frac{h_X r_A}{h_Y}.$$

Selection on the basis of a genetical correlation will thus only be more effective than direct selection if both h_X and r_A are high relative to h_Y. Of course, in cases of enforced indirect selection, as in forest trees, where selection may be on the basis of correlation between traits expressed in seedlings and traits of mature trees, the criterion of efficiency takes second place to the criterion of necessity.

Correlated response to selection may also be taken to mean the effect of selection on a metric trait on particular distinguishable gene loci. Although many long-term selection experiments have been carried out, they have tended not to investigate specific gene loci, so that few data are available to bear on this point. The well-established changes over time in Composite Cross V, an experimental population of barley (Suneson 1956), have to some extent been correlated with changes in gene frequency at four esterase loci (Allard *et al.* 1972), but there is no evidence to show that yield changes result from such single gene changes. Similarly, in the Illinois maize selection experiment previously mentioned, it seemed unlikely that the temporal gene frequency changes at six isozyme loci reflected directional selection for any alleles of these genes, despite overall evidence of non-neutrality (Brown 1971). Contrasting results arose in one study of reciprocal recurrent and full-sib selection for increased yield in maize (Stuber *et al.* 1980). As shown in Table 7.7, allele frequencies altered consistently at three unlinked loci as yield changed. Equivalent, but less marked responses were seen at three other loci. Further investigations appear to confirm these results (Frei *et al.* 1985b). If such associations are widespread, they have important implications for prediction by means of readily distinguishable markers (Goodman and Stuber 1983). It should however, be noted that strong

Table 7.7

Response to selection for yield in maize (FS = full sib selection, RRS = reciprocal recurrent selection). (From Stuber *et al.* 1980)

Cycle of selection	Response to selection (g/plant)		Allele frequencies					
			Acph1		β-glu1-k		Phi1e	
	FS	RRS	FS	RRS	FS	RRS	FS	RRS
0	—	—	0.20	0.20	0.57	0.57	0.98	0.98
6	55.3	78.0	0.54	0.58	0.82	0.87	0.74	0.83
9	62.1	59.0	0.43	0.58	0.91	0.86	0.71	0.87

but irrelevant association between linked genes may arise through crosses between plants from very small, highly selected populations.

7.4.2.3 Selection in more than one environment In a plant breeding programme of any magnitude, assessment of genotypes will normally be carried out in a range of environments. Analysis of genotype-environment interaction in such circumstances is discussed in section 6.3. Here, we consider the problem of correlation between responses in different environments.

Falconer (1952) showed that the response in a trait expected in the kth environment to selection in the jth is

$$CR_{kj} = ih_k h_j \surd(V_{P_k} r^2_{G_{jk}})$$
$$= i\,\mathrm{Cov}_{G_{jk}}/\surd V_{P_j}$$

where $r_{G_{jk}}$ and $\mathrm{Cov}_{G_{jk}}$ are the genotypic correlation and covariance between values in the two environments. Similar relationships will allow use of a selection index incorporating information from more than one environment.

Wright (1976b) has emphasized that the correlated response in mean performance over all relevant environments, termed by him 'general adaptation', is the most important case. Consider the simple model for the performance of m genotypes in n environments.

$$y_{ij} = \mu + g_i + e_j + (ge)_{ij} + \varepsilon_{ij}$$

where μ is the overall mean, g_i the additive effect of the ith genotype, e_j that of the jth environment, $(ge)_{ij}$ their interaction and ε_{ij} the error.

In Chapter 6, it was shown how regression on mean performance in environments could be used to describe genotype–environment interactions in a tractable way. Wright (1977a) has shown how response to selection in differing environments may be treated by this approach.

Average response to selection in single environments can be shown to be $i\surd(V_A + V_{AE} - V_{\surd V_P})$, where V_{AE} is the variance of genotype–environment interaction effects and $V_{\surd V_P}$ is the variance of the phenotypic standard deviation. ($V_{\surd V_P}$ diminishes the response by reducing the accuracy of prediction.) Selection for a single variety with general adaptation for all environments, using a sample of one environment and the same value of i, will yield a correlated response of $iV_A/\surd(V_A + V_{AE} - V_{\surd V_P})$. Thus, response in this latter case is diminished in the proportion $V_A/(V_A + V_{AE} - V_{\surd V_P})$; the superiority of independent selection derives from the assessment of far more experimental results. If a sample of n environments is used in the assessment of general adaptation, the relative disadvantage becomes

$$V_A/\surd[(V_A + V_{AE} - V_{\surd V_P})\{V_A + V_{AE} - V_{\surd V_P})/n\}].$$

No matter what environments are used, the upper limit to response for general adaptation will be $i\sqrt{V_A}$, but as n increases this limit is approached and the relative disadvantage of the method tends to the value $\sqrt{\{V_A/(V_A + V_{AE} - V_{\sqrt{V_P}})\}}$. Wright has suggested that this and the value for one sample represent bounds which will guide the experimenter. If they are low in practice, then selection for general adaptation must be inefficient, and selection within homogeneous subgroups of environments should be practised.

The approach outlined above may be extended (cf. Comstock and Moll 1963; Wright 1977a) to cover the models discussed in Section 6.3.3.

7.4.2.4 Selection index The problem of truncation selection for more than one trait may be approached in a number of ways (see e.g. Gallais 1973). The question of selection in early generations for yield as against qualitative traits has already been considered; in most cases, the approach advocated amounts to a combination of independent culling levels for the different traits, sometimes practised as tandem selection, i.e. the traits are selected for in turn, in successive generations rather than simultaneously. The merits of such systems have been shown, in general, to be less than those of selection by means of an index (Young 1961; Finney 1962).

The selection index, as commonly understood, means a linear combination of a number of traits, introduced by Smith (1936), following Fisher (1936), as an example of a discriminant function; other early work was that of Hazel (1943), who formalized the genetics of selection indexes. Although an index could be any linear or non-linear combination of traits (see for example, Compton and Bahadur 1977, who carried out selection in maize on the product total grain yield × proportion of standing plants × proportion of undropped ears, described as 'yield of standing plants'), the theory has been properly developed mainly for Smith's (1936) index, which will be discussed here.

Consider a set of traits with phenotypic values $P_1, P_2, P_3, \ldots, P_n$ made up as

$$P_i = G_i + E_i$$

as usual. Then consider that the net economic merit of an individual is

$$X = \sum_i a_i G_i$$

where the a_i, $i = 1, \ldots n$, are economic weights attached to the different traits. (These weights will be determined from the contributions of the different traits to efficiency of production, price, etc., and have no genetical content; see Andrus and McGilliard 1975; Hogsett and Nordskog 1958.) It might be thought that selection for X would be most appropriate, but this neglects the genetic and phenotypic relationships between the traits. Consider any arbitrary function

$$B = \sum_i b_i P_i.$$

The magnitude of correlation between B and X will be a rational criterion for choosing among possible Bs. It was shown by Smith (1936) and others that the optimal b_is are obtained from the simultaneous solution of the set of equations

$$\begin{aligned}
b_1 w_{11} + b_2 w_{12} + \ldots + b_n w_{1n} &= m_1 \\
b_1 w_{21} + b_2 w_{22} + \ldots + b_n w_{2n} &= m_2 \\
&\cdot \\
&\cdot \\
&\cdot \\
b_1 w_{n1} + b_2 w_{n2} + \ldots + b_n w_{nn} &= m_n
\end{aligned}$$

where w_{ii} is the phenotypic variance of the ith trait $= V_{P_i}$; w_{ij} is the phenotypic covariance of the ith and jth traits $(i \neq j)$ or Cov_P; and m_i = covariance of P_i with X. This may be expressed more simply in matrix notation, which also allows description of other aspects of selection indices.

Let $\mathbf{W}$ be the matrix of phenotypic variances and covariances, $\mathbf{b}$ the column vector of arbitrary weights, and $\mathbf{m}$ the column vector of genetical covariances of index traits with the aggregate breeding value. Then the optimum value of $\mathbf{b}$ is

$$\mathbf{b}_I = \mathbf{W}^{-1}\mathbf{m}$$

yielding the optimal selection index

$$B_I = \mathbf{m}'\mathbf{W}^{-1}\mathbf{p}$$

and expected gain in X as

$$\Delta X_I = i(\mathbf{m}'\mathbf{W}^{-1}\mathbf{m})^{\frac{1}{2}}$$

for an intensity of selection i (in units of the standard deviation of B_I, i.e. $i = S/\sqrt{V_{B_I}}$).

An important extra requirement in an index may be that some traits, especially yield, should be increased without limit while others, e.g. height and heading date, may have optimal values, i.e. if μ_i is the population mean of P_i before selection, then for $i = 1, \ldots r$, the breeder wishes to change μ_i to $\mu_i + k_i$, while for $i = r + 1, \ldots n$, the aim is to increase μ_i as much as possible. This means maximizing ΔX with constraint

$$m_i S/V_{B_1} = \alpha k_i, \qquad i = 1, \ldots r$$

where α is a constant of proportionality.

This problem has been solved by Tallis (1962), and the optimal value of **b** may be written as

$$\mathbf{b} = \{\mathbf{I} - \mathbf{W}^{-1}\mathbf{U}'_r(\mathbf{U}_r\mathbf{W}^{-1}\mathbf{U}'_r)^{-1}\mathbf{U}_r\}\mathbf{W}^{-1}\mathbf{m} + \mathbf{W}^{-1}\mathbf{U}'_r(\mathbf{U}_r\mathbf{W}^{-1}\mathbf{U}'_r)^{-1}\mathbf{k}$$

where **U** is the genetic variance–covariance matrix (elements u_{ij}), and $\mathbf{U}_r$ is the $r \times n$ matrix formed from **U** by removing the last $n - r$ rows, and **k** is the column vector of k_is. If $r = 0$, the solution becomes

$$\mathbf{b}_I = \mathbf{W}^{-1}\mathbf{m}$$

as before. Cunningham *et al.* (1970) have presented another method for obtaining a restricted index, but though simpler it is essentially similar. An index of the same kind which will achieve more rapid progress has been developed by Harville (1975). It is also simpler, the optimal **b*** being essentially part of **b**:

$$\mathbf{b}^* = \mathbf{W}^{-1}\mathbf{U}'_r(\mathbf{U}\mathbf{W}^{-1}\mathbf{U}'_r)^{-1}\mathbf{k}$$

After $1/\alpha$ cycles of selection by Tallis's method, where $\alpha = S/V_{B_I}$, the traits P_i, $i = 1, \dots r$ should be at their optima, and traits P_i, $1 = r+1, \dots n$ should have shown maximum genetical advance, though this is unlikely in practice since **U** and **W** will change with time, and are subject to large errors of measurement (Tallis 1960; Finney 1962).

Both these points may be illustrated in data of Manning (1956), who used a selection index for improvement of yield in cotton. He used the traits (1) bolls per plant, (2) seeds per boll, (3) lint per seed, and (4) yield per plant, but as the first three traits should determine the fourth, yield per plant was not incorporated in the index. Tables 7.8 and 7.9 show the estimates of genetic and phenotypic variances and covariances over 7

Table 7.8

Estimates of genetical variances and covariances for three traits in cotton (from Manning 1956). See text for description of traits

Season	u_{11}	u_{22}	u_{33}	u_{12}	u_{13}	u_{23}
1948–9	4.619	0.082	0.0278	−0.058	−0.074	−0.060
1949–50	0	0.436	0.0561	−0.088	0.081	−0.051
1950–1	0.273	1.542	0.0552	0.659	−0.110	−0.101
1951–2	3.606	0.452	0.1054	−0.362	−0.240	−0.073
1952–3	0	0.142	0.0250	−0.192	−0.078	−0.040
1953–4	0.218	0.536	0.0119	0.334	−0.067	−0.116
1954–5	0.118	0.338	0.0139	0.149	−0.023	−0.068
Means	1.262	0.504	0.0422	0.063	−0.073	−0.074

Table 7.9
Estimates of phenotypic variances and covariances for three traits in cotton (from Manning 1956). See text for description of traits

Season	w_{11}	w_{22}	w_{33}	w_{12}	w_{13}	w_{23}
1948–9	7.298	0.927	0.0480	−0.121	0.090	−0.087
1949–50	4.590	1.259	0.0753	0.124	−0.020	−0.077
1950–1	1.028	2.157	0.0659	0.566	−0.083	−0.120
1951–2	6.717	1.253	0.1228	−0.183	−0.183	−0.182
1952–3	0.785	0.689	0.0374	−0.040	−0.078	−0.054
1953–4	1.801	0.903	0.0322	−0.176	−0.070	−0.007
1954–5	0.582	1.315	0.0404	0.004	−0.015	−0.077
Means	3.257	1.215	0.0602	0.023	−0.051	−0.072

years. Only in a limited proportion of cases were the estimates significantly different from zero, and in some cases, e.g. phenotypic covariance of bolls per plant and seeds per boll, the variation between years was vast. Nonetheless, selection was practised using an index, with coefficients (calculated as outlined above) shown in Table 7.10. Over the course of the experiment, despite the inaccuracy in **U** and **W**, selection achieved an improvement of over 30 per cent, relative to the original material, a rate of improvement of about 4 per cent per annum. Annual estimation of index weights would of course in many cases be prohibitively expensive.

Many applications such as the one described have shown the merits of the index method of selection, though its use in early generation selection (where it might be particularly effective) seems to have been limited (Matzinger *et al.* 1977). Restricted indexes like that of Tallis have been less widely used, though as Rosielle *et al.* (1977) showed, in the case of oats where both straw and grain have economic value but tall, late plants

Table 7.10
Coefficients used in the selection index (from Manning 1956)

Season	b_1	b_2	b_3
1948–9	0.872	1.179	6.958
1949–50	−0.009	0.145	0.232
1950–1	0.170	0.305	1.254
1951–2	0.599	−0.057	2.481
1952–3	−2.475	−1.122	−10.253
1953–4	0.188	0.290	0.444
1954–5	0.236	0.392	0.015

would have agronomic disadvantages, they will probably be more effective than unrestricted indexes in maximizing gains in economic value in most circumstances, since they require no more estimations of parameters than the unrestricted indexes, and little more calculation.

Considered as a single trait, an index can be examined for its correlated response to selection for one of the traits in the index. Such an index is called a performance index, and has been useful in assessing the causes of a cessation in response to selection, a problem which has been important in poultry and other animals (Nordskog 1978), but not yet in plants.

There are situations in which the selection programme determines that several traits must be selected for in different stages, in which case a selection index for simultaneous selection cannot be used. One such two-stage scheme, of practical value in tree-breeding, has been described by Cotterill and James (1981).

Suppose that there are three traits with phenotypic values in any individual P_1, P_2, P_3, means μ_1, μ_2, μ_3, and phenotypic variances and correlations V_{P_1}, V_{P_2}, V_{P_3} and $r_{P_{12}}$, $r_{P_{13}}$, $r_{P_{23}}$.

Stage (1) of selection is for individual performance, stage (2) is for individual performance, mean progeny performance or an index of both. Then at stage (1), P_1 is truncated at a distance x_1 from the mean, a proportion p_1 being retained, and at stage (2) P_2 is truncated to retain a proportion p_2 of those previously retained. Intensity of selection at stage (1) is i_1 and at stage (2), i_2.

Then Cotterill and James show that the response in P_3 is

$$R = i_1 r_{P_{13}} + i_2\{r_{P_{23}} - r_{12} r_{13} i_1(i_1 - x_1)\}/\{1 - r^2_{P_{12}} i_1(i_1 - x_1)\}^{\frac{1}{2}}$$

If selection is on the basis of an index of individual and family performance, then

$$R = i_1 r_{P_{13}} + i\{(r_{P_{13}} + r_{P_{23}})^2/(1 - r^2_{P_{12}}) - 2 r_{P_{13}} r_{P_{23}}/(1 - r_{P_{12}}) - r^2_{P_{12}} i_1(i_1 - x_1)\}^{\frac{1}{2}}$$

If stage (1) selection is on individual P_1, stage (2) on progeny performance for P_2, measured as the mean of n progeny, and P_3 is the breeding objective, and if heritabilities and genetical correlations are h_1^2, h_2^2, and $r_{G_{12}}$, $r_{G_{13}}$, $r_{G_{23}}$, and t is the intraclass correlation among family members in the progeny test, then

$$r_{12} = r_{G_{12}} h_1 h_2/[4\{1 + (n-1)t\}/n]^{\frac{1}{2}}$$

$$r_{13} = r_{G_{13}} h_1$$

$$r_{23} = r_{G_{23}} h_2/[4\{1 + (n-1)t\}/n]^{\frac{1}{2}}$$

Such a situation may arise in the initial stages of a tree-breeding programme where parents are chosen on their attributes as mature trees, but their progeny must be chosen on nursery or early traits which can be used to predict mature tree performance (Section 15.2; see also Cotterill 1984).

7.5 Competition

The predictions of selection response described in the previous sections were all developed on the basis that plants of different genotype do not interact with one another. This is manifestly not the case, as shown on many occasions. In two related experiments, to take an example mentioned briefly in Section 5.2.2, Phung and Rathjen (1976) first grew F_1 hybrid seeds of the wheat crosses Warimek × Halberd and Wariquam × Halberd in a stand of Halberd at five frequencies: 1/25, 1/16, 1/9, 1/4, and 1/2, and then subsequently grew F_4 lines from the crosses Warimek × Halberd and Wariquam × Gabo in stands of Wariquam and Halberd, respectively, at frequencies 1/16 and 3/16.

In the first experiment, yield of individual F_1 plants declined linearly with frequency, the coefficients of regression on frequency being 2.4 for Warimek × Halberd and 3.1 for Wariquam × Halberd. Since yield per plant was about 3.5 g at a frequency of 1/25, this meant a decline of 30–50 per cent over the range 1/25 to 1/2. In the second experiment, reductions in plot yields for the hybrids were 38 and 28 per cent for the crosses grown in Wariquam and Halberd, respectively. In all cases, therefore, substantial frequency-dependent advantage accrued to the rarer genotypes. It was observed in two seasons, with different plant spacing (the first experiment using spaced plants in small plots, the second, large contiguous plots). In addition, in the first experiment the hybrids were taller than the background, while in the second experiment Halberd was taller than the relevant hybrids, Wariquam shorter. Plant maturity also seemed to be irrelevant. Thus, the phenomenon of frequency-dependent advantage may be expected to be common and it is not the only possible kind of interaction between genotypes.

Given that such interactions between genotypes may be common, it is clear that response to selection in the presence of interaction should be investigated, to see how the predictions may need to be modified. Theoretical and practical studies are needed. Griffing (1967, 1968a,b, 1969, 1976a,b) has made considerable progress with the former. Gallais (1976b) has suggested that the whole of the theory of selection response in plant breeding should be reformulated to take account of the interaction between plants.

7.5.1 Selection among groups

Griffing (1967) began by considering groups of size 2, and distinguishing between the direct additive effect of a gene and the associate additive effect of a gene as it influences the other member of the group according to the genotype of the second individual, and similarly for the dominance effects. It is then possible to obtain the covariance between direct and associate additive effects, say $\mathrm{Cov}_A(\delta, \alpha)$, and to show that the effect of one cycle of selection of intensity i is to produce a response

$$R = i\{V_A(\delta) + \mathrm{Cov}_A(\delta, \alpha)\}\surd V_P$$

(Here, $V_A(\delta)$ relates only to the direct additive effects.)

Since $\mathrm{Cov}_A(\delta, \alpha)$ can be negative, R may be positive, zero, or negative according to the relationship between $V_A(\delta)$ and $\mathrm{Cov}_A(\delta, \alpha)$. If the two plants are in direct competition, $\mathrm{Cov}_A(\delta, \alpha)$ may well be negative, so that response will be less than in the case of independent genotypes.

Griffing was able to suggest a criterion for group selection which would ensure that $R > 0$, though not that it would be greater than the response for independent genotypes in all cases. This is to select on the basis of the group mean, when response will be

$$R = \tfrac{1}{2}i\{V_A(\delta) + 2\mathrm{Cov}_A(\delta, \alpha) + V_A(\alpha)\}/\surd V_P(\text{group})$$

Here, $V_A(\alpha)$ is the variance of associate additive effects, and $V_P(\text{group})$ is the group phenotypic variance. Because of the relationship between variances and covariances,

$$V_A(\delta) + 2\mathrm{Cov}_A(\delta, \alpha) + V_A(\alpha) \geqq 0,$$

so that $R \geqq 0$. This result may be extended to groups of size n. Group selection corresponds to family selection in animal breeding (cf. Falconer 1982), with the differences that in animals improvement is sought in individual performance while in plants it is aggregate performance and that the logical basis for choosing groups need not be relationship by descent.

In Section 6.3.1, two applications of interaction between neighbouring plants to estimation of genetical parameters are considered; these are essentially particular cases of the general model of Griffing, and show how it may be useful, though the forest-tree work was in fact developed as a consequence of earlier work of Sakai (1955), and the other method was developed because of the problems in estimating $V_A(\alpha)$ and $V_A(\delta)$ from simple experiments. Nonetheless, the implications of Griffing's work are still very general. Rotili *et al.* (1976) have presented some evidence that direct and associative effects may be antagonistic.

Griffing (1968a) showed that group selection, although always acting so as to increase the mean of the variable of interest, could on occasion be inefficient. Individual selection applied within groups of size n yields a response

$$R_{\text{ind}} = i\{V_A(\delta) + (n-1)\,\mathrm{Cov}_A(\delta, \alpha)\}/\surd V_P(\text{ind}).$$

Selection among groups of size n yields

$$R_{\text{group}} = i\{V_A(\delta) + 2(n-1)\text{Cov}_A(\delta, \alpha) + (n-1)^2 V_A(\alpha)\}/\{n\surd V_P(\text{group})\}.$$

In the special case where individuals do not interact within groups,

$$R_{\text{group}}/R_{\text{ind}} \cong 1/\surd n.$$

Although this analysis showed, for example, how selection of pasture plants on an individual basis might yield no improvement in sward productivity, yet there was still the problem of potential inefficiency. Griffing (1969) accordingly showed how a selection index, developed as in Section 7.4.2.4, might be applied to direct and associate additive effects on genes, a process corresponding to an index based on individual and within family values in animal breeding (cf. Falconer 1982).

In later papers, Griffing (1976a,b) developed these ideas further, to deal with selection among full-sib and other non-random groups, and showed that the more intense the competition between plants, the more non-random should be the groups. Inbred clonal groups as produced by doubling haploids obtained from multiply heterozygous parents may be ideal in certain cases. Table 7.11 shows some comparative results, with

$$B_1 = V_A(\delta) + 2(n-1)\text{Cov}_A(\delta, \alpha) + (n-1)^2 V_A(\alpha)$$

(summed over all loci) and

$$B_2 = V_E + (n-1)\text{Cov}_E.$$

(Cov_E is the covariance between the environmental effects associated with different phenotypes within groups.) Here, $B_1 \neq 0$ and $B_2 \cong 0$ implies high heritability, $B_1 = B_2$ moderate heritability and $B_1 \ll B_2$ low heritability. Thus, the lower the heritability the greater the advantage of non-random groups, as might be expected from a knowledge of individual and within family selection responses.

Table 7.11

Efficiency, in terms of change in gene frequency for a given intensity of selection, of different modes of selection relative to random group selection (from Griffing 1967)

Group	Relationship of B_1 and B_2		
	$B_1 \neq 0, B_2 \cong 0$	$B_1 = B_2$	$B_1 \ll B_2$
Random	1	1	1
Full-sib	$\surd\{(n+1)/2\}$	$\surd(n+3)\left(\dfrac{n+1}{n+3}\right)$	$(n+1)/2$
Inbred clonal	$\surd 2n$	$n\sqrt{\dfrac{2}{n+1}}$	$2n$

7.5.2 Genotype–environment interaction

As has already been described, competition between genotypes may be treated as a form of genotype–environment interaction. Use of the models described in Section 7.4.2.3 allows prediction of response of a population of genotypes to truncation selection based on individual yield, whereas the theory outlined in the previous section may require complex estimation procedures (as noted by Hamblin and Rosielle 1978), though what follows is very closely analogous to Griffing's work.

Consider the model

$$y_{ij} = \mu + a_i(\delta) + a_j(\alpha) + \gamma_j a_i(\delta) + s_{ij}$$

for the determination of a trait y in individuals of genotype i associated with individuals of genotype j, where $a_i(\delta)$ is the average direct effect of the genotype, $a_j(\alpha)$ the average effect of genotype j as a competitor (an associate effect), γ_j is the coefficient of regression of associate effects on direct effects, and s_{ij} is a residual linearly independent of $a_i(\delta)$. Wright (1977a) has shown that selection for y based on values of $a_i(\delta)$ will yield

$$R = \Delta a(\delta)\{1 + b_{a(\alpha)|\Delta a(\delta)} + b_{\gamma|\Delta a(\delta)} . \Delta a(\delta)\}$$

where $\Delta a(\delta)$ is the change in $a(\delta)$ and $b_{a(\delta)|\Delta a(\delta)}$ and $b_{\gamma|\Delta a(\delta)}$ are the coefficients of regression of $a(\alpha)$ and γ on $\Delta a(\delta)$. Now $b_{a(\alpha)|\Delta a(\delta)}$ will generally be negative, for the reasons already outlined, so that as Griffing showed more generally, response will in such cases be retarded by selection in inappropriate environments. The relationship with γ is less certain, though if $b_{\gamma|\Delta a(\delta)}$ is substantial, non-linear selection responses will result.

Wright (1977a) has examined a large number of published experiments, and found estimates of $b_{a(\alpha)|\Delta a(\delta)}$ less than zero in all cases, and only 7 of 42 estimates of $b_{\gamma|\Delta a(\delta)}$ positive. Potential responses to direct selection vary greatly, being maximal for mixtures among unrelated species, and least for related individuals of a single species. Thus, Wright's analysis may be regarded as showing in practice what Griffing has predicted.

8

Reproductive systems

8.1 Classification

As has already been noted, the breeding system of a plant will to some extent determine the strategy of the plant breeder. The possible range of reproductive systems is very large (Table 8.1), and presents both problems and opportunities for the breeder.

To determine that a plant is self-incompatible is relatively simple; this is the case even where the species is protandrous or protogynous, and has relatively synchronized flowering, since plants of identical or very similar genotype may usually be obtained and tested by crossing. Recognition of heteromorphic self-incompatibility, monoecy, or dioecy is even simpler. However, given that a plant seems to be self-fertilizing, assessment of the degree of out-crossing may be vital.

A simple method of assessing the degree of out-crossing in apparently self-fertilized plants is given by Marshall and Broué (1973). It requires only that a plant population be polymorphic for a diallelic locus with alleles say A_1, A_2 and that the probability, $1-f$, of random out-crossing be constant, with a consequent probability of self-fertilization of f. Then the progeny of a random sample of recessive homozygotes in the case of a recessive gene, or either homozygote in the case of codominance, will yield an estimate of the degree of out-crossing. Observed and expected numbers of homozygous and heterozygous progeny are as follows:

	Observed	Expected
Homozygotes (say A_1A_1)	n_1	$n\{f+(1-f)q\}$
Heterozygotes (A_1A_2)	n_2	$n(1-f)p$
Total	n	n

Here, p is the relative frequency of A_1 and q the relative frequency of A_2. Then the maximum likelihood estimate of f is given by $f=(np-n_2)/(np)$ for known p, and the variance of f is the maximum likelihood value $(1-f)\{1-(1-f)p\}/(np)$. If, however, p is estimated from the same

Table 8.1
Systems of reproduction [mainly from Stebbins (1941) and Brown (1972)]

Asexual	Vegetative apomixis, yielding no seeds	
		Fairly common in some groups, e.g. bulbils in onion, cormils in Gladiolus, leaf plantlets in Kalanchoe
	Agamospermy, yielding seeds	
		Adventitious embryony, i.e. a type of sporophytic budding
		Parthenogenesis, i.e. female gamete yields a sporophyte without syngamy, e.g. in maize, wheat, tobacco
		Non-recurrent apomixis, i.e. normal meiosis yielding a haploid sporophyte via a haploid gametophyte, e.g. in maize
		Agamogony, i.e. a daughter sporophyte develops from a diploid gametophyte
		Apomeiosis, i.e. diploid gametophyte functioning in meiosis
		Apospory, i.e. gametophyte formed by meiotic division
		Pseudogamy, i.e. maternally derived offspring after induction by pollination but no gametic fusion, e.g. in potato, strawberry
		Semigamy, i.e. maternally and/or paternally derived offspring after induction by pollination but no gametic fusion, e.g. in cotton
		Apogamety, i.e. embryo develops from non-egg nucleus
		Polyembryony, i.e. occurrence in the seed of more than one embryo. Common in *Citrus*, also in Mango, some *Cassia*
Sexual	Autogamy, yielding at least 90% self-pollinated progeny	
		Cleistogamy, i.e. pollination within the unopened flower. Occasional in many plant families, common in grasses.
	Allogamy, yielding at least 50% outcrossed progeny	
		Protandry, stamens ripening before stigmas, common in Campanulaceae, Goodeniaceae, Asteraceae (Compositae)
		Protogyny, stigmas ripening before stamens
		Monoecy, having male and female flowers physically separate on the plant. Common, e.g. most cucurbits, Chenopodiaceae
		Dioecy, having male and female flowers on separate plants sometimes with chromosomal sex mechanism. Moderately common, e.g. hops, hemp
		Andromonoecy, having some hermaphrodite and some male flowers
		Androdioecy, some hermaphrodite, some male plants. Rare
		Gynomonoecy, having some hermaphrodite and some female flowers
		Gynodioecy, having some hermaphrodite and some female plants
		Self-incompatibility, i.e. plants unable to be fertilized by pollen of particular phenotype which is perfectly functional elsewhere. May be heteromorphic, e.g. *Primula sinensis* or homomorphic, if so may be sporophytic, e.g. *Iberis amara* or gametophytic, e.g. rye, cherry

sample, then the variance of f is approximately $(1-f)\{1-(1-f)p\}/(np)+(1-f)^2\mathrm{Var}(p)/p^2$. Using this method, Marshall and Broué estimated the degree of out-crossing as approximately 0.0015 ± 0.0007 in subterranean clover *Trifolium subterraneum.* (Symon (1954) had earlier drawn attention to the level of heterozygosity in field populations of this plant.) More extensive computational methods are given by Brown (1975), and Brown *et al.* (1975), who found the remarkable value of 76 per cent out-crossing in the large forest tree *Eucalyptus obliqua.* Cheliak *et al.* (1985) found a rate of 88 per cent out-crossing in *Pinus banksiana* and strong evidence that this was declining by several per cent annually over a four year period. From the work of Fisher (1941) and others later, it is not clear how species which allow such high levels of self-fertilization will not evolve complete self-compatibility. Less than 10 per cent selfing, as found by Müller (1977) in *Pinus sylvestris,* is a more likely value in predominantly outcrossing species, just as less than 10 per cent out-crossing, as found in *T. subterraneum* or *Hordeum spontaneum* (Brown *et al.* 1978), appears reasonable for species which highly evolved self-fertilizing mechanisms. The proportion of selfing and out-crossing maintained by different breeding systems and the consequences for fitness have been considered by many workers from Fisher (1941) to Lloyd (1975).

Since the most recent theoretical investigations of the evolution of breeding systems continue to suggest that almost complete selfing or self incompatibility are the only stable systems (Lande and Schemske 1985; Schemske and Lande 1985), I have tried to determine by simulation whether there are conditions under which genes allowing self-fertilization will not spread in a self-incompatible species. The models examined allow the proportions of incompatible and compatible pollen to influence the probability of self-fertilization (Visser and Marcucci 1984), but the only factor which significantly delays fixation of a gene which permits incompatible pollination including selfing, is lethality on selfing. Thus, intermediate proportions of selfing remain anomalous.

While the equilibrial frequencies of deleterious recessives may be different under different reproductive systems, highly developed crop plants are so far from their relevant equilibria that these calculations are irrelevant in the present context (though of evolutionary interest). Table 9.1 illustrates this. However, as Brown *et al.* (1975) noted, much of the genetical analysis of forest trees depends upon assumptions about random outcrossing which may well not be met, so their findings may be important for the genetics of natural populations which are cropped, but not highly selected, or in the early stages of a breeding programme for such species. If out-crossing is thought to be almost complete, for example, but there is in fact a substantial degree of close inbreeding, V_A

and gca estimation by out-crossing methods will be substantially in error (Namkoong 1966; Eldridge 1978a,b).

8.2 Asexual reproduction

Asexually reproducing species have two obvious advantages from the breeder's point of view: 'sports', i.e. mutants of unspecified and possibly genetically unstable nature may be indefinitely propagated vegetatively. Genotypes which are used for crossing may also be maintained for comparative and other purposes, whereas plants solely propagated sexually must in general have their genes reassorted each generation. Thus, asexual reproduction is very widely used in horticulture.

The techniques for selection among clones are discussed in Section 7.2.2; no genetics is involved. If crosses are made and favourable genotypes obtained in highly heterozygous form, these may then be maintained vegetatively; otherwise, breeding of plants which are only facultatively asexual is as for other sexually reproducing species, inbred or outbred. The techniques are discussed in Chapters 9 and 10.

Matromorphy, such as occurs through pseudogamy, appears to offer the advantage that homozygous progeny may be produced where there is chromosomal doubling after at least the first stage of meiosis, thereby shortening the production of true-breeding lines by many generations. Unfortunately, this does not seem to be the case in either *Nicotiana rustica* (Virk *et al.* 1977) or *Brassica* (Mackay 1972). It is possible, but unlikely, that irradiated pollen may yet provide a means of achieving these results.

8.3 Sexual reproduction

Perennial plants may almost invariably be maintained and increased in number vegetatively, so that the techniques alluded to in the previous section may be used for them, in addition to those which only apply to plants reproducing sexually. For annual plants, where the only widely available techniques involve the sexual cycle, the reproductive system is more critical in determining the breeding approach.

Selection methods for inbreeding species are described in Chapters 7 and 10, those for outbreeding species in Chapters 7 and 9. Here, we consider a few matters relating directly to the breeding system itself.

8.3.1 Autogamous species

Once it has been established that a particular plant is largely self-fertilizing, it may be necessary to generate male sterility for hybrid breeding, to resort to mutation breeding to generate variability for specific traits, or to make wide crosses to ancestral populations or related species, but in general breeding techniques are well-established.

8.3.2 Allogamous species

If plants maintain a high level of out-crossing through physical separation of the flowers (monoecy; the same remarks apply to protandry and protogyny also), selfing techniques for hybrid breeding may need to be established; this has been accomplished for most major crop species where it is necessary.

If plants are outbred through dioecy, then if the crop is obtained from one sex only (as with asparagus under intensive conditions, and perhaps textile hemp), 'sexing' methods or somatic cell genetics may be needed to eliminate the unwanted sex. Lloyd and Webb (1977) have shown that secondary sexual characters that could be used for sexing are unfortunately rare in plants.

Self-incompatibility is the area of greatest genetical interest. Since it essentially ensures that selfing is impossible, inbreeding must be by sib-crossing (or parent–offspring crossing for perennial crops) unless self-compatibility is inducible. In many cases, breakdown of self-incompatibility may be achieved by environmental modification, e.g. pollination in the bud (before anthesis), delayed pollination, pollination at the end of the breeding season, or radiation treatment of pollen. De Nettancourt (1977) has summarized these non-genetical techniques. Genetical methods are also available, e.g. artificial mutagenesis directed at the incompatibility genes or modification of the female parent's genetical background. An example of the latter technique is provided by Busbice *et al.* (1975). They were able to modify the degree of self-incompatibility very readily in lucerne by simple selection for or against self-sterility. Starting with only 100 randomly chosen clones of one cultivar, Cherokee, they were able to double self-fertility to 12.6 per cent ovules developing into seed (and increase cross-fertility slightly) and reduce it to 1.4 per cent (and decrease cross-fertility substantially) from the original level of 6.2 per cent, by only two cycles of selection. If the self-incompatibility system is more precise than this, progress will probably be slower unless artificial mutagenesis is used, as discussed in Chapter 12, though natural mutants lacking pollen specificity but retaining stylar specificity are well known in several species and could be used where available (Lundqvist 1958).

Inter-specific incompatibility is also a problem in much breeding work, but the evidence summarized by de Nettancourt (1977) suggests that most cases of inter-specific incompatibility are unilateral and are really an extension of intra-specific incompatibility, in that self-compatible species will accept pollen from closely related self-incompatible species, but the reciprocal cross will fail. In such cases, it is frequently found that application of compatible pollen made inviable by irradiation or other such treatment (so-called 'mentor' or 'recognition' pollination) with viable incompatible pollen of the chosen male parent allows satisfactory levels of fertilization. This has proved important in interspecific crosses in tree breeding (see for example Knox *et al.* 1972).

9 Heterosis

9.1 Introduction

Knowledge of the related phenomena of inbreeding depression and hybrid vigour is at least as old as recorded history; see Darlington (1971) for a provocative (and widely contradicted; see for example Thoday 1971) view of the role of these phenomena in the recent evolutionary history of man. Systematic investigation of inbreeding depression began at least with Kölreuter in the eighteenth century, and was given great stimulus in the following century by the work of Darwin (1876).

Although breeders had begun to plan for hybrid vigour in the late part of the last century, it was not until the early years of this century that widespread utilization of heterosis began. Zirkle (1952) and Hayes (1952) should be consulted for an account of the history of the use of hybrids.

Claims for heterosis breeding include the following: it provides maximum performance under optimal growing conditions; it provides phenotypic stability under stress; it allows the joint improvement of traits (for example, components of yield) which are negatively associated through linkage or pleiotropy; and it allows the combination in one individual of harmonious sets of traits from very diverse parents without necessarily disturbing the existing harmony (Griffing and Langridge 1963).

The greatest long-term success so far with heterotic breeding has of course been hybrid maize in the United States (see Hayes 1952). However, as Table 9.1 shows, it is now very widespread indeed. Surprisingly, in view of this very great success, the genetical theory underlying the manifestation of heterosis is not well-developed. This is despite a vast and to some extent very successful elaboration of the theory of inbreeding. Much work remains to be done, and the theory outlined in Sections 9.2 and 9.3 below will emphasize how many gaps there are. Section 9.4 gives a brief account of the limited extent of heterosis theory underlying practice.

First, however, we should consider the definition of heterosis. It is essentially a phenotypic phenomenon, and is defined operationally in

Table 9.1

Past and potential introduction of commercial hybrid varieties in food crops, largely relating to North American experience (modified from Wittwer 1974)

Before 1955		1955–74		1975–	
Crop	Year	Crop	Year	Crop	Year
Field corn (*Zea mays*)[m]	1921	Sorghum (*Sorghum bicolor*)[s]	1955	Asparagus (*Asparagus officinalis*)[d]	1975
Sweet corn (*Z. mays*)[m]	1933	Sugar beet (*Beta vulgaris*)[i]	1957	Celery (*Apium graveolens*)[c]	1975
Eggplant (*Solanum melongena*)[s, a]	1939	Broccoli (*Brassica oleracea*)[i]	1961	Oats (*Avena* spp.)[s]	1980
Summer squash (*Cucurbita pepo*)[m]	1941	Spinach (*Spinacea oleracea*)[d]	1961	Rye (*Secale cereale*)[i]	1980
Tomato (*Lycopersicon esculentum*)[s]	1943	Beetroot (*Beta vulgaris*)[p]	1962	Potato (*Solanum tuberosum*)[c]	1980
Slicing cucumber (*Cucumis sativus*)[m]	1945	Bussels sprouts (*Brassica oleracea*)[i]	1963	Soyabean (*Glycine max*)[s]	1980
Onion (*Allium cepa*)[c]	1948	Carrot (*Daucus carota*)[p]	1964	Haricot bean (*Phaseolus vulgaris*)[s]	1985
Watermelon (*Citrullus lanatus*)[m]	1949	Pearl millet (*Pennisetum typhoides*)[q]	1965	Field bean (*Vicia faba*)[s]	1985
Winter squash (*Cucurbita* spp.)[m]	1950	Coconut (*Cocos nucifera*)[c]	1965	Peas (*pisum sativum*)[s]	1985
Pepper (*Capsicum annuum*)[s]	1954	Cauliflower (*Brassica oleracea*)[i]	1966	Lettuce (*Lactuca sativa*)[s]	1990
Muskmelon (*Cucumis melo*)[m]	1954	Lucerne (*Medicago sativa*)[i]	1968		
Pickling cucumber (*Cucumis sativus*)[m]	1954	Barley (*Hordeum vulgare*)[s]	1968		
Cabbage (*Brassica oleracea*)[i]	1954	Wheat (*Triticum aestivum*)[s]	1969		
		Rice (*Oryza sativa*)[s]	1972		
		Sunflower (*Helianthus annuus*)[i]	1972		

a Andromonoecious, c cross-fertilizing but self-compatible, d dioecious, i self-incompatible, m monoecious, p protandrous, q protogynous, s self-fertilizing.

terms of the performance of progeny generations, but because one of the models proposed for its causation is the same as the model of balanced polymorphism (see Section 5.2.1), it has frequently been confused with over-dominance. Over-dominance or heterozygous advantage, where the heterozygote at a diallelic locus is at an advantage to both homozygotes (see Li (1967) for the extension to multiple alleles), has been proposed as a widespread phenomenon by Ford (1945), but the evidence is still limited. In particular, evidence of 'single locus heterosis', i.e. an over-dominant effect of a single locus on a metric trait, is sparse. The enzyme activities of heterozygotes, where gene products as such are being assayed, generally (though not invariably) lie between those of the relevant homozygotes. Berger (1976) has reviewed the evidence for heterosis at the gene product level. To support his thesis that greater catalytic efficiency of enzymes from heterozygotes may account for much of the observed polymorphism (cf. Haldane 1954; Abel 1972), he has given a number of relatively clear-cut examples. One is alcohol dehydrogenase (ADH) in maize, discussed in Section 6.2.2. Schwartz and Laughner (1969) bred plants heterozygous for alleles whose products had high stability and low activity and low stability and high activity, respectively, and showed that these heterozygotes had greater activity (assessed ordinally only) and stability than either parent, a form of complementary heterosis. Brown *et al.* (1976), however, failed to detect heterosis for enzyme activity at an ADH locus in soft brome (*Bromus mollis* L.). Griffing and Langridge (1963), arguing from results of experiments with *Arabidopsis thaliana*, had earlier suggested that much heterosis, particularly in species which are naturally self-fertilized, might be explicable as the result of complementation between alleles with differentially temperature-sensitive products.

At the next level, that of the specific metabolite, the simple animal example shown in Table 9.2 illustrates very clearly the phenomenon of single locus heterosis. Here, heterosis may be measured as

$$H_p = \text{heterozygous value} - \text{mean of homozygous values} = 5.75$$

or

$$H_p = \text{heterozygous value} \times 100/\text{mean of homozygous values} = 153.5 \text{ per cent.}$$

These different values correspond, for crosses between strains or inbred lines, to

$$H_p = \mathrm{F}_1 \text{ mean} - \text{mean of parental lines} = \overline{\mathrm{F}}_1 - (\overline{\mathrm{P}}_1 + \overline{\mathrm{P}}_2)/2$$

and

$$H_p = \overline{\mathrm{F}}_1 \times 100/\{(\overline{\mathrm{P}}_1 + \overline{\mathrm{P}}_2)/2\}.$$

Table 9.2

Phosphoglucomutase$_2$ genotypes and liver glycogen content (in mg/g) in fasted fieldmice (*Apodemus sylvaticus*) (from Leigh Brown 1977)

Genotype	*aa*	*ac*	*cc*
Mean	6.3	16.5	15.2
S.D.	1.66	4.02	2.25
Sample size	14	18	16

These expressions constitute simple alternative definitions of the level of heterosis.

At a third level of complexity or difficulty of resolution lies a trait such as yield, and here examples of single gene heterosis have long been regarded as extremely rare (Schuler and Sprague 1956), but this may not be the case. For example, Gottschalk (1976) described a small group of fasciated pea mutants which caused a high degree of heterosis for seed production and other traits in the heterozygote of crosses between the homozygous fasciated plant, and other inbred lines and cultivars. (Since fasciation is recessive, the improved yield of the F_1 over the better parent was not due to fasciation.) The plants were not grown at commercial density, so the practical merits of using the gene to improve yield remain unknown, but the yield increases were very substantial. Using the Albina 7 mutant (a) in barley, well known to show unifactorial heterosis for yield and its components (Gustafsson 1946), Ellerström and Hagberg (1967) were able to show that such effects can carry over to autotetraploids, where the rank order of genotypes for several components of yield was $AAaa \geqq AAAa \geqq AAAA > Aaaa$. (*A* is the non-albino allele of the locus.) Explanation of either of these examples is still to come.

Crossing of inbred lines, or more especially different populations, will of course not necessarily result in heterosis (cf., for example, Sved 1976) nor will the heterosis necessarily be useful (cf., for example, Dobzhansky 1952), but since such cases will readily be recognized during a breeding programme and since theory does not yet allow their general prediction the discussion will deal almost entirely with useful heterosis. In essence, this requires that the F_1 be greater than both P_1 and P_2.

9.2 The genetical basis of heterosis

9.2.1 Inbreeding depression

In Chapter 5, we have already examined in some detail the effect of self-fertilization on genetical variation within and between plants. Here, we shall be considering the effect of such inbreeding on the genotypic value of any particular trait of interest. However, it will be of value first to list

Table 9.3
Inbreeding of polysomics (from Parsons 1959)

Number of chromosome sets	Time to attain the same degree of homozygosity as the diploid	
	under self-fertilization	under sib-mating
2	1	1
4	3.80	2.67
6	6.58	4.35
8	9.35	6.05
10	12.12	7.74

some further relevant results on increasing homozygosity within plants on inbreeding.

The result of self-fertilization in diploids, the simplest kind of inbreeding, is to halve the heterozygosity at any particular locus in each generation. This means that between seven and eight generations are needed to achieve better than 99 per cent homozygosity, compared with 27 to 28 for tetrasomics, while progress under full-sib crossing is much slower, the respective ranges being 23–24 and 62–63 generations (Fisher 1949). Table 9.3 shows the relative time to attain a given degree of homozygosity for higher ploidy, using the diploid as the standard in each case. Double reduction of frequency α increases the rate of approach to homozygosity in the proportion

$$\ln\left(\frac{5-2\alpha}{6}\right)\bigg/\ln\left(\frac{5}{6}\right).$$

(These results apply to autopolyploids only.) It is clear that progress towards homozygosity is greatly delayed by polyploidy, though this effect is moderated slightly by the presence of a significant degree of double reduction. This will mean that the production of homozygous lines of autopolyploids will be a slow process, unless cytogenetical manipulations (Section 14.5) can be used, and of course if self-incompatibility is present, then with sib-mating the process will be very slow indeed.

To determine the effect of self-fertilization on the genotypic values for a trait, consider the following simple one-locus model:

Genotype	A_1A_1	A_1A_2	A_2A_2
Genotypic value	g_{11}	g_{12}	g_{22}
Panmictic equlibrial frequencies	p^2	$2pq$	q^2

The mean genotypic value initially will be

$$M_0 = p^2 g_{11} + 2pqg_{12} + q^2 g_{22}.$$

After n generations of selfing, this will become

$$M_n = M_0 - \{1 - (\tfrac{1}{2})^n\}pq\{2g_{12} - g_{11} - g_{22}\}.$$

Hence, there will be inbreeding depression, unless the heterozygote is inferior to the average of the homozygotes in genotypic value, which is unlikely, on the basis of extensive experience. Setting $g_{11} = a$, $g_{12} = d$, and $g_{22} = -a$, we see that inbreeding depression simply depends, phrasing it another way, on the level of dominance at the locus. For complete additivity, there is no inbreeding depression. In the absence of interactions between loci, these results may be generalized to many loci by addition. However, such absence will be rare.

The case of autotetraploids, which has been described by Bennett (1976), presents interesting contrasts with the diploid case. Consider the following simple model:

Genotype	$A_1A_1A_1A_1$	$A_1A_1A_1A_2$	$A_1A_1A_2A_2$	$A_1A_2A_2A_2$	$A_2A_2A_2A_2$
Genotypic value	g_{1111}	g_{1112}	g_{1122}	g_{1222}	g_{2222}
Panmictic equilibrial frequencies	p^4	$4p^3q$	$6p^2q^2$	$4pq^2$	q^4

This yields the initial mean genotypic value

$$M_0 = p^4 g_{1111} + 4p^3 q g_{1112} + 6p^2q^2 g_{1122} + 4pq^3 g_{1222} + q^4 g_{2222}.$$

The genotypic mean in the nth generation of selfing will then be:

$$\begin{aligned} M_n = M_0 &- \{1 - (5/6)^n\}\tfrac{1}{4}pqI' \\ &- \{1 - (\tfrac{1}{2})^n\}pq(q-p)I'' \\ &- \{1 - (1/6)^n\}\tfrac{1}{4}pq(p-q)^2 I''' \end{aligned}$$

where

$$\begin{aligned} I' &= -7g_{1111} + 4g_{1112} + 6g_{1122} + 4g_{1222} - 7g_{2222} \\ I'' &= g_{1111} - 2g_{1112} + 2g_{1222} - g_{2222} \\ I''' &= -g_{1111} + 4g_{1112} - 6g_{1122} + 4g_{1222} - g_{2222}. \end{aligned}$$

Here, there are three interaction effects I', I'', and I''', which are the generalization for a diallelic tetrasomic locus of the dominance deviation for the disomic case. It can be seen from the form of I'' that even with complete additivity of allelic effects, inbreeding depression is still possible. If p and q are approximately equal, I'' and I''' will have minimal effect on inbreeding depression, while I' will affect inbreeding only

Table 9.4
Green forage yield (g/plant) of seven clones of crested wheatgrass *Agropyron cristatum* (L.) Gaertn. (from Dewey 1969)

	Diploid			Tetraploid		
	Clone	Selfed	Open-pollinated	Clone	Selfed	Open-pollinated
	412	192	503	488	417	536
	588	362	592	642	534	606
	612	365	608	508	477	623
	705	241	508	442	293	465
	375	192	540	528	440	746
	822	452	598	795	633	746
	292	81	295	388	341	571
Mean	544	269	521	542	448	613

slowly. Bennett has pointed out that this would explain the results of Dewey (1969) and others, that newly synthesized autotetraploids show a much lower inbreeding depression than older natural autotetraploids. Table 9.4 shows Dewey's results for the selfing and outpollination for seven clones of diploid and newly synthesized autotetraploid crested wheatgrass. While the effect of self-fertilization is extremely variable, particularly in diploids, we see that overall the selfed diploids yield only 52 per cent of the open-pollinated value, while for tetraploids the selfed plants yield 73 per cent of the open-pollinated value. These results may be contrasted with earlier work of Dewey (1966), where the corresponding values were 64 and 50 per cent, the tetraploid here being naturally occurring.

The greater inbreeding depression of long-established autotetraploids as compared with diploids is to be expected, as Bennett has pointed out, on account of the well-known (e.g. Mayo 1971) fact that the equilibrial gene frequency for deleterious recessives is higher in tetraploids than in diploids. Accordingly, selfing long-established tetraploids should yield inbreeding depression more rapidly than in diploids, on account of deleterious recessives. Newly synthesized tetraploids should not contain many more such rare deleterious recessives at higher frequency than in the diploid, since equilibrium will not have been reached. Indeed, in such cases the rate of inbreeding depression has been shown to be slower in tetraploids than in diploids (Lundqvist 1966, 1969), though heterosis was greater in tetraploids for yield and most of its components. Gallais (1984) has suggested that a zero level of inbreeding may not be optimal in outbreeding tetraploids, an idea which requires much more work.

Overall, inbreeding depression may arise both as a result of homozygosity *per se* where dominant phenotypes are not necessarily at an advantage to recessive phenotypes, and also where deleterious recessives are uncovered by homozygosity. Lerner (1954) summarized many studies showing that phenotypic variability in the sense of environmental plasticity was directly dependent upon the degree of inbreeding, providing thereby the main evidence for the deleterious effect of homozygosity *per se.* Mitton (1978) has further shown that in natural populations of the killifish *Fundulus heteroclitus* heterozygosity at identifiable gene loci is associated with lower phenotypic variability. Other studies have confirmed Mitton's results. Frei *et al.* (1986a) have suggested that isozyme dissimilarity, a measure related to heterozygosity, may be a valid predictor of performance in crosses of inbred lines of maize. There may therefore be grounds for re-examining the role of homozygosity *per se* with these newer techniques, but in general this explanation has not yielded the hoped-for return.

Inbreeding depression may also result from the breakdown of balanced linkage combinations, as suggested by Mather and others (Mather and Jinks 1971). Other specific models will be considered in the next section.

It can be seen from the results shown in Table 9.4, which are representative of a wide range of inbreeding studies, that yield depression following intense inbreeding can be very rapid and severe. Selfing represents the maximum possible rate of inbreeding, but in many studies it has been shown that the effect of inbreeding is approximately linear over a wide range of values of the inbreeding coefficient, F, regardless of the system of inbreeding used. The rate of decline of yield in maize with increasing F is shown in Table 9.5. If much of the inbreeding depression is due to the elimination of deleterious recessives, then any lines which persist for more than a few generations of very close inbreeding will

Table 9.5

Yield depression in maize per unit initial yield per 0.01 increase in F (modified from Burton *et al.* 1978)

Method of inbreeding	Sing *et al.* (1967)	Hallauer and Sears (1973)	Cornelius and Dudley (1974)	Burton *et al.* (1978)
Selfing	—	−0.007	−0.007	—
Sib-mating	—	—	−0.006	—
Double-double cross pedigrees	−0.004	—	—	−0.005
Balanced pedigrees	—	—	—	−0.006

contain very few such deleterious recessives, and may contain a higher than usual proportion of advantageous genes.

If we simply consider the effect of inbreeding on fitness, then following Crow (1952), we can set up the pattern of fitnesses at a single diallelic locus given in Table 9.6.

Here, the homozygous values refer to the same gene frequencies as in the panmictic case, so that one has to imagine that inbreeding to homozygosity has been rapid enough for no change to occur in gene frequency, as would be the case with doubled haploids from gametophytic culture. For the recessive and incomplete dominant case, u represents the rate of mutation from allelic class A_1 to allelic class A_2.

Suppose that we let $s_1 = s_2 = 0.1$, $h = 0.5$, and $u = 10^{-5}$. Then we obtain the following ratios of variances:

$$[\{s_1s_2/(s_1 + s_2)\}^2]/(us_1) = 500$$
$$\{s_1s_2/(s_1 + s_2)\}^2/(2uhs_1) = 2500$$
$$\{s_1s_2(s_1 - s_2)^2/(s_1 + s_2)^2\}/[\{s_1s_2/(s_1 + s_2)\}^2] = 0$$
$$\{2\surd(u/s_1)\}/(us_1) = 2000$$
$$(us_1/h)/(2uhs_1) = 2$$

Given rough equivalence in the numbers of loci displaying the different kinds of variation, in the panmictic case variation in fitness will be mostly a reflection of over-dominant loci, while under inbreeding, partly and completely recessive deleterious alleles will contribute far more to the variance. Under panmixia, 90 per cent of the variance might be of the overdominant kind, 10 per cent of the other, while with complete selfing the reverse might be the case. These considerations led Crow (1952) to suggest that inbreeding depression and recovery of vigour on crossing would largely be the result of rare recessive defects, while the vigour of non-inbreds and hybrid vigour over equilibrial populations would be a reflection of over-dominance. In the next section this will be considered further. It should, however, be noted that for Crow's conclusions to apply to hybrid vigour for economically important traits, these must be strongly positive correlated with fitness; this will be true for some, such as grain number, but not necessarily for others, such as grain weight.

A more extensive analysis by Jinks (1981), of height, flowering time and yield in two Nicotiana species, in barley and in maize, found values of $(2V_D/V_A)$ to lie between zero and unity in all cases except yield in maize. He therefore concluded that heterosis arises in general from dominance of alleles of widely dispersed genes, so that it should be possible to extract pure-breeding lines equalling or out-performing the best F_1 hybrids, provided that linkage combinations could be broken down, which might take many generations.

Table 9.6
Pattern of fitness at a single diallelic locus

				Panmictic equilibrium with mutation and selection		Complete homozygosity	
Fitness pattern	A_1A_1	A_1A_2	A_2A_2	Mean	Variance	Mean	Variance
Recessive	1	1	$1 - s_1$	$1 - u$	us_1	$1 - \sqrt{(us_1)} - \sqrt{(u/s_1)}$	$2\sqrt{(u/s_1)}$
Incomplete dominant	1	$1 - hs_1$	$1 - s_1$	$1 - 2u$	$2uhs_1$	$1 - u/h$	us_1/h
Heterozygous advantage	$1 - s_2$	1	$1 - s_1$	$1 - s_1s_2/(s_1 + s_2)$	$\{s_1s_2/(s_1 + s_2)\}^2$	$1 - 2s_1s_2/(s_1 + s_2)$	$\dfrac{s_1s_2(s_1 - s_2)^2}{(s_1 + s_2)^2}$

9.2.2 Simple models of heterosis

In order to develop a full account of heterosis in genetical terms, one must first obtain an understanding of multiple locus polymorphism. Although single locus polymorphism is reasonably well understood, it is not a sufficient description of multiple locus polymorphism to regard the latter as simply the sum or product of a set of single locus polymorphisms (Ewens and Thomson 1977).

Many investigations have been made of two-locus polymorphisms in diploids; see in particular Bodmer and Felsenstein (1967), and Karlin and Feldman (1970). Little of the theoretical investigation of polyploidy has been relevant to heterosis (see for example Gallais 1976d). Most investigations have been made of models having symmetric viability in diploids, with Darwinian fitnesses as set out below:

	B_1B_1	B_1B_2	B_2B_2
A_1A_1	$1-\sigma$	$1-\beta$	$1-\alpha$
A_1A_2	$1-\gamma$	1	$1-\gamma$
A_2A_2	$1-\alpha$	$1-\beta$	$1-\sigma$

with gamete frequencies as follows:

A_1B_1	A_1B_2	A_2B_1	B_2B_2
x_1	x_2	x_3	x_4

Then general solutions exist of the form $\hat{x}_1 = \hat{x}_4$, $\hat{x}_2 = \hat{x}_3$ and these are stable for sufficiently small θ, where θ is the recombination frequency between loci A and B. Other equilibria exist with

$$\hat{x}_1 \neq \hat{x}_4,\ \hat{x}_2 = \hat{x}_3$$
$$\hat{x}_1 = \hat{s}_4,\ \hat{x}_2 \neq \hat{x}_3$$
$$\hat{x}_1 \neq \hat{x}_4,\ \hat{x}_2 \neq \hat{x}_3$$

These are stable only if the symmetric equilibria are unstable.

As the number of loci increases, the complexity of the system increases far more than proportionately; Feldman *et al.* (1974) give an account of the three-locus symmetric viability model, where the marginal two-locus viabilities are all of the type shown above, so that all triple heterozygotes have fitness $1-\delta$. For tight linkage, there may be four stable equilibria, each having one pair of complementary chromosomes high in frequency, all others low. For looser linkage, the only stable symmetric equilibrium is that with complete linkage equilibrium; i.e. $\Delta = x_1x_4 - x_2x_3 = 0$. Intermediate linkage values may lead to both types of equilibria in some cases, and indeed equilibria may exist with all pairwise linkage disequilibria zero, but non-zero disequilibrium between all three loci. Such an equilibrium may be stable with very tight linkage.

This brief sketch of the problems of investigating multilocus polymorphism is merely intended to indicate their complexity, and illustrate how far we are from understanding such polymorphism sufficiently well to allow us to interpret heterosis in terms of the interactions between a specified set of loci. (It is not even known, for example, whether genes affect fitness additively or multiplicatively, to a first approximation.) Attempts have, however, been made to consider models yielding heterosis from the interaction of several loci. Taking a simple case, consider a pair of loci with genotypic effects as given by Mather and Jinks (1971):

	B_1B_1	B_1B_2	B_2B_2
A_1A_1	$a_A + a_B + i_{AB}$	$a_A + d_B + j_{AB}$	$a_A - a_B - i_{AB}$
A_1A_2	$d_A + a_B + j_{BA}$	$d_A + d_B + l_{AB}$	$d_A - a_B - j_{BA}$
A_2A_2	$-a_A + a_B + i_{AB}$	$-a_A + d_B - j_{AB}$	$-a_A - a_B + i_{AB}$

Now if we seek to determine the heterosis arising from a cross of two populations segregating for these loci, then suppose that gamete frequencies are as follows:

	A_1B_1	A_1B_2	A_2B_1	A_2B_2
Population 1	x_1	x_2	x_3	x_4
Population 2	x'_1	x'_2	x'_3	x'_4

and gene frequencies are as follows:

	A_1	B_1
Population 1	p_A	p_B
Population 2	p'_A	p'_B

Then Arunachalam (1977) has shown that in the cross between these two populations, heterosis, measured as $H_p = \overline{F}_1 - (\overline{P}_1 + \overline{P}_2)/2$, amounts to

$$\begin{aligned} H_p = {} & (p'_A - p_A)^2(a_A + j_{BA}) + (p'_B - p_B)^2(a_B - j_{AB}) \\ & - 2\{p'_A - p_A - (x'_1 - x_1)\}^2 l_{AB} \\ & + (p'_A - p_A)(p'_B - p_B)(l_{AB} - i_{AB} - j_{BA} + j_{AB}) \\ & + 2(p'_A - p_A)\{p'_A - p_A - (x'_1 - x_1)\}(l_{AB} - j_{BA}) \\ & - 2(p'_B - p_B)\{p'_A - p_A - (x'_1 - x_1)\}(l_{AB} + j_{AB}) \end{aligned}$$

This means that even if gene frequencies in both populations are equal, linkage of loci A and B, and linkage disequilibrium such that $x'_1 \neq x_1$ will ensure $H_p \neq 0$.

In the absence of interaction between the loci, i.e.

$$i_{AB} = j_{AB} = j_{BA} = l_{AB} = 0,$$

Arunachalam has also pointed out that

$$H_p = (p'_A - p_A)^2 d_A + (p'_B - p_B)^2 d_B$$

which is the two-locus extension of the single-locus case

$$H_p = (p'_A - p_A)^2 d_A$$

given by Falconer (1982).

If all the interactions are zero except the additive × additive component, i.e. $i_{AB} \neq 0$,

$$H_p = -(p'_A - p_A)(p'_B - p_B) i_{AB}.$$

Arunachalam has emphasized that dominance is therefore not necessary for heterosis to occur, though the reality of such a pattern of interactions has yet to be demonstrated.

Sved (1972) had earlier shown that gametic association (linkage disequilibrium) might be important in generating heterotic effects associated with chromosome segments, especially associated overdominance (i.e. apparent over-dominance arising from non-random association between intrinsically non-over-dominant loci and other loci with dominance or over-dominance; cf. also Ohta and Kimura (1970), and Ohta (1971)). Associated over-dominance is unlikely to be a persistent phenomenon (Sved 1971, 1972), but in population or inbred line crosses may be quite frequent. Wright (1983) attempted to investigate gametic association in mass selection as against selection based on means of half-sib families and found that, for pairs of genes, breakdown of unfavourable associations would in general be aided more by mass selection than by half-sib selection. However, precise prediction of such effects would require knowledge of the values of most of the variables just discussed, e.g. genotypic values and level of association, and these will virtually never be known for all the genes influencing yield, the trait of greatest importance.

From these results it is clear that evidence of epistatic heterosis should be sought, using methods more orientated towards the effects of specific gene loci than those of Jinks (1955, 1981). This is not to deny the potential utility of the prediction of inbred line performance of Jinks and Pooni (1976) (see Section 9.4.2.3), but rather to emphasize that in many specific cases a small number of genes will be important and the causation of epistasis therefore of interest. It is noteworthy that Seyffert and Forkmann have in a model system demonstrated that heterosis can arise from additive × additive and additive × dominance effects of a few loci (see Section 4.7).

9.3 The physiological basis of heterosis

It will be evident from the analyses of Crow (1952) and Jinks (1981) that elimination of deleterious recessives by inbreeding is most unlikely to account entirely for heterosis in outbreeding species. This was earlier noted by Fisher (1949), but has occasionally been disputed, e.g. Hayman (1960). In normally self-pollinating species, however, the major role of such recessives appears to have been established.

In addition to the accumulation of favourable dominants, over-dominance, interaction between non-allelic genes and heterozygosity *per se* have been suggested as genetical mechanisms for heterosis. Genetical complementation may be a manifestation of one or more of these phenomena, and is a valid alternative description of some aspects of heterosis. For example, any history of maize breeding will point out that most early hybrids resulted from crosses involving Northern Flint and Southern Dent types of corn, a case of the two basic types each providing something the other lacks, at some unspecified level (Anderson and Brown 1952). At a more precisely defined level, Sage and Hobson (1973) and others have suggested that mitochondrial complementation might be used as an indicator of heterosis for yield in breeding hybrid wheat, a conclusion contradicted by more recent work (Sen 1981).

Even where such sub-cellular investigation has not yet proved practical, it may be possible to partition a complex trait like yield into components whose genetical variation can be investigated, leading to a more meaningful interpretation of heterosis for yield. It was recognized very early in the investigation of hybrid vigour that this might arise in a complex trait through simple additive or dominant genetical interaction at different loci for components of the complex trait. For example, Keeble and Pellew (1910) showed an example of multiplicative interaction for height in peas, where in one parent there was dominance for length of internode, and in the other dominance for thickness of stem, leading to heterosis for height in the F_1. Williams (1959) termed this phenomenon somatic multiplication, and an example is set out in Table 9.7. From this model, and consideration of experiments on yield in

Table 9.7

Fruit yield as determined by two components, each with (a) additive gene action and (b) dominant gene action (modified from Williams 1959)

	Fruit number (x)		Fruit weight (y)		Yield (xy)	
	(a)	(b)	(a)	(b)	(a)	(b)
P_1	3	2	1	1	3	2
F_1	2	2	2	2	4	4
P_2	1	1	3	2	3	2

Table 9.8

Yield components of sorghum hybrids and their parents (from Sinha and Khanna 1975)

	Number of whorls/ panicle	Number of branches/ panicle	Number of grains/ panicle	Number of grains/ branch	1000-grain weight (g)	Grain weight/ panicle (g)
msCK60B	9.3	54.8	1321	24.10	21.25	30.9
IS3691	13.0	94.3	1206	13.35	34.90	25.2
msCK60BxIS3691	11.0	89.5	2453	27.49	29.70	58.2

tomatoes, Williams concluded that analysis of yield as a single trait was bound to lead to a non-additive genetical interpretation which would be inappropriate, since there was in fact genetical additivity for components which were multiplied to give yield. This approach was criticized by Hayman (1960) on the grounds, first, that it did not contribute anything new, secondly, that certain examples existed which could not be explained on these grounds, and thirdly, that transformation of scale (in this case by use of logarithms) would change the perceived mode of gene action to additivity, leaving heterosis determined by additive × additive interaction between favourable dominants for different traits. Despite such criticism, the concept was revived by Sinha and Khanna (1975), possibly because it gives a simple unified description of many experimental results. Table 9.8 shows how it has been applied to yield in sorghum by Sinha and Khanna.

The yield of sorghum was computed by Sinha and Khanna as Grain yield = number of rachis × number of grains per rachis × kernel weight. The three components are inherited dominantly or with partial dominance, and determine yield multiplicatively. Many other similar examples may be adduced.

While this approach is attractive for its simplicity and generality, it fails to encompass a great deal of well-established plant breeding data. Apart from cases where there is negligible evidence of multiplicative interaction (e.g. Geiger and Wahle 1978), it is not readily applicable to analyses of genetical variability such as the results of Moll *et al.* (1960) on variability in an advanced generation of a cross of two open-pollinated varieties of corn. In the F_3, V_A was greater than in the parents, and the F_3 mean was greater than either parental population mean (though less than the F_1 mean). The increase in V_A is not a prediction of the somatic multiplication model, though the changes in mean may be. It is more generally inappropriate to explain single gene heterosis. Its application to self-pollinated crops has been limited, so that this area of hybrid theory (cf. for example Hayes and Foster 1976), perhaps the most obscure despite the work of Jinks (1981), may benefit from such simplification. (Table 9.1 shows how hybrid varieties of self-fertilizing species were utilized early in the course of hybrid breeding, despite the lack of any theoretical basis for the work.)

9.4 Response to selection for heterosis

From Table 9.1, it is clear that a theoretical account for breeding for heterosis cannot encompass details of breeding plans for any particular crop, though the general ideas are very simple. For crosses between widely differing populations, varieties or species, the choice may depend upon an understanding of complementary traits in the different classes. If such crosses are between perennial or vegetatively propagable species, only the level of heterosis if the F_1 will be important, but for all other heterotic breeding plans, the cost of producing seed annually will be critical. This will determine the form of the breeding plan, given normal market forces preventing supernormal profits. For established crops, the technique will involve the development of inbred lines having desirable traits obtainable in homozygotes. As shown in Chapter 5, V_A will always increase on inbreeding in diploids, with all the variance being between lines on attainment of complete homozygosity, thereby allowing better discrimination between lines. However, this will not always be the case in autopolyploids; moreover an intermediate optimum may exist for the level of inbreeding (Gallais 1976d; see also Zali and Allard 1976). Thus, polyploids may require different treatment, quite apart from the extra time needed for a given level of homozygosity to be achieved. Once the breeding approach has been decided, the later steps will involve crosses between the established inbred lines to find the best combinations, and multiplication crosses to produce commercial quantities of seed. What methods are appropriate for any particular plant species will depend upon the relevance of these general principles to the plant in question; for example, even with established crops, the breeding history may mean that an apparently inappropriate method may in fact achieve significant advance. Thus, heterosis for yield in hybrids between the potato subspecies *tuberosum* and *andigena* may reach 15 per cent or more (Cubillos and Plaisted 1976).

Methods for detection and estimation of general and specific combining abilities are the most fully developed, and have been described in Chapter 4. Because of the essentially non-additive nature of the variance utilized in hybrid breeding, simple prediction of hybrid performance from parental performance has never been successful, though simple correlations between production traits of inbred lines and their single cross hybrids are generally positive (but small in absolute magnitude; see, e.g., e Gama and Hallauer 1977).

Despite the widespread and increasing utilization of heterosis, little of its theoretical development has related to response to selection for heterosis; rather the emphasis has been on developing the most efficient ways of obtaining crosses between inbred lines, generally by obtaining factors, nuclear or cytoplasmic, determining male sterility (thereby obviating the need for large-scale physical emasculation of plants) and restoring fertility in the resulting F_1 plants for commercial open pollination. The short sections which follow therefore indicate briefly where some progress has been made in the development of the genetical theory.

9.4.1 Types of cross

Cockerham (1961) has considered in some detail the implications of genetical variances for a hybrid breeding plan. The key points which he has identified for a practical plan are as follows: the amount of inbreeding needed in the inbred lines; the type of cross to be used, i.e., single, three-way, or double; the order of combination of lines; estimation and interpretation of variances between crosses; and prediction of the performance of one kind of cross from that of another.

Consider four lines, A, B, C, and D. Then AB represents a single cross from the mating of line A with line B. A·BC represents a three-way cross from the mating of line A with the single cross BC. AB·CD represents a double cross from the mating of the two single crosses AB and CD. Then A·BC and B·AC are different three-way crosses.

In order to examine the response to selection within such crosses, Cockerham considered the partition of variance in the different kinds of cross:

Source	Degrees of freedom	Expected mean squares
Replicates	$r-1$	
Crosses	$c-1$	$V_E + nV_G(\mathrm{R}) + nrV_G(\mathrm{C})$
Interaction	$(r-1)(c-1)$	$V_E + nV_G(\mathrm{R})$
Residual	$rc(n-1)$	V_E

where $V_G(\mathrm{C})$ is the genetical variance attributable to the c unrelated crosses, and $V_G(\mathrm{R})$ is residual genetical variance. In terms of this analysis, the responses to one generation of selection among single (SC) three-way (3W) or double (DC) crosses are:

$$R_{\mathrm{SC}} = \frac{i\mathrm{Cov}_{\mathrm{AB,AB}}}{\sqrt{[\{V_E(\mathrm{SC})/nr\} + \{V_G(\mathrm{R})(\mathrm{SC})/r\} + \mathrm{Cov}_{\mathrm{AB,AB}}]}}$$

$$R_{3\mathrm{W}} = \frac{i\mathrm{Cov}_{\mathrm{A\cdot BC,A\cdot BC}}}{\sqrt{[\{V_E(3\mathrm{W})/nr\} + \{V_G(\mathrm{R})(3\mathrm{W})/r\} + \mathrm{Cov}_{\mathrm{A\cdot BC,A\cdot BC}}]}}$$

$$R_{\mathrm{DC}} = \frac{i\mathrm{Cov}_{\mathrm{AB\cdot CD,AB\cdot CD}}}{\sqrt{[\{V_E(\mathrm{DC})/nr\} + \{V_G(\mathrm{R})(\mathrm{DC})/r\} + \mathrm{Cov}_{\mathrm{AB\cdot CD,AB\cdot CD}}]}}$$

It can be shown that $R_{\mathrm{SC}} > R_{3\mathrm{W}} > R_{\mathrm{DC}}$. The ratio will be 4:3:2 when $V_G = V_A$ and will increase the advantage of the single cross as V_D, V_I, etc., become more important. For autotetraploids, however, Gallais (1968) has shown theoretically that double crosses may be better than single crosses.

Other results related to inbreeding and the order in which crosses should be made all tend to indicate that more gain is expected from selection among single than three-way crosses, and among three-way than double crosses. As Cockerham has pointed out, the single cross is the type which is the least compatible with established hybrid corn procedures. There are two justifications for using double crosses, however: the first is the improved production of seed, and the second is that genotype by environment interactions may be suppressed by the greater mixture of genotypes that is achieved. This latter point does not of course take into account that there is no guarantee that the most appropriate combinations will be obtained by the double cross nor that particular widely adapted genotypes will not be as frequently represented in single cross progenies (though this is less likely; see for example Allard 1961). Cockerham has suggested that single crosses should always be used at the start of a hybrid breeding plan, on account of the greater gains to be expected. The more non-additive genetical variance that is present, the more important it will be to adopt some hybrid breeding plan. In the case of autotetraploids, there is some evidence to support Gallais's (1968) contention that DC populations will out-perform SC populations (Dunbier and Bingham 1975). This may be a result of allelic interactions different in kind from those possible in diploids (Gallais 1967, 1968), or may merely reflect the different response to inbreeding which we have already discussed.

One problem of double crosses is the large number of possibilities which must be evaluated in a breeding programme. Rawlings and Cockerham (1962) presented the analysis of the complete set of $3p!/\{(p-4)!4!\}$ double cross hybrids from p lines, allowing one to assess epistatic effects, which is not possible with the diallel cross. In fact, such an extension of the diallel analysis allows estimation of gca and different types of sca: effects arising from particular pairs, triplets or quartets of lines. Since an evaluation of only ten lines requires 630 crosses, and since one must assume that linkage effects are unimportant, which is unlikely in the progeny of a cross between two F_1's, results of this method have rarely been published, thought it may have been used commercially.

9.4.2 Recurrent selection schemes

Any breeder seeking to make use of heterosis will want not simply to choose the best available parents on the basis of specific combining ability, cross them and expect the resultant cross to be the best possible genotype for all purposes. Selection must be practised continuously. At least four schemes of such selection have been suggested.

First, selection may be based only on performance in particular pure lines (PLS), i.e. lines will be inbred as closely as possible using selfing or sibbing, depending on the plant's breeding system, and crosses made among the best lines. In general, this scheme cannot take advantage of most forms of epistatic or non-allelic interaction, so that selection progress among the pure lines will depend mainly on V_A and to a lesser extent on V_{AA}.

After identification of the crosses which are best in some sense, the breeder may seek to introduce advantageous genes from lines less good than the best cross, $P_1 \times P_2$, say. Dudley (1984) has presented a very simple method for identifying favourable dominant alleles in inferior lines, P_i, $i = 1, 2$, for substitution into the F_1 of $P_1 \times P_2$. Dudley showed that the expression

$$k\bar{p}a = [\{(\overline{P_1 \times P_i}) - \overline{P}_1\}(\overline{F}_1 - \overline{P}_2) - \{(\overline{P_2 \times P_i}) - \overline{P}_2\}(\overline{F}_1 - \overline{P}_1)]/\{2(\overline{P}_1 - \overline{P}_2)\}$$

estimates the product of the relative number of genes (ka) not carrying at least one favourable allele in $P_1 \times P_2$ and the mean frequency $\bar{p}$ of favourable alleles in the population P_i. To evaluate 100 lines completely, one requires 303 populations: P_1, P_2, $P_1 \times P_2$, $P_1 \times P_i$, $P_2 \times P_i$ and P_i, a substantial number of progenies to be raised. Ranking the P_i for their potential contributions will require the P_i crosses only in addition to P_1, P_2 and $P_1 \times P_2$. Such a procedure will allow comparison of different untried or inferior populations, through their values of $k\bar{p}a$, but it will not yield much insight into the underlying genetics of the apparent dominants detected, since gametic association and epistasis have been assumed absent in deriving the relationship shown above. The procedure illustrates the difficulty of improving advanced hybrids by simple means.

A second type of selection scheme makes use of selection in one population on the basis of crosses to another population. The second population can either be the parental lines, so that the test is a topcross (Jenkins 1940), or it can be the basis of crossing to a particular inbred tester line. This second kind of scheme is called recurrent selection to a tester (RST; Hull 1945).

Both of the recurrent selection to tester schemes have the disadvantage that selection is only practised in half the material used for breeding. Comstock *et al.* (1949) suggested a further development now called reciprocal recurrent selection (RRS), which takes account of this objection to the other schemes. Here, two populations are maintained, and selection is practised within both populations on the basis of performance in crosses between the populations. Thus, a sketch of the plan would be as follows. Suppose that the two populations are called A and B. Then in generation one, self A and outcross the same plants to a number (4 or 5) of random B plants. Make the reciprocal crosses. In generation two, evaluate the crosses. In generation three, plant the selfed seed from the plants with good test crosses and make all possible single crosses between lines within the two populations. This will provide the source for the next generation.

A number of theoretical and practical investigations have been made of the different recurrent selection schemes. (For details, see Cress 1966,

1967; Darrah *et al.* 1978; Hill 1970.) For a single locus, with genotypic values as follows:

	A_1A_1	A_1A_2	A_2A_2
Genotypic value	a	d	$-a$

the response to one cycle of selection may be written (Comstock *et al.* 1949)

$$R = \{p_1(1-p_1)(a+d-2dp_2)^2 + p_2(1-p_2)(a+d-2dp_1)^2\}\frac{S}{2\sqrt{V_P}}.$$

Here, p_1 and p_2 are the frequencies of gene A_1 in the two populations, S is the selection differential, and V_P is the phenotypic variance for the trait, generally yield. From this simple expression, it may be seen that response to selection depends, as would be expected, both on gene frequency in the two different populations and on the genotypic values. Cress (1966) showed that the PLS method, which may be expected to use mostly general combining ability, can in some cases be as effective as the RRS method, for high frequencies of advantageous dominants. Otherwise the RRS method is better. In a case where there were many advantageous dominants, any method which combined them in heterozygous form as simply as possible would be the best.

An increase in heterosis as RRS (or other hybrid selection) continues has often been interpreted as demonstrating overdominance. However, Moll *et al.* (1978) have shown that this is not so. Consider the simplest case, of two populations with a single diallelic locus affecting the trait of interest, with frequencies of the advantageous allele p_1 and p_2 in the two populations ($p_1 > p_2$) and metric values as above. Then heterosis depends upon $\Delta \propto p_1(1-p_1)^2 - p_2(1-p_2)^2$ for selection within cultivars (such as full-sib family selection) and to $\Delta \propto (1-p_1)(1-p_2)(p_1-p_2)$ for selection between cultivars (such as RRS). Now

$$H_{\mathrm{p}} = \Delta\{2(p_1-p_2)+\Delta\}d$$

Hence, $\Delta > 0$ and $H_{\mathrm{p}} > 0$ for RRS, while for intracultivar selection, initial gene frequencies in the two populations determine the outcome, with a likely initial decrease in H_{p} and a slower subsequent increase. Moll *et al.* (1978) found RRS gave greater heterosis than full-sib family selection, in accordance with their approximate theoretical analysis.

Hill (1970) has carried out the most intensive theoretical investigation of the PLS, RST, and RRS methods, but did not consider epistasis, linkage, or multiple allelism. Since epistasis may be a critically important contributor to heterosis in certain cases, his analysis will not be completely general, but he has shown that RRS > RST for his assumptions.

This will be particularly the case where there are dominants with low gene frequency. With partial dominance, RST will be the worst method, and the other two will be approximately equivalent. On Hill's analysis, PLS was not useful with over-dominance, but as noted already this single locus analysis is not completely general. Hill considered limits to response under the different methods, and it may be generally concluded that RRS is overall somewhat superior, though with particular exceptions. As emphasized by Cockerham (1961) as well as Hill (1970), the selection scheme to be used will depend crucially upon the nature of the genetical variation available for selection.

Darrah *et al.* (1978) showed over 6 years in three different maize populations that RRS would out-perform ear-to-row selection, i.e. planting out of half-sibs from chosen plants, followed by selection within families (Lonnquist 1964), by about 30–50 per cent per cycle of 2 years. However, where two similar crops could be grown in one year the simpler method would be better.

Paterniani and Vencovsky (1977) have noted that RRS has not been used as widely as might have been expected, despite its general soundness from a theoretical point of view, and suggested that this was on the following grounds: it is labour-intensive, since selfing and out-crossing simultaneously involves much work, thereby reducing the number of genotypes testable; the testers may be unrepresentative, since only a few plants are used; there is no scope for recombination among selected selfed progenies, as this might require an extra generation, which will slow the procedure even further; and the procedure is very slow already, with tests in only one year of about five. To overcome these difficulties, Paterniani and Vencovsky suggested RRS based on half-sib family evaluation. This, they pointed out, can be carried out so as to complete one cycle of selection in three years, generations or seasons. The procedure is as follows. In generation one, a number of open-pollinated ears (i.e. half-sib families) from population A are planted ear-to-row as females. Male rows are planted with seeds of population B. In a separate block, the reciprocal planting is made, i.e. population B as females, population A as males. Limitations of experimental size will determine how many half-sib families may be planted out with culling on the basis of undesirable traits or poor performance. In generation two, the progenies (A half-sibs × B and B half-sibs × A) may be evaluated. In generation three, the best half-sib families of population A may be recombined, with the remaining half-sib seed, to obtain the next generation of A. The reciprocal crosses will be made to obtain the next generation of B. This experimental design is essentially based on the same principles as the original RRS, and Paterniani and Vencovsky obtained an increase of more than 5 per cent in yield in one cycle of selection, though as they did not conduct a straightforward RRS experiment in parallel, the interpretation of this result is difficult. Nonetheless, the revised method appears promising.

Another approach to improvement of RRS is that of Moreno-Gonzalez and Grossman (1976). As we have seen, single-locus overdominance is not always important in heterosis (cf. also Horner *et al.* 1973), while RRS was developed specifically to ensure that such overdominance was utilized. Moreno-Gonzalez and Grossman therefore suggested two modifications which would use variation having other patterns of dominance.

The first, designated RRS-I, would test population 1 with L2, a line derived from population 2 by family selection for low yield, while 2 would be tested with 1 as before. The second, RRS-II, would test 1 with L2, and 2 with L1, a line derived from 1 by family selection for lower yield. Consider the customary single locus model:

Genotype	A_1A_1	A_1A_2	A_2A_2
Genotypic effect	a	d	$-a$

α_j = average effect of a gene substitution in population ($j = a, b, la$, or lb)
p_j = frequency of A_1 in population$_j$
V_P = phenotypic variance of half-sib families.

Then the responses to one cycle of selection are:

$$R_{\text{RRS}} = \frac{i}{2\sqrt{V_P}}\left[p_1(1-p_1)\alpha_2^2 + p_2(1-p_2)\alpha_1^2 - i\frac{dp_1(1-p_1)p_2(1-p_2)\alpha_1\alpha_2}{\sqrt{V_P}}\right]$$

$$R_{\text{RRS-I}} = \frac{i}{2\sqrt{V_P}}\left[p_1(1-p_1)\alpha_2\alpha_{l2} + p_2(1-p_2)\alpha_1^2 - i\frac{dp_1(1-p_1)p_2(1-p_2)\alpha_1\alpha_{l2}}{\sqrt{V_P}}\right]$$

$$R_{\text{RRS-II}} = \frac{i}{2\sqrt{V_P}}\left[p_1(1-p_1)\alpha_2\alpha_{l2} + p_2(1-p_2)\alpha_1\alpha_1 - i\frac{dp_1(1-p_1)p_2(1-p_2)\alpha_{l1}\alpha_{l2}}{\sqrt{V_P}}\right]$$

From these expressions, for partial or complete dominance,

$$R_{\text{RRS-II}} > R_{\text{RRS-I}} > R_{\text{RRS}}.$$

For some over-dominance, RRS-II will be least effective, and RRS will show increasing advantage over RRS-I as over-dominance becomes more important. With a mixture of patterns of dominance, it may be best to use RRS-II first to increase the frequency of favourable alleles at loci showing partial or complete dominance, later using RRS or RRS-I to maximize over-dominant response.

None of these modifications of RRS is designed to utilize interactions between alleles at different loci, as will be seen more clearly in the next section.

9.4.2.1 General combining ability selection Griffing (1962a) showed how selection for one or both of individual phenotypic value and gca would yield after n generations the values shown in Table 9.9. This is probably as close as possible to the use of a selection index of individual value and gca for simultaneous selection. If selection for heterosis is desired, then testing in three populations, the parentals, A and B say, and the hybrid AB, will be necessary.

In subsequent work, Griffing (1962b, 1963) considered and formalized the general problem for gca selection with two populations. As may be seen from Table 9.10, prediction of the response to n cycles of such selection requires estimates of many parameters which may not ordinarily be obtained. Moreover, as emphasized by Griffing (1962a), his general approach deals only with infinite populations in initial genetical equilibrium, and with small gene effects. However, such an approach is necessary to advance our understanding of selection for heterosis. As an example of our defective understanding of such selection, we should note that, as pointed out by Schnell (1961), RRS schemes do not generally use epistatic genetical variance, despite the aim of RRS. Griffing's (1962b) formulation allows us to see (in Table 9.10) that only additive × additive effects of genes at different loci from the same population contribute to the response in selection. Griffing (1962b), however, emphasized that RRS would eventually allow isolation and incorporation of rare over-dominant gene combinations in the hybrid population.

These results have not yet been extended to selection in more than one environment. Combining abilities are as subject to environment interactions as other measurable genetical effects, but while such interactions have been used in taxonomic classification of the Mexican races of maize (Cervantes *et al.* 1978), they do not appear to have been widely regarded as relevant to practical breeding problems. As their magnitude may be very substantial relative to actual gca values (Gordon 1979), this is perhaps unfortunate. Possibly a combination of the approach just outlined with that of Wright (1977a) described in Section 7.4.2.3 could prove fruitful.

Table 9.9

Incremental changes in population means due to n cycles of selection in a population with genetical variation arising from two linked loci manifesting epistasis (from Griffing 1962a)

		Females (1) No selection	Females (1) Individual selection	Females (1) gca selection
Males (2)	No selection	0	$\frac{i_1(\text{ind.})}{\sqrt{V_{P_1}(\text{ind.})}} X$	$\frac{i_1(\text{half-sibs})}{\sqrt{V_{P_1}(\text{half-sibs})}} Y_1$
	Individual selection	$\frac{i_2(\text{ind.})}{\sqrt{V_{P_2}(\text{ind.})}} X$	$\frac{i_2(\text{ind.})}{\sqrt{V_{P_2}(\text{ind.})}} X + \frac{i_1(\text{ind.})}{\sqrt{V_{P_1}(\text{ind.})}} X$	$\frac{i_2(\text{ind.})}{\sqrt{V_{P_2}(\text{ind.})}} X + \frac{i_1(\text{half-sibs})}{V_{P_1}(\text{half-sibs})} Y_1$
	gca selection	$\frac{i_2(\text{half-sibs})}{V_{P_2}(\text{half-sibs})} Y_2$	$\frac{i_2(\text{half-sibs})}{V_{P_2}(\text{half-sibs})} Y_2 + \frac{i_1(\text{ind.})}{V_{P_1}(\text{ind.})} X$	$\frac{i_2(\text{half-sibs})}{V_{P_2}(\text{half-sibs})} Y_2 + \frac{i_1(\text{half-sibs})}{V_{P_1}(\text{half-sibs})} Y_1$

where

$$X = \left\{\tfrac{1}{2} n V_A + \left[\sum_{r=1}^{n} (1-\bar{\theta})^{r-1}\right] \tfrac{1}{4} V_{AA}\right\}$$

and

$$Y_j = \left\{\tfrac{1}{4} n V_A + \tfrac{1}{4}[1-2\theta_j]^2 \left[\sum_{r=1}^{n} (1-\theta_j)^{r-1}\right] \tfrac{1}{4} V_{AA}\right\}$$

with recombination fraction θ_j between the two loci in sex j (1 = female, 2 = male) and $\bar{\theta} = (\theta_1 + \theta_2)/2$.

Table 9.10

Response to n cycles of selection for the testing schemes shown, where T_{ij} means i tested in j, and all variation arises from alleles at two linked loci (from Griffing 1962b)

	T_{11}	T_{12}	T_{21}	T_{22}
T_{11}	$2U_1$	$U_1 + U_5$	$U_5 + U_3$	$U_5 + U_6$
T_{12}		$2U_5$	$U_2 + U_3$	$U_2 + U_4$
T_{21}			$2U_6$	$U_6 + U_4$
T_{22}				$2U_4$

i subscript = genes from population i
(i) = measured in population i

$$U_1 = \frac{S}{V_{\text{gca}}(12)}\left\{\tfrac{1}{2}nV_{A_1}(1) + \tfrac{1}{4}[1+(1-2\theta)^2]\left[\sum_{r=1}^{n}(1-\theta)^{r-1}\right]V_{AA_1}(1)\right\}$$

$$U_2 = \frac{S}{V_{\text{gca}}(12)}\left\{\tfrac{1}{2}nV_{A_1}(12) + \tfrac{1}{4}[1+(1-2\theta)^2]\left[\sum_{r=1}^{n}(1-\theta)^{r-1}\right]V_{AA_1}(12)\right\}$$

$$U_3 = \frac{S}{V_{\text{gca}}(12)}\left\{\tfrac{1}{2}nV_{A_2}(12) + \tfrac{1}{4}[1+(1-2\theta)^2]\left[\sum_{r=1}^{n}(1-\theta)^{r-1}\right]V_{AA_2}(12)\right\}$$

$$U_4 = \frac{S}{V_{\text{gca}}(12)}\left\{\tfrac{1}{2}nV_{A_2}(2) + \tfrac{1}{4}[1+(1-2\theta)^2]\left[\sum_{r=1}^{n}(1-\theta)^{r-1}\right]V_{AA_2}(2)\right\}$$

$$U_5 = \frac{S}{V_{\text{gca}}(12)}\left\{\tfrac{1}{2}n\text{Cov}_1 + \tfrac{1}{4}[1+(1-2\theta)^2]\left[\sum_{r=1}^{n}(1-\theta)^{r-1}\right]\text{Cov}_2\right\}$$

$$U_6 = \frac{S}{V_{\text{gca}}(12)}\left\{\tfrac{1}{2}n\text{Cov}_3 + \tfrac{1}{4}[1+(1-2\theta)^2]\left[\sum_{r=1}^{n}(1-\theta)^{r-1}\right]Cov_4\right\}$$

Cov_1 = covariance between additive effects of alleles from population 1 measured in 1 and 12 respectively.
Cov_3 = Cov_1 with 2 substituted for 1 and vice versa.
Cov_2 = covariance between additive × additive interaction effects of alleles for population 1 measured in 1 and 12, respectively.
Cov_4 = Cov_2 with 2 substituted for 1 and vice versa.

9.4.2.2 Synthetic varieties These are produced essentially by inter-crossing a number of genotypes selected for good gca in all possible combinations, with the resulting populations being maintained simply by out-crossing. They were suggested by Hayes and Garber (1919) as a relatively simple means of utilizing heterosis. Because less generations of selfing and testing are needed and because the seed does not need to be generated anew for each sowing, there are a number of advantages in principle over straightforward hybrid production.

First, the gene pool is enlarged, since many parents may be used; secondly, the cheapness means that the varieties may be grown on marginal land or smaller farms; thirdly, the more broadly based populations should be more rapidly adaptable; fourthly, it allows heterosis to be used in open-pollinated crops with difficult floral morphology, such as many forage crops; and fifthly, it is particularly suited to plants such as perennial forage crops where genotype–environment interaction may be critical.

In 1922, Wright pointed out that between the F_1 and F_2 a drop in performance could be expected, the magnitude of which would depend upon the number n of parents used; this was formalized by Kinman and Sprague (1945) as

$$\hat{F}_2 = \overline{F}_1 - \frac{1}{n}(\overline{F}_1 - \overline{P})$$

where $\hat{F}_2$ is the predicted F_2 mean, $\overline{F}_1$ the mean of all possible crosses, and $\overline{P}$ the mean of all parents. For ploidy of $2k$, the equivalent value is

$$\hat{F}_2 = \overline{F}_1 - \frac{2k-1}{kn}(\overline{F}_1 - \overline{P}),$$

i.e. the effect is greater than for diploids (Gallais 1974, 1975).

In the absence of linkage and epistasis, the equilibrium level of heterosis is reached rapidly in a completely panmictic diploid population, but both epistasis and linkage (as well as polyploidy) delay the approach to equilibrium. In forage crops, whether annual or perennial, synthetics are widely used, having particular advantages for self-incompatible crops such as lucerne. Thus, it is of great importance to know the rate of approach to equilibrium, i.e. the decline from the F_1 level of heterosis. Considering only additive × additive epistatic interaction I between pairs of loci, Gallais (1975) has shown that the expected level of performance in the mth generation is

$$\hat{F}_m = \text{equilibrium value} + 2(1-\theta)^{m-1}I$$

where θ is the frequency of recombination between loci. Thus, linkage is of crucial importance in determining the rate of this decline.

Following Hill (1966) and Wright (1973), Gallais (1975) introduced the concepts of general and specific synthetic abilities as extensions of combining abilities, but as shown that given only $\overline{P}$ and $\overline{F}_1$ and mean gca values, a selection index combining $\overline{P}$ and mean gca will give the best prediction of synthetic performance. Busbice (1969, 1970) had earlier shown how to take into account inbreeding in synthetic varieties, and Wright (1977b) had shown in *Vicia faba* L. that the level of inbreeding rapidly approached an equilibrium value depending largely on the degree of outcrossing, so that Busbice's method of yield prediction, which essentially requires knowledge only of performance in two generations, together with n and the degree of out-crossing, could be applied with confidence. However, development of a new synthetic may follow a more heuristic path in many cases; Gallais (1977) suggested the outline breeding plan shown in Fig. 9.1 for cocksfoot (*Dactylis glomerata*).

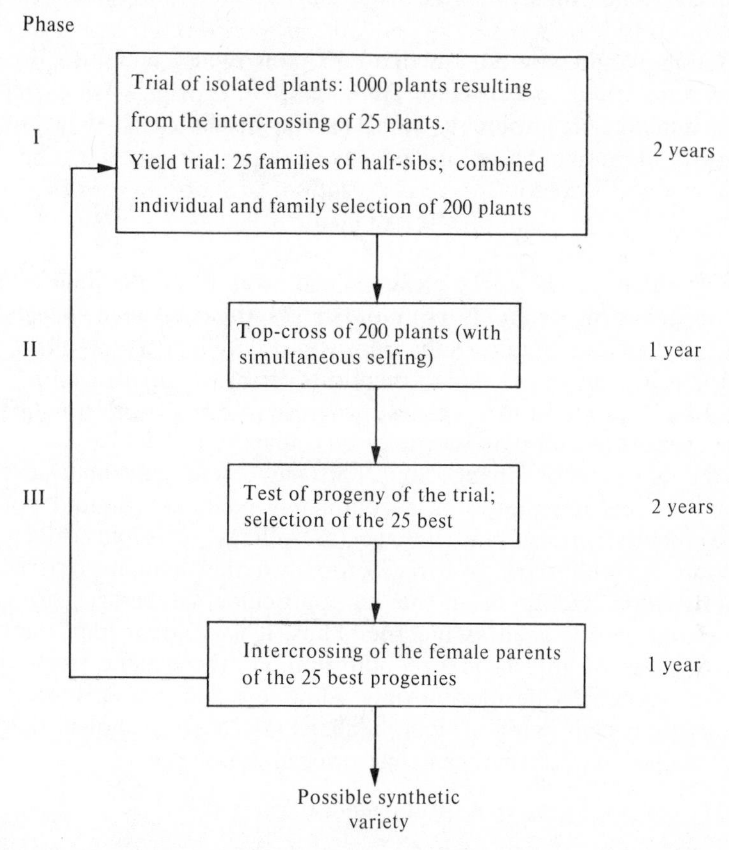

Fig. 9.1. A breeding plan for producing synthetic varieties of a forage crop, *Dactylis glomerata* (Gallais 1977). Redrawn with permission of the Editor, *Annales de l'Amélioration des Plantes.*

9.4.2.3 Prediction of the performance of inbred lines Jinks and Pooni (1976) used the distribution of breeding values, assumed normal with variance depending on $V_A + V_E$, to predict the proportion of inbred lines derived by single seed descent which would exceed any given value. If the F_1 were heterotic, the distribution of dominance deviations would be relevant. It was recommended that progenies of the triple test cross, where the F_2 is back-crossed to P_1, P_2, and F_1, should be used for the most precise estimation of V_A. Both genotype–environment interaction and epistasis could influence the outcome, but in most cases one would expect these to be of minor importance.

Snape (1982), using this method in wheat, noted that single seed descent to F_6 was necessary. Wheat, of course, is almost completely self-fertilized, so that heterosis is unlikely *a priori*, but growing the necessary generations of selfing is simple and will form part of any wheat breeding programme. Snape considered that there was relatively good agreement between observed and expected frequencies of transgressive segregants, supporting the conclusion of Jinks (1981) discussed in Section 9.2.1. However, more than half of the possible comparisons showed significant departure of observation from expectation, so that much work remains before the method is completely validated.

Apart from what has just been discussed, the main problem with the method is that for the very large variances of variance components to be measured with precision, large numbers of progenies need to be raised from crosses which may only infrequently form part of a breeding programme. Thus, while replacement of hybrids by superior inbred lines should be possible in many cases, progress in this direction is likely to remain very slow.

10
Selection methods for self-fertilizing crops

10.1 Introduction

Most selection methods designed for largely self-fertilizing plants depend upon some adaptation of the ideas of Johannsen (1903, 1909), in that it has been assumed that optimal homozygous genotypes are obtainable, and may be obtained by selection among completely homozygous plants or lines obtained by rigorous inbreeding. (The diallel cross was introduced originally by Schmidt (1919) as a logical outcome of Johannsen's approach.) The value of breeding for heterosis in self-fertilizing crops, demonstrated many years ago, has displayed the inadequacy of any simple pure line method. This is despite the continuing absense of a completely satisfactory theory of heterosis, as noted in the previous chapter.

Selection may or may not follow crossing, so that selection schemes may be divided into those not involving crossing (i.e. selection among differing homozygotes) and those involving crossing (i.e. selection among mixtures of related individuals). Within the general framework of selection among homozygotes, two primary approaches may be distinguished (see for example Andrus 1963). These are pure line breeding, where a new variety is the progeny of a single pure line, i.e. maximum selection intensity is combined with maximum inbreeding, and 'mass' selection, where a new variety is obtained by combining the progenies of many pure lines.

Selection among the progeny of crosses between parents possessing desirable attributes constitutes the bulk of breeding activity for self-fertilized plants. Generally, however, the segregating generations are regarded as problematical for selection, as has already been mentioned in Chapter 7.

A third distinct class of breeding system may be distinguished, the back-cross system, whereby a desirable trait (e.g. earliness or disease resistance) is incorporated into an otherwise good variety by repeated back-crossing. While such a system is not necessarily applicable only to self-fertilizing crops, it is most readily used in such cases, and therefore will be treated in this chapter.

Before these systems are described in more detail (in Sections 10.3, 10.4, and 10.6), I shall examine in Section 10.2 some aspects of selection criteria which are especially relevant to self-fertilized plants. In Section 10.5, some theoretical and experimental comparisons of different selection methods are appraised.

10.2 Selection criteria

If the trait for which selection is to be practised is discrete rather than continuous in its distribution, then the approach to selection will be that described in Section 7.4.1, with the addition of the back-cross methods of Section 10.6. If several traits are to be combined, then the criterion may well be a selection index of some kind, used within whatever breeding system is chosen. However, much of the theory of selection so far described in this book relates to individual or between family selection. At times, selection within families will be more appropriate, particularly when there is a large proportion of environmental variance common to members of a family, i.e. familial effects are partially confounded with environmental effects (as may be the case for initial selection of parental material for breeding a tree crop). Falconer (1960) summarized the theory of family selection as it relates to outbreeding organisms, and it was later emphasized that much of the theory may be applied directly to inbreeding species (Pederson 1969a). Because Pederson's treatment of response to selection in self-fertilized species had been extended directly to family selection, his approach is outlined below. Response to individual and between family selection was described in Section 7.3.2. Using the same notation, response to t_2 cycles of within family selection may be shown to be

$$R = \overline{X}_{t_2}\,(\text{selected}) - \overline{X}_{t_2}\,(\text{unselected})$$

$$= i \sum_{t=0}^{t_2-1} \frac{\text{Cov}_{G_{t,t_2}}\,(\text{within family})}{\sqrt{V_{P_t}}\,(\text{within family})}\,.$$

As for individual selection, generalization from this basic result is very difficult, and application would depend upon the estimation of many unknown parameters. Accordingly, Pederson (1969b) analysed a population with genetical variation determined by a single locus, first the initial random mating case and secondly the case of only two alleles. In the former situation, in the absence of dominance,

$$\text{Cov}_{G_{t,t_2}}\,(\text{between families}) = 2\{1 - (\tfrac{1}{2})^t\}V_A$$

$$\text{Cov}_{G_{t,t_2}}\,(\text{within family}) = (\tfrac{1}{2})^t V_A.$$

In the latter (diallelic) case, with initial genotypic frequencies

$$\begin{array}{ccc} A_1A_1 & A_1A_2 & A_2A_2 \\ P & Q & R \end{array}$$

$$\text{Cov}_G\,(\text{between families}) = \frac{1 - 4(\frac{1}{2})^t Q - (R-P)^2}{R + P - (R-P)^2}\, V_A$$

$$+\, 2\left[\frac{\{2 - (\frac{1}{2})^t\}\{1 - (R-P)^2\} - 6(\frac{1}{2})^t Q - (\frac{1}{2})^{t_2}\{R + P - (R-P)^2\}}{R + P - (R-P)^2}\right] V_{AD}$$

$$+ \left[2(R-P)^2 Q \left[\frac{\{1 - (\frac{1}{2})^t\}\{1 - (R-P)^2\} - 2(\frac{1}{2})^t Q}{\{(R+P) - (R-P)^2\}4RP}\right]\right.$$

$$\left. +\, (\tfrac{1}{2})^{t_2+1}\left[\frac{\{1 - 4(\frac{1}{2})^t Q\}\{(R+P) - (R-P)^2\} - 4(R - P^2 Q)}{4RP}\right]\right] V_D$$

$\text{Cov}_{G_{t_1 t_2}}$ (within family)

$$= \left[\frac{2(\frac{1}{2})^t Q}{(R+P) - (R-P)^2}\, V_A + \frac{8(\frac{1}{2})^t Q}{(R+P) - (R-P)^2}\, V_{AD}\right]$$

$$+ \left[\frac{4(\frac{1}{2})^t (R-P)^2 Q}{\{(R+P) - (R-P)^2\}4RP} + \frac{(\frac{1}{2})^{t_2+1}\{(R+P) - (R-P)^2\}}{4RP}\right] V_D$$

In the special case where initial gene frequencies are equal, V_{AD} is zero by definition, whence

$$\text{Cov}_{G_{t_1 t_2}}\,(\text{between families}) = \frac{1 - 4(\frac{1}{2})^t Q}{R + P}\{V_A + (\tfrac{1}{2})^{t+1} V_D\}$$

$$\text{Cov}_{G_{t_1 t_2}}\,(\text{within family}) = \frac{2(\frac{1}{2})^t Q}{R + P}\, V_A + \frac{(\frac{1}{2})^{t+1}}{R + P}\, V_D.$$

As would be expected, within family selection rapidly becomes ineffective, as the variances decline with inbreeding, but the analyses of Lush (1947) and Falconer (1960) show how combined selection (i.e. an index based on individual and family values) may be used depending on the intra-class correlation between members of families. The lower this correlation the higher the relative efficiency of combined selection as against individual selection, the disadvantage of selection within families declining as family size increases, but increasing with t.

Considering, for example, selection of the F_3 stage, individual selection may be more efficient than selection between families if h^2 is high. A selection index combining individual and family records will rarely be helpful, provided that F_3 families are of sufficient size ($\geqq 25$). Selection within families will be relatively more effective for small family sizes and large common environmental effects, but rarely otherwise.

General comparisons of family and individual selection have also been made by England (1977), with an initial design for the estimation of genetical parameters which may be Griffing's diallel or North Carolina (Comstock and Robinson 1952). England's approach allows considerations of experimental design to be evaluated. If families are of size xy, where x is the number of replications and y the number of plants per plot, then the response of family selection relative to mass selection is

$$\frac{1+(xy-1)r}{\sqrt{xy\{1+(xy-1)t\}}}$$

where r, the correlation between breeding values for a given relationship, is $\frac{1}{2}$ for full sibs, $\frac{1}{4}$ for half sibs, and 1 for clonal propagules, and t is the intra-class correlation coefficient, i.e.

$$t_{\mathrm{FS}} = \frac{\frac{1}{2}V_A + \frac{1}{4}V_D}{V_A + V_D + V_{E'}}$$

$$t_{\mathrm{HS}} = \frac{\frac{1}{4}V_A}{V_A + V_D + V_{E'}}$$

$$t_{\mathrm{clone}} = \frac{V_A + V_D}{V_A + V_D + V_{E'}}$$

Here, $V_{E'}$ includes competitive as well as environmental effects. From England's analysis, family selection using single plant plots with several replications will be best for low values of h^2 (analogous to the animal breeding case (Falconer 1960)), while plots larger than about eight plants will yield negligible extra information.

Jinks and Pooni (1981a) suggested, on the basis of the results of extensive experiments with *Nicotiana rustica*, that family selection will be more effective than individual selection in the early generations following crossing. Individual selection, on the other hand, will be more successful in the F_7 or F_8 generations than in the F_2. This would be expected if much of the genetical variance was non-additive, since by the F_7 individual heterozygosity would be rather low.

Progeny testing, where the criterion of selection is the mean value of an individual's progeny, is a form of family selection which is essentially suitable only for perennial plants, where it may be very advantageous (Owino 1977). (It should be noted that progeny testing strictly means that parents are chosen for further breeding on the basis of their progenies' performance which is generally impossible for annual plants. Thus, methods such as the F_2 progeny scheme (Section 10.4.1) are not progeny tests as such.) It also suffers from the extra time necessary for the progeny to be evaluated. Choo and Kannenberg (1979) have recommended progeny testing in maize breeding on account of the greater gains to be expected from this method than from modified row-to-row selection (Section 9.4.2), noting that the delay inherent in the method may be cut by growing two generations a year. However, the other methods can also be accelerated in the same way, so that the delay remains real. Nonetheless, because the mean of an individual's offspring gives a reliable estimate of the individual's breeding value, *A*, progeny testing has occasionally been used successfully, for example with the improvement of Upland cotton variety BP52 in Uganda (Manning 1963). It is a very old method, having been used successfully in the development of the sugar beet in the early nineteenth century.

10.3 Selection among homozygotes

Two approaches to selection among existing homozygous genotypes have evolved, largely through trial and error: pure line breeding, and mass selection. Both methods may still have value in choosing among exotic lines, either to found a breeding programme based on crossing, or simply to select the best variety for a new environment.

Pure line breeding depends upon the existence or creation of a large population made up of many distinct, but largely homozygous genotypes. Large numbers of individual plants, especially of small-grained seed crops, may thus be grown, and selection then is on an individual plant basis. These plants are then multiplied, say ear-to-row, and much selection, for 'type' or discrete characters such as disease resistance, may then be conducted. Finally, multi-site yield trials will be carried out over several years.

It has been suggested, for example by Byth and Caldwell (1970), that superior F_2-derived heterogeneous lines of soyabean (*Glycine max* (L.) Merr.) would out-perform pure lines selected from within the heterogeneous lines through loss in such pure lines of genetically determined non-additive co-operative effects. Thus, heterogeneous populations may be used for superior yield or stability. However, Boerma and Cooper (1975b) found in four crosses of soyabeans involving high-yielding and large-seeded cultivars that pure lines could be isolated from superior heterogeneous lines which would equal or out-perform the relevant heterogeneous lines. Thus, pure line breeding may have a role in changing from genetically heterogeneous varieties or cultivars to homogeneous ones, should this prove desirable.

The alternative to pure line selection is so-called 'mass selection,' but it is not the same as the mass or truncation selection discussed in detail in Chapter 7. In inbreds, the primary variable population is set up in the same manner, but initial selection amounts to culling, i.e. selection of fairly low intensity on the basis of defects, so that it has had much use, in the past, in 'purifying' commercial seed of existing varieties. Its use in highly developed crop plants seems likely to remain minor.

10.4 Selection following crossing

The basic plan of any pure line system must be to obtain a set of homozygous genotypes which can then be multiplied and grown in some suitable trial. It is most appropriate for annual plants. Selection can occur at any stage of the process. The initial population can be a set of F_1s, a set of existing varieties (which are unlikely to be completely homozygous, since many self-fertilized species manifest a small proportion of outcrossing) or any other desired grouping, though generally it will be a set of F_1s.

10.4.1 Pedigree selection

In the pedigree method, which is generally applied to the segregating generations of the cross or set of crosses, a record is maintained of the ancestry of each plant chosen, and initially (F_2 to F_4 approximately) selection is carried out within families as well as between families. In later generations, selection is largely between related families, so that from about the F_7, the pedigree allows the elimination of all but the best in each group of families.

The constraints of Finney (1966) discussed in Chapter 7 roughly determine the pattern of selection. Broadly, the F_1 must be large enough to yield an adequate F_2, while the F_2 must be sufficiently large to yield an adequate set of F_3 families, i.e. 100 or more. The ratio of F_2 individuals to F_3 families will be between 10 to 1 and 100 to 1. Each F_3 family must be large enough to give an indication of remaining variability, i.e. the extent of expressed genetical segregation, as well as of basic agronomic characteristics. The F_4 will be much as the F_3, while by the F_5, commercial planting rate or spacing should be achieved. In subsequent generations, multi-site yield trials begin.

The pedigree method was claimed by Allard (1960) to allow the greatest exercise of skill of any method used for self-fertilizing crops. This is because the choice of parents is critical, and also because the intensity and type of selection to be practised in early generations are determined by the breeder rather than by the application of standard genetical and statistical techniques. However, this perhaps allows decision-making to be excessively arbitrary, and to some extent reflects our ignorance of the genetics of yield.

In Section 10.5, some comparisons are made between the methods;

here we merely note that the computer has taken much of the drudgery out of record-keeping and has thereby allowed small groups of plant breeders to handle far more material far more precisely. Rathjen and Lamacraft (1972) described a computer system for cereal breeding which provides a record of the experiments conducted, analyses the data as they are collected, and generates appropriate further experiments on the basis of these data. This system, which was one of the first to be made available to other breeders and which is currently undergoing extensive redevelopment, used the pedigree trial method of Lupton and Whitehouse (1957), a method especially appropriate for computer development, since its main advantage is that quantitative estimates of yield and product quality may be obtained early in the breeding programme, thereby making the breeder's experience and skill less critical requirements.

In the pedigree trial method, the F_2 and F_3 are straightforwardly pedigreed, but in the F_4 single plants are chosen from the most promising families for continued pedigree selection, while the rest of these chosen F_4 families are bulked for a yield trial in the next year. This is repeated in the F_5 and F_6, the seed for each subsequent trial being from progeny rows, not the previous trial. While this modification of the pedigree method has the advantage already given, as well as allowing more intense or more precise early selection, it requires more attention at each stage. A variant of this method, developed by Breakwell and Hutton (1939), was in use in South Australia in the 1930s. It is shown in Fig. 10.1.

Lupton and Whitehouse (1957) suggested another method, obviating to some extent the extra space and labour requirements at each stage, which they call the F_2 progeny method (essentially a development of Harrington's (1937) mass pedigree method). Here, progenies of selected (pedigreed) F_2 plants are grown in yield trials without selection in F_4, F_5, or F_6, after which single plant selections are made. This method also allows early assessment of yield or quality, with less labour (from the F_4 onward) than the other method, though with a loss of precision in identifying individual promising genotypes. It is used with success in Australia (Rathjen and Pugsley 1978), but is less relevant to European practice because of the need for homozygosity for Plant Variety Rights.

Another closely related method applicable to winter cereals is accelerated pedigree selection (Valentine 1984). Here, single F_1 plants are grown after crossing, then single rows from these plants, then multiple rows, followed by 2 years of yield trials. Because of PVR requirements, the second year of yield trials is also used for assertainment of uniformity, as required for registration.

Thus, modifications of the simple pedigree method may be made for any particular purpose, incorporating aspects of mass selection as necessary.

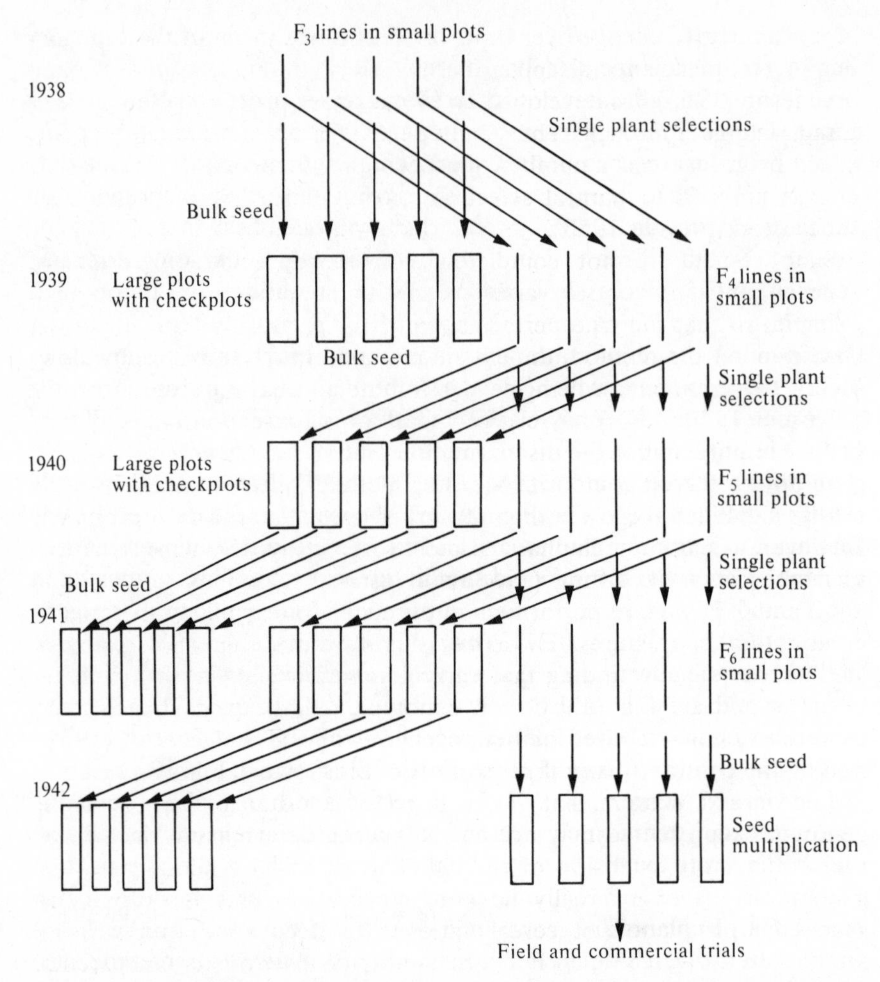

Fig. 10.1. The modified pedigree method of Breakwell and Hutton (1939).

10.4.2 Natural selection

We have seen that two of the key problems of pedigree breeding are that natural selection (for viability, etc.) and competition between plants are minimized, and that various modifications have been suggested to take account of these difficulties. Natural selection may be directional or stabilizing. Directional selection as discussed earlier will be important in any mixture of genotypes (such as that formed by the segregating generations from a cross). Bulk population breeding requires that the F_2 generation be planted as a set of very large plots, preferably at commercial densities. Natural selection and segregation both act to change genotypic and hence gene frequencies. Many studies have shown how varietal mixtures change over time, poorer competitors usually reaching very low frequencies or disappearing.

Suneson (1956) has developed an evolutionary plant breeding method based on this approach. Thus, bulk seed of a cross or set of crosses would be grown over a number of generations, increases in yield occurring as a result of natural selection, as has been observed (see also Suneson and Stevens 1953).

Such results are not completely convincing, since, for example, Suneson's (1956) control variety seems to have been unsatisfactorily variable, so that the true performance of the crosses is hard to assess. Furthermore, the method, though simple and cheap, is inherently slow, several field generations being needed for effective natural selection.

Jensen (1970, 1978) has claimed that natural selection among genotypes will not hamper the discrimination among such genotypes in early generations, so that composite breeding methods such as Suneson's need not be significantly slower than the other methods already considered. However, as has been emphasized most forcefully by Donald (1968) on physiological and Griffing (1968a) on genetical grounds (cf. sections 2.4.2 and 7.5.1), good performers in pure stand are unlikely to be good competitors in mixtures. Evidence is conflicting, Byth and Caldwell (1970) for example finding that early generation yield testing in soyabeans would be biased by inter-genotypic competition (hence their preference, noted above, for heterogeneous lines); Gedge *et al.* (1978) finding the contrary. Even if general conclusions may be drawn, and it is by no means certain that in each specific instance the particular genotypes represented may not be the crucial determinant, the advantages of a short extension of the crossing or heterogeneous genotype generations have not really been demonstrated. This is despite the successful production of cereal varieties by Jensen's diallel crossing method. In this method, an F_1 is obtained from a set of suitable parents, crossed in partial diallel, then composited and crosses made of selected plants for two or more generations, selection among lines beginning in the F_5. The parents used include specific gene donors, e.g. lines carrying single genes for disease resistance, existing cultivars, advanced breeding material, and single crosses, thereby allowing new single crosses, double crosses, three-way crosses, and back-crosses. The method's main advantage, therefore, may lie in the choice of parents rather than in the extra segregating generations.

In principle, however, extension of this period allows two changes to occur: consistently deleterious alleles will be eliminated; and a balance may be struck at loci manifesting heterozygous advantage or frequency dependence. As we have seen in Chapter 5, heterozygous advantage will rarely be strong enough to be relevant to self-fertilized plants, but frequency dependence may be important. It has certainly been shown to exist in a number of cases (Phung and Rathjen 1977). The topic is reviewed by Ayala and Campbell (1974), but because in plant populations frequency and density dependence may be closely related, the treatment of density-dependent selection of Clarke (1972) will be described for its insights into behaviour in a heterogeneous population of plants. It is based on panmixia, but the lessons are more general, since the segregating generations of a breeding programme at least begin with the maximum possible heterozygosity.

Consider a given area containing N breeding plants. Suppose that a single locus A, with alleles A, a at frequencies p and $1-p$ is acted upon by natural selection. There may be density-independent selective mortality:

Genotypes	AA	Aa	aa
Selective values	v_1		v_2

Since there is an excess of offspring in the proportion ε, assumed to be constant for all genotypes, there will also be density-dependent selective mortality:

Genotypes	AA Aa	aa
Selective values	$x_1 = \dfrac{k_1}{k_1 + w_1Np(2-p) + \alpha w_2N(1-p)^2}$	$x_2 = \dfrac{k_2}{k_2 + w_2N(1-p)^2 + \beta w_1Np(1-p)}$

where $w_1 = \varepsilon v_1$, $w_2 = \varepsilon v_2$, and k_1 and k_2 measure the extent to which the environment can support each phenotype in the absence of the other. The constants α and β $(0 \leqq \alpha, \beta \leqq 1)$ represent the degree of competition between the phenotypes. Only if $\alpha = \beta = 0$ is there no competition.

Then the two types of selection may be combined:

Genotype	AA Aa	aa
Overall selective value	$y_1 = w_1x_1$	$y_2 = w_2x_2$

Then the selective change in gene frequency is

$$\Delta p = \frac{pq^2(y_1 - y_2)}{y_1 + q^2(y_2 - y_1)}.$$

If there is an equilibrium, this must arise when the phenotypes have overall selective values equal, i.e. $y_1 = y_2$, and this yields

$$\hat{p} = 1 - \sqrt{\frac{w_1N(w_2k_2 - \beta w_1k_1) + k_1k_2(w_2 - w_1)}{N\{w_1k_1(w_2 - \beta w_1) + w_2k_2(w_1 - \alpha w_2)\}}}.$$

N will generally change from generation to generation, even when $p = \hat{p}$, so that $\hat{p}$ is not usually a true equilibrium. (Only if $w_1 = w_2$ is the equilibrium independent of N.)

It is also possible to investigate the change in number within the given area:

$$\Delta N = N\{(1-p)^2(y_2 - y_1) + y_1 - 1\}.$$

Joint equilibrium values for p and N may now be obtained, where these exist. If $w_1 > 1$ and $w_2 > 1$, as implied in the model, then a stable equilibrium will exist for

$$\alpha < \frac{k_1(w_1 - 1)}{k_2(w_2 - 1)}$$

and

$$\beta < \frac{k_2(w_2 - 1)}{k_1(w_1 - 1)}$$

Thus, if there is substantial heterogeneity in the environment, it seems likely that plant competition (plus the other selective forces summarized in w_1 and w_2) will maintain the polymorphism, and in addition allow N to be larger than for either monomorphic population.

Because the locus A showed complete dominance, the analysis allows us to note that a similar situation could occur with only two genotypes, AA and aa, and while the equilibrium (if any) would be at a different point, it should be possible for a mixture of homozygotes to have higher mean fitness and hence productivity than either single homozygote. In view of our ignorance of the forces involved in such competition, especially how these may be determined by specific gene loci, further experimentation would seem more immediately necessary than further elaboration of over-simplified theory.

10.5 Comparison of methods

Many distinctive aspects of the different methods have already been discussed. Pedigree breeding allows the most intense selection and inbreeding, while bulk selection allows minimum levels of both but much greater natural selection and perhaps greater recombination during the segregating generations. Thus, if early generation selection for a quantitative trait like yield is regarded as vital, strict pedigree breeding may not be practised. As noted in Chapter 7, the evidence for the value of early generation selection is ambiguous, possibly because of the variation in V_A among experiments and because of similar variation in genotype–environment interaction (Weber 1984). The roles of linkage and non-allelic interaction, in particular, have also not been properly elucidated, though epistasis as such will probably be unimportant. Provided that extensive recombination is not generally critical, the widespread use of doubled haploids from the F_1 generation would allow homozygosity to be rapidly achieved and hence to some extent resolve this conflict.

At the practical level, a number of studies have been conducted comparing the results of different selection practices on the same initial material. One such is an investigation, by selection in tomato and by computer simulation, of pedigree, single seed descent and bulk methods (Casali and Tigchelaar 1975a,b).

Single seed descent (SSD) is a modification of pedigree selection suggested by Goulden (1939) to overcome the slowness of pedigree breeding. (It also resembles the bulk method, to some extent, in the absence of selection in early generations.) He pointed out that the breeder of self-fertilized plants had two aims in setting up a cross: to obtain homozygous (or true breeding) lines and to select among these lines for agronomically desirable characters. If both aims are pursued simultaneously this requires large numbers from the F_2 onwards, as we have seen, and only one generation a year may be grown. Goulden therefore suggested that the aims be separated, and the first objective be sought by growing substantial numbers of F_2 plants and cutting down the number of progeny from each to at most two, and then repeating this to the F_6, which might thereby be reached in two years rather than five, with two generations grown in the glasshouse each winter, and one in the field each summer. Selection would begin in the F_6. Snape and Riggs (1975) have shown by simulation techniques that the distributions of the F_6 lines after three generations of SSD will be similar whatever the pattern of inheritance of the trait of interest, but that with substantial heterosis the F_2 will be a poor predictor of F_6 line performance.

Casali and Tigchelaar (1975a) tested SSD against straightforward pedigree selection in tomatoes. Two crosses (of a larger number made) were taken through the F_2 and down to F_7 by pedigree selection, but in addition 12 F_5 lines were obtained from the F_2 by SSD, 12 F_6 from the F_3, and 12 F_7 from the F_4. The general conclusion was that pedigree selection was not very effective overall, but did increase early yield. Where pedigree selection did work, it was superior to SSD, and vice versa. They claimed further that it would be best (for effectiveness, speed, and economy) to combine pedigree and SSD methods, with strong selection in the F_2 and F_3 for characters of high heritability, then SSD to obtain F_6 or later generations rapidly. This conclusion is stronger than their data would allow, though it is often advocated on certain theoretical grounds and criticized on others, as noted earlier (Section 7.3.3).

This experimental comparison was extended by Casali and Tigchelaar

(1975b) to include bulk selection, using computer simulation. They used a model having 20 gene loci, with no linkage, dominance, or epistasis, so that $V_G = V_A$. From 400 F_2 plants, pedigree selection was practised to obtain two F_6 lines, with the proportion selected increasing by doubling from 0.5 to 8 per cent from F_2 to F_5. Bulk selection was carried out by choosing the best 8 per cent in each generation, to obtain 32 F_6 lines also. SSD was conducted at three levels of random selection, from 8 per cent upwards, yielding 32 or more F_6 SSD lines. Pedigree and SSD were combined, as suggested in their earlier work, as were bulk and SSD.

The results were not clear cut, but in general for $h^2 > 0.2$, pedigree and bulk selection were less ineffective than the two SSD methods in increasing yield, as might be expected. For low heritability, SSD was most successful, especially at maintaining V_G; this latter advantage was to some extent carried over to higher heritabilities. Mass selection appeared to perform better at intermediate heritabilities (0.25, 0.5) than at low (0.1) or high (0.75) values.

In another experimental study, Yap *et al.* (1977) compared visual and index (combining pod length and number) pedigree selection and visual mass selection for yield in an F_2 from ten crosses of long bean *Vigna sesquipedalis* Fruw. In the F_2, 20 plants were selected for the F_3 of each method, and taken through to 300 F_5 lines. The three methods did not yield significantly different sets of F_5 lines, though there was some evidence that the pedigree methods were better than their bulk method. Pod yield showed $h^2 = 0.08$, so that this result would seem to be in concordance with the other work discussed above.

In an apparently more decisive experiment, Knott and Kumar (1975) have unequivocally claimed that SSD is superior to early generation yield testing. They made two crosses of spring wheat (Wisconsin 261 × Manitou, Wisconsin 261 × Pitic 62). The F_2 generations were selected for leaf and stem rust resistance and culled for height, maturity and other traits, then the same material was used for both selection procedures. For SSD, single F_2 plants gave the F_3 and so to the F_5 by the same method, then F_5 lines from individual F_4 plants were used to obtain the F_6 for yield testing. For the yield testing method (YT), the F_2 gave F_3 lines which were yield tested, then F_4 and F_5 by pedigree, the F_4 being bulked for the F_5 yield test. Thus, the SSD F_6 and the YT F_5 were compared for yield. Overall, the YT F_5 out-performed the SSD F_6, but only because there were fewer very poor YT lines; the top 20 per cent of SSD lines were as good as the top 20 per cent of YT lines in both cases. Heritability for yield was moderate in one cross (0.30 for Wisconsin 261 × Manitou) and low in the other (0.13 for Wisconsin 261 × Pitic 62). SSD was held by Knott and Kumar to be superior on the basis of efficiency, as it was faster, cheaper and at least as effective as early generation yield testing. Boerma and Cooper (1975a) also held SSD to be superior on the basis of similar results on soyabean, though they also obtained by the pure line selection process mentioned in Section 10.3, lines equal to their best SSD derived lines. Peirce (1977) found SSD to be inferior to pedigree selection or a combination of the two in tomato, but the differences were once again modest, and some progress was achieved with all three methods.

These comparative studies illustrate two important problems: experiments must be both large and lengthy to give unequivocal results; and there is not sufficient theory on which to base the comparisons, especially with regard to the scale of experiments. This will usually be based on scaled-down commercial practice or the boundaries of resources available, and may thereby induce a bias against early generation yield testing. Nonetheless, the genetical arguments against such testing remain strong: heterosis and genotype–environment interaction will both make results misleading if the aim is to predict homozygous, pure-stand performance.

10.6 Back-crossing

Back-cross breeding is designed to incorporate a desirable trait into an otherwise acceptable variety of a crop plant; an example is given in section 7.2.1. Thus, the aim is to transfer only the gene or genes of interest from the donor parent B to the recurrent parent A. The criteria then must be that a suitable variety A exists (generally a current commercially successful variety), that the trait from B is readily recoverable (which usually means that it must be determined by very few genes), and that the desirable features of A may be restored by a reasonable number of backcrosses. Disease resistance, plant height, and earliness are traits which are usually suitable for incorporation by back-cross breeding.

Bassett and Woods (1978) have suggested a procedure for incorporating into the early generations of a selection programme an assessment of the mode of inheritance of a desirable genotype where this is unknown, by using variance among F_3 lines. This has been advanced because of the dilemma of starting a breeding programme first or determining the genetics of the trait first. However, by the F_3 it should in any case be clear if usable phenotypes are reliably available among segregants, thereby fulfilling the second criterion above, or not. Thus, the genetical analysis may be supererogatory, initially at least.

Bartlett and Haldane (1934, 1935) provided the basic infinite population theory which allows us to see the effects of the back-crossing procedure on elimination of unwanted genes from B.

In the simplest case, where the trait is determined by a single locus *C* and is dominant, B is crossed to A and the progeny (carrying the allele *C*) are back-crossed to A. The progeny of this cross are then culled (the genotype cc is removed) and the remainder back-crossed to B. This process, repeated for n generations, reduces the mean segment length of B genome about C to $\frac{2^n - 1}{n2^n}$, which is approximately $1/n$ for large n.

In the case of a recessive, the breeding procedure is as follows:

Parent B				Parent A
cc	$\times$			CC
	$\downarrow$			
	Cc	$\otimes$		
		$\downarrow$		Parent A
		cc	$\times$	CC

Repeated m times, i.e. $2m+1$ generations involving $n = 2m$ generations of crossing, this procedure yields a mean segment length of B genome about C of $\frac{2^{m+1} - 1}{(m+1)2^{m+1}}$, which is approximately $2/n$ for large n.

The actual breeding procedure used has the following general outline. A and B are crossed. The F_1 is back-crossed to A, and selection is carried out for the desirable phenotype from B. The resulting plants are selfed and their seed is grown as a large F_2-like generation which is selected for attributes of A among those plants bearing the trait from B. These selected plants may be selfed to produce an F_3-like generation in which further selection is practised as in the F_2-like generation, or back-crossed to A. If the F_3 is produced, it is back-crossed to A instead.

This procedure is repeated for up to six back-crosses, after which further selfings are made to ensure homozygosity for the gene or genes determining the desirable trait for B. For the simple transfer of a dominant, less than 100 plants are needed (in principle) in any back-cross (over a set of six) or F_2 and less than 2000 in any F_3, to be confident at a very high level of probability of at least one homozygous CC in the F_3. Homozygous recessives or incomplete dominants can be recognized in the F_2, so that even smaller numbers can be used. However, these small numbers would preclude rigorous selection for attributes of A, so that larger numbers should be used where the mechanics of the crosses are not a problem.

The theory of Bartlett and Haldane therefore indicates that the length of segment introduced with any desired single gene will be in the range $1/n$ to $2/n$. The proportion of chromosomes remaining from B will be most $(\frac{1}{2})^n$ and probably much less, given worthwhile selection for the attributes of parent A. Overall, for k chromosomes, the proportion from parent B will be less than $\frac{1}{k}\left\{\frac{2}{n} + (k-1)(\frac{1}{2})^n\right\}$. Thus, if $n = 6$ and $k = 21$, this limiting value will be about 0.03. If a trait is determined by more than one gene, there will be more residual genetical material from B, but the principle remains the same.

The fact that $n \geqq 6$ for reasonable assurance of incorporation of most of the genes of the better parent in a back-crossing programme with $k = 21$, highlights the extreme duration of such a programme in a crop such as wheat. If the aim is to incorporate desired genes from several different parents into one otherwise satisfactory genotype, the problem is compounded. However, as complex properties become identified with gene properties, early generation identification of the desirable phenotypes may abbreviate the number of back-crosses necessary for each part of the programme. For example, some aspects of bread quality in wheat are associated with particular storage proteins and may therefore be identified well before sufficient grain is available for dough-making. As more such major genes are identified, mapped, and characterized biochemically, back-cross breeding, aided by the techniques discussed in Chapter 11, will again become a major part of large-scale breeding programmes.

11

Molecular genetics

11.1 Introduction

The justification for presenting fundamental aspects of gene function and regulation after a consideration of selection on the basis of Mendelian inheritance is only partly historical, in that this knowledge has been recently won. More importantly, the methods subsumed under the heading 'genetic engineering' do not replace existing genetical methods of crossing and selection: they add to what exists, extend its scope and will bring certain previously unattainable goals within our reach. They have as yet, however, contributed little to what is grown for survival or for sale (Borlaug 1983).

I give a brief account of gene translation and the regulation of protein synthesis in order to show how and where the contributions to plant improvement of 'recombinant DNA technology' will be made.

11.2 The genetical mechanism

In all plants, the genetical information is carried in DNA double helices, with the long-elucidated universal set of triplet codons of four bases in chromosomal sequences and a few minor variations in the mitochrondrial but not the chloroplast genome. Green plants differ from other organisms in having this third genome, concerned with photosynthetic carbon reduction. Both cytoplasmic genomes are of importance. Inheritance of traits coded for by non-nuclear genes is non-Mendelian and often maternal, since the cytoplasm of a new zygote is supplied by the female parent. Maternal effects can bias standard techniques. For example, consider a cross between P_1 and P_2, mid-parent value $\overline{P}$, with two different F_1 values for a quantitative trait, $F_1\ (1 \times 2)$ and $F_1\ (2 \times 1)$, from the crosses $P_1 \times P_2$ and $P_2 \times P_1$:

$$F_1\ (1 \times 2) = d + \overline{P} + (P_1 - \overline{P} - d)m$$
$$F_1\ (2 \times 1) = d + \overline{P} + (P_2 - \overline{P} - d)m$$

Here, d is the dominance effect (Section 4.2) and m is the maternal effect, $0 < m < 1$. Then the heritability of the trait will be biased upwards, to $V_A/\{V_A + V_E/(1-m)^2\}$, when estimated from F_1 and F_2 variances (Chandraratna and Sakai 1960). However, when investigated by molecular techniques, non-nuclear genes present opportunities rather than problems, as will become clear.

The processes of gene transcription and translation are very similar for all three genomes.

In transcription, which is the initial process in the synthesis of a protein coded for by a particular structural gene, the DNA helix is locally unwound from the start of that gene and the coding strand, not its complement, is copied to produce a messenger RNA (mRNA) template. The mRNA template moves into the cytoplasm where ribosomes assemble on it. Here translation occurs. Transfer RNA (tRNA) molecules carry amino-acids to the template in the ribosome. A particular sequence of bases specifies for each tRNA the site where the amino-acid is attached. There is a specific tRNA molecule which recognizes each mRNA codon, the matching being by mRNA codon to tRNA anticodon (complementary codon).

There is much ribosomal RNA (rRNA), made on the large and variable number of rRNA genes in the nucleoli, structures associated with the chromosomes carrying the rRNA genes. These genes were initially classified by size of gene product on the basis of rate of sedimentation on ultracentrifugation as 40S, 60S, etc., and later have been associated with the ribosomal structures, aggregates of rRNA and many proteins (binding sites), in which protein synthesis occurs. The ribosomes themselves are aggregated into complex structures called polysomes (polyribosomes). Evidence from selection experiments in *Drosophila melanogaster* at one stage suggested that an increase in the number of rRNA genes might be associated with an increase in certain quantitative traits, opening the way to direct gene selection, but this unfortunately has not proved to be the case (Frankham 1980).

Initiation of the synthesis of a polypeptide chain, in chloroplasts as in bacteria, is always at the codon AUG, the codon for methionine, the methionine being modified to Nformylmethionine first. In all three genomes, termination is via three codons, UAA, UGA, and UAG. It has frequently been suggested, for example by James (1984), that many plant seeds could become food crops through the removal of plant toxins and an obvious route to this might be the modification of reading initiation or termination in the gene coding for the toxin; one target would be *Lathyrus sativa*, regular ingestion of which brings about a progressive paralysis called lathyrism.

The regulation of these processes in eukaryotes is not completely elucidated. The present state of understanding of prokaryotic gene regulation is one of the triumphs of molecular biology and has been the starting point for gaining similar knowledge of eukaryotes. One of the most important results has been the elucidation of the major differences between systems. In bacteria, specific repressor molecules regulate the rate of mRNA synthesis. Each repressor is a small polypeptide coded for by a specific gene. Specific inducers transmit signals for synthesis to begin. For example, lactose is the inducer for the synthesis of β-galactosidase. The repressor molecule binds near the beginning of the β-galactosidase gene at a specific site called the operator. Lactose binds to the repressor, thereby preventing the repressor from binding to the operator, so that mRNA synthesis begins, RNA polymerase binding to the promoter which is close to the operator. In bacteria, functionally related genes are often organized into clusters called operons.

In higher organisms, the differences are striking. Functionally related genes are not organized into operons, nor are regulatory elements necessarily located adjacent to the structural genes. In maize, aspects of the regulation of the amount formed of particular gene products have been clearly elucidated (Scandalios and Baum 1982). Three catalase (superoxide dismutase) genes have been identified, and the rate of synthesis of one of them shown to be determined by an unlinked gene whose alleles act additively. One form of alcohol dehydrogenase, on the other hand, has been shown to have its rate of synthesis regulated by a gene with dominant and recessive phenotypes, and by an inhibitor, a proteolytic enzyme coded for by another gene.

Furthermore, differentiation of cell types cannot occur in prokaryotes, but is important in eukaryotes. For example, in the system of genes which control segmentation in *Drosophila*, there are at least eight similar genes involved, about half of which have regulatory elements located on the same chromosome, but at least one of which has a repressor coded for by a physically remote gene (Lewis 1978). These genes specify the developmental pathways for the body segments of the fly, determining each segment uniquely, and have sequences in common, sequences also found conserved in vertebrates (McGinnis *et al.* 1984*a*; Carrasco 1984; McGinnis *et al.* 1984*b*). Thus, there is evidence that regulatory mechanisms may have much in common across the animal kingdom, so that a complete understanding of one system such as these homoeotic genes in *Drosophila* may provide a general understanding of much of regulation. Sequential synthesis of components of reiterated structures is in any case a fundamental characteristic of plant growth. Many genes in plants are also tissue-specific in action, and in at least one case, a gene for aleurone pigment in maize, the controlling element appears to be located on the same chromosome close to the structural gene. It does not determine the signal product which activates the pigment-synthesizing gene, but rather in which tissue the pigment is made (Dooner 1979).

Organization of cells into nucleus and cytoplasm also determines differences from prokaryotes in regulatory mechanisms. Thus, steroid hormones may bind to specific receptors in the cytoplasm, forming a complex which is transported into the nucleus to initiate or repress synthesis. Hence, more complex mechanisms are necessary to ensure membrane transport, an example of which is discussed in Section 11.3.

In this brief account, I have omitted many complexities. One of the most notable is the presence, within genes, of introns, apparently meaningless DNA sequences which are transcribed but then excised before translation. For example, the leghaemoglobin genes of soybean contain three introns, two corresponding precisely with those found in vertebrates (Jensen *et al.* 1981). However, at least one insect globin gene lacks introns entirely (Antoine and Niessing 1984). If introns prove to be necessary for protein synthesis or gene regulation in some cases, this will influence at least one of the methods of DNA manipulation to be described below, the use of complementary DNA.

In general, one can conclude that the main methods for plant improvement will be the introduction of new genes using the existing regulatory mechanisms, however imperfectly understood, and the improvement of understood processes. Alteration and addition of genes are so precise that traits such as yield will be influenced indirectly, until regulation is much more completely elucidated, through removal or modification of limitations to yield such as disease susceptibility.

11.3 Nucleic acid manipulation

This includes the alteration or removal of existing genes in a genome, the addition of genes, the induction of protein synthesis by genes, and the modification of regulation of existing or added genes. After a brief account of current methodology, I consider actual and potential applications of these methods.

The central problems are to obtain the incorporation of alien structural genes into the host genome and then to ensure their expression (Cohen 1979). For the first step, the requirements are a method for obtaining the DNA to be inserted, a cloning vehicle or vector (a replicon) which is self-replicating in the host, a method of joining the DNA to the vector DNA, a method of introducing the modified vector into host cells, and a method for identifying the cells containing the modified vector.

Initial identification of the DNA to be inserted requires either knowledge of the protein sequence so that an artificial DNA can be constructed corresponding to part of the protein (this DNA being used to find the mRNA for the whole protein by hybridization) or of the mRNA sequence so that a complementary DNA sequence can be made, or by other knowledge of the relevant genes, such as their number and location. One purpose of the analysis of the genome of a potential host crop plant will in fact be to determine whether there are sequences having sufficient homology with a given probe to allow incorporation of new genes by recombination. In many cases, there will be some knowledge of what relevant genes the genome of a potential host holds, but this will be incomplete. In this case a process like the following may be carried out.

Whole genomic DNA can be obtained by various methods outside the range of this discussion. Test sequences (probes) which are to be sought in the genomic DNA are incorporated into plasmids (vectors) by methods which have very rapidly become standard (Maniatis *et al.* 1982). (The plasmids are taken up by bacteria, and a colony cloned from a single bacterium will contain many copies of the DNA sequence inserted into a particular plasmid.) The genomic DNA is digested by particular restriction endonucleases, enzymes which cut the DNA at particular palindromic sequences wherever these occur. The fragmented DNA is then separated by electrophoresis and made available for hybridization with the probe (Southern 1975). Radioactive labelling of the probe then allows identification of the relevant sequence in the genome. Other techniques permit identification of the chromosomal location of the sequence.

Another approach is to reverse the order of events and clone the whole genome (i.e. make a genome 'library') and select the clone containing the particular sequence desired. To be 99% confident that the whole genome will be expressed in such a library, one needs 1500 cloned fragments for *E. coli*, 4600 for yeast, 4800 for *Drosophila* and 800,000 for mammals. Thus, in plants progress will occur through specific examination of individual systems until much cheaper, faster procedures are available (Flavell 1984).

The vectors for multiplication of DNA are already well established: they are bacterial plasmids which can be used reliably from extensive basic molecular biological development over three decades. Vectors for insertion of alien DNA into crop plants are more problematical. First, there are bacterial plasmids. Of these, the most promising is the Ti plasmid of *Agrobacterium tumefaciens* to be described below. Secondly, there are in some species, for example yeasts, endogenous plasmids. Thirdly, it may be possible to alter mitochondrial or plasmid DNA. The genes for cytoplasmic male sterility used in hybrid maize breeding (Section 14.4) are mitochrondrial, for example, and were exploited before their nature was elucidated. Transposition of DNA between mitochondrion, chloroplast and nucleus appears possible, given that sequences have been found in common between nucleus and chloroplast in spinach (Timmis and Steele Scott 1983) and nucleus and mitochondrion in maize (Kemble *et al.* 1983). Fourthly, there are double-stranded DNA caulimoviruses (the best studied is cauliflower mosaic virus, which can easily be introduced systematically into healthy plants; see Hohn *et al.* 1982). Finally, there are the transposable elements first elucidated in maize by McClintock (1951, 1956). These are systems in which DNA elements which are autonomous in respect to movement can insert themselves or other related pieces of DNA at particular chromosomal sites. They have probably been important in evolution, in generating variability through haphazard alteration of sequences and propinquity, and will, when they can be systematically used, allow planned addition of new genes in a manner to which host organisms have already become adapted (Peacock *et al.* 1983; Howard and Dennis 1984; Flavell 1984). They can cause undirected heritable variation which can be used to increase response to selection (Mackay 1984). Like microinjection into ova of multiple copies of a gene, transposable elements have been used first in animal systems (Scholnick *et al.* 1983; Hammer *et al.* 1984).

The Ti plasmid deserves special mention. *Agrobacterium tumefaciens* causes crown gall, one of very few true plant cancers, in a wide range of dicotyledonous host plants. The tumour-inducing (Ti) plasmid is responsible for the bacterium's ability to infect host plants and cause unregulated cell proliferation together with the production of novel metabolites. Part of the Ti plasmid, the T-DNA, is transferred to a plant cell and inserted into the plant genome. The T-DNA can be cloned in *E. coli* and modified by insertion of alien DNA before its use as a vector in *A. tumefaciens.* It has already been used to insert genes for a maize storage protein (a zein), soybean leghaemoglobin and yeast alcohol dehydrogenase into various plants, though some genes were not expressed (Shaw 1984) presumably because suitable promoters were not used. Many genes have been expressed; chimaeric genes are formed between the structural gene of interest and an active promoter. It is possible that the plasmid may also be used to modify monocotyledonous plants (Hooykaas-Van Slogteren *et al.* 1984). At least one non-pathogenic strain produces a bacteriocin, a product which allows biological control of the disease (Kerr and Htay 1974; Kerr 1980). Thus, there is scope both for use of the plasmid as a DNA vector and also for manipulation of its environment; the plasmid itself is not involved in bacteriocin synthesis. Modified vectors based on the Ti plasmid can now be readily constructed and are coming into widespread experimental use (An *et al.* 1985).

One example of successful gene transfer is the secretion of a wheat α-amylase in yeast (*Saccharomyces cerevisiae*) after transfer by means of a yeast plasmid (Rothstein 1984). Here, not only was the cDNA sequence of the α-amylase transcribed, the protein itself was translocated across the endoplasmic reticulum by use of a signal peptide which must be recognized and cleaved from the protein during this process. While there might be application for a yeast modified in this way in brewing, the success of the experiment is of greater interest in suggesting how the regulatory mechanism of one eukaryote may be used successfully with genes from another. Since animal genes have also been expressed in yeast, it is obviously a very favourable host; much work on gene regulation is still needed (Barton and Brill 1983). What is by now clear is that DNA from virtually any source can be introduced successfully into plants, and may be expressed if attached to an active plant promoter.

11.4 Applications

As well as the examples already discussed, DNA manipulation has already allowed the isolation and sequencing of many nuclear genes, the detection of variation in gene number and structure, the physical mapping of the chloroplast genome, the mapping and identification of nuclear genes, extensive investigation of highly reiterated sequences, analysis of cytoplasmic variation and investigation of the role of mitochondrial DNA in pollen development (Flavell 1981, 1984). This and much of the work on vectors remain basic research, though with direct application in some cases.

Applications can arise from work carried out for other purposes. For example, the restriction endonucleases, which cleave DNA at specific palindromic sites, have revealed widespread variation in the number and location of these sites. This variation has been termed restriction fragment length polymorphism (RFLP) and is the subject of much investigation (Weir 1983). However, no matter what its origin or evolutionary significance, it may have direct application in plant breeding, for example, through strain identification for Plant Variety Right Protection, for marking parental genes in new lines established after crossing and in mapping genes influencing quantitative traits (Burr *et al.* 1983). The reason for this possible direct utility is that the RFLP variants are present at many sites throughout the genome, thereby making it likely that the mapping techniques discussed in Chapter 4 can be efficiently used.

In the future, however, much of the DNA manipulation may be expected to be aimed towards the direct modification of genes underlying vitally important metabolic processes.

11.4.1 Single gene systems

The two most important systems are photoreduction of atmospheric carbon dioxide and fixation of atmospheric nitrogen (Hollaender 1977), but much work will also be done on gene-for-gene systems of resistance to pathogens, to be discussed in Chapter 13.

11.4.1.1 Photosynthesis and photorespiration Photosynthetic carbon reduction and oxidation cycles are sets of reactions integrated between the chloroplast, where carbon is reduced (photosynthesis), and the peroxisome and mitochondrion, where it is oxidized (photorespiration). Both sets of reactions are initiated by ribulose-1,5-biphosphate carboxylase/oxygenase, which consists of a large subunit coded for by chloroplast genes and a small subunit coded for by nuclear genes, thereby allowing its modification through protoplast fusion (Section 11.5.3). This enzyme has two functions, carboxylating ribulose-1,5-biphosphate to 3-phosphoglycerate, and oxygenating it to 2-phosphoglycolate and 3-phosphoglycerate. Thus, enrichment of the atmosphere with carbon dioxide in the presence of excess oxygen will suppress the former. This and other relatively simple manipulations have allowed the identification of mutants of at least eight enzymes involved in these pathways, all mutants requiring enrichment with carbon dioxide for growth (Somerville 1984).

In principle, it should be possible to obtain mutants which either improve carbon dioxide fixation or diminish photorespiration, but none has as yet been identified. Further, given that the relevant genes have been identified, incorporation of additional copies will be possible, where this can be shown to be advantageous.

11.4.1.2 Nitrogen fixation The bacterium-legume symbiosis which fixes atmospheric nitrogen is, after photosynthesis, probably the most frequently considered target for genetical manipulation, to improve the system in legumes, to incorporate the system into non-legumes, and to reconstruct it as a non-symbiotic system. Much of the work has been basic investigation of nitrogen metabolism, especially the central reactions involving nitrogenases. Lim *et al.* (1979) have identified five general areas where the symbiosis might be improved: photosynthetic carbon dioxide fixation and photorespiration in the host plant; gene regulation and energy efficiency in the bacterium; and better matching of bacterium and host. To many, energy efficiency seems the most promising target, since the energy involved in the reduction of one nitrogen molecule requires the reduction of a little more than one molecule of glucose. Also, some strains of the symbiotic *Rhizobium*, which appear to be more efficient than others in nitrogen fixation, carry a hydrogenase which regenerates the energy source adenosine triphosphate (Schubert *et al.* 1978; Barton and Brill 1983; Brewin 1984).

The rewards of genetical engineering of this system will be very great when they are finally won. Meanwhile, there is great scope for improving its efficiency in many species by conventional selection techniques (Iruthayathas *et al.* 1985).

11.5 Somatic cell genetics

Somatic cell genetics, which may be defined for the present purposes as the culture and manipulation of plant cells *in vitro* (cf. Pontecorvo 1975), offers many advantages to the plant breeder. First, there is the possibility of regeneration of whole plants from single cells, showing the totipotency of somatic cells, which may be regarded as a special form of asexual reproduction, but which may be applied to haploid tissue (Section 14.2) and to polyploid tissue such as endosperm, which has been shown to be totipotent in wheat, rice, apple and grape (Mu *et al.* 1977). (Thus, large numbers of superior plants may be grown more rapidly than by conventional asexual reproduction via cuttings.) Secondly, given that haploid culture is possible, rapid homozygosity may be achieved through chromosome doubling of haploid plants. Thirdly, there is the possibility of making wide crosses, not just avoiding incompatibility barriers, but also obviating the reproductive barriers between completely isolated taxa. Fourthly, cell culture allows very rapid mass selection procedures to be applied, for pathogen resistance and other important attributes. Fifthly, there is the possibility of the rapid multiplication of good genotypes, another modification of ordinary asexual reproduction (clonal propagation). Sixthly, as implied above, new techniques for modifying ploidy are made available. Finally, cells in culture may provide the ideal medium for genetical manipulations of the kinds just described.

Virtually any part of a plant may be used to grow callus though young tissue may be preferred. Whole plant regeneration is not always possible, but it has rapidly become much easier than when it was first attempted. Techniques for plant cell tissue culture, however, are outside the scope of this book. (See, for example, Scowcroft 1977; Brettell and Ingram 1979; Chaleff 1983; Chaleff and Carlson 1975.) Tissue culture has been successful on an industrial scale in crops as far apart as the oil palm (James 1984) and pyrethrum *Chrysanthemum cinerariaefolia* (Levy 1981).

11.5.1 Protoplasts

Protoplasts, that is plant cells which have had their cell walls removed, have been widely used in plant somatic cell genetics since the first regeneration of whole plants from protoplasts in about 1970 (see Takebe *et al.* 1971). Successful culture and regeneration of plants from protoplasts was rapidly achieved for the following species: *Nicotiana tabacum, Nicotiana sylvestris, Daucus carota, Petunia hybrida, Petunia parodii, Asparagus officinalis, Brassica napus, Brassica campestris, Antirrhinum majus, Datura innoxia, Ranunculus sceleratus, Solanum tuberosum, Lycopersicon esculentum, Arabidopsis thaliana, Atropa belladonna,* and

Citrus sinensis (Scowcroft 1977; Tsai *et al.* 1977; Shepard *et al.* 1983). Techniques have been improved to such an extent that protoplasts from almost any crop species may now be cultured. Apart from the regeneration of whole plants from protoplasts, these cells have the advantage that with no cell wall, they may be treated by many of the techniques of mammalian somatic cell genetics (Carlson 1973).

11.5.2 Mutant selection and detection

The key to detection and selection of mutants in cell culture is the very large number of physically homogenous cells which may be manipulated. Screening of very large numbers of plants is not easy, though it has been successful on many occasions. An illustration of the possibilities of screening is given by Table 7.1, and many commercial cereal varieties have also been obtained by screening advanced generations of crosses for disease resistance. For example, Warigo, an Australian spring wheat, was selected for yield and stem rust resistance at about the F_7 generation, and was found to be resistant also to leaf rust, flag smut, loose smut, and mildew (Phipps *et al.* 1943). However, screening for artificial mutants, even among plants initially identical genetically, requires far greater numbers and hence will be vastly more difficult than screening a larger number of cells in culture, using the techniques of microbial genetics. Thus, all the methods of artificial mutagenesis to be discussed in Chapter 12 may be appropriate to cell culture, and the results correspondingly may be expected to be better than using whole plants or seeds or other appropriate tissues of whole plants, provided that the trait of interest may be assayed in single cells.

In addition, selective media may be used to screen for natural mutants, since cell numbers are large enough for the frequency of such mutants to be useful (see, for example, Kandra and Maliga 1977). Thus, hypoxanthine phosphoribosyl transferase deficiency can be selected for in plant cell culture (Bright and Northcott 1975) as was earlier the case in mammalian cell culture (Velazquez 1975). More particularly, however, economically important traits such as disease resistance may be screened for in cell culture, provided that cell culture response to the toxins correlates strongly with plant response to the pathogen (Buiatti *et al.* 1985). Thus, Gengebach *et al.* (1977) were able to select in cell culture for resistance to the pathotoxin of *Helminthosporium maydis* race T. They used a culture of cells from a plant having Texas male-sterile cytoplasm (cms-T). This cytoplasm determines susceptibility to race T of the leaf blight, so that susceptibility to the disease and male sterility are jointly maternally inherited; the genes are mitochondrial. After five cycles of selection, 65 resistant plants were regenerated from resistant cells. Of these, 52 were male-fertile, and 13 male-sterile, but not cms-T in phenotype. Resistance was maternally inherited, hence selection had also involved the mitochondrial equivalent of segregation or recombination or mutation. In another study, Carlson (1973) achieved the isolation of mutant clones of tobacco cells resistant to the pathogen *Pseudomonas tabaci*, inheritance nuclear rather than cytoplasmic. Matern *et al.* (1978), by selecting for early blight (*Alternaria solani*) toxin insensitivity in cultures of protoplasts of 'Russet Burbank' potatoes, were able to regenerate plants resistant to the blight through two generations of vegetative reproduction. A similar process has achieved resistance for late blight of potatoes through cell selection in the presence of *Phytophthora infestans* pathotoxins (Behnke 1980). (It should, however, be noted that Sanford *et al.* (1984) have pointed out that natural mutants termed 'bolters' occur with reasonable frequency in Russett Burbank, and that the protoplast-derived clones are very similar to the bolters, so that protoplast-derived lines may not have the utility originally expected.)

Resistance to herbicides may also be usefully selected for, allowing different strategies of weed control (Polacco and Polacco 1977). For example, selection in cell culture of *Nicotiana tabacum* for picloram resistance led in many cases not only to resistance to this herbicide but to hydroxyurea as well (Chaleff 1980, 1983).

Other important physiological responses such as drought tolerance, as they become biochemically characterized, will similarly be amenable to rapid selection in cell culture. Aluminium tolerant carrots have, for example, been produced by reconstituting plants from protoplasts grown in a medium containing an excess of aluminium chloride (reviewed by King 1984). Sodium chloride tolerance, an important trait, since there are about 400 million hectares of salt-affected land worldwide, has also been selected for successfully in rice, capsicum, lucerne, and tobacco (Croughan *et al.* 1981). Liu and Chen (1978) and others have claimed that mutant lines of sugarcane (*Saccharum* sp. hybrid) may be obtained, by reconstituting plants from callus culture without selection, which are superior in cane yield, sugar yield and stalk number. Given the genetical heterogeneity of interspecific hybrids, these results are not surprising, though it has not proved easy to incorporate the resulting plants into breeding programmes. However, long-term culture would certainly lead to genetical change, though the proportion of useful changes might not be very substantial.

The variability which arises in protoplast culture is not always wanted. If it is desired to regenerate a number of identical plants under aseptic conditions (e.g. for investigation of nitrogen-fixation), other techniques must be used. One is the culture of immature zygotic embryoes, which has been successfully carried out for several forage legumes (Maheswaran and Williams 1984).

11.5.3 Somatic hybridization

As noted earlier, one of the advantages of protoplasts is that, having no cell wall, they may be treated by some of the techniques of mammalian cell hybridization. Since protoplast fusion was first achieved (see Carlson *et al.* 1972), it has been used to attempt many different crosses. For example, Kao *et al.* (1974) were able by means of polyethylene glycol induction to obtain substantial proportions of heterokaryocytes between soyabean, two *Vicia* species, the pea (*Pisum sativa*), barley, maize, and two *Nicotiana* species. Carlson *et al.* (1972) showed that the somatic hybrid *Nicotiana glauca* × *N. langsdorfeae* yielded, on regeneration of whole plants, some which appeared the same as the sexually derived amphiploid. Shepard *et al.* (1983) have fused protoplasts from species both closely related and unrelated, e.g. potato and tomato, and *Arabidopsis thaliana* and *Brassica campestris.* Given such successes, it is of potential utility that systematic loss of particular chromosome sets occurs in interspecific somatic hybrids in plants (e.g. *Hordeum vulgare* × *H. bulbosum*; see Kasha 1975), as had earlier been observed in mammalian hybrids (Velazquez 1975; Puck 1981). (Where this does not occur, and in cases where wide crosses are not wanted, chromosomal loss may be induced by treatment with specific agents such as carbamates (Wood 1982; Roth and Lark 1984).)

Somatic hybridization should be an important alternative to wide crosses achieved by other means, but the techniques and results as yet remain largely experimental. Hybridization of male sterile and fertile strains of one species may provide a more rapid method of fertility restoration than that available by ordinary crosses (Belliard *et al.* 1977; Izhar *et al.* 1984), but this too is for the future.

12

Induced mutation

12.1 Introduction

Natural rates of mutation are low. Kahler *et al.* (1984) obtained upper limit estimates of 4×10^{-6} for isozyme genes and 10^{-6} for genes affecting morphological traits in barley, in agreement with many earlier results. It is not surprising, therefore, that, not long after the discovery of artificial, radiation-induced mutagenesis by Muller (1927), plant breeders began to use this technique in a search for both novel mutations and additional variability in quantitative traits. The earliest economically successful induction of mutations was by Tollenaar (1934, 1938) in tobacco. Soon after this, chemical mutagenesis was discovered by Auerbach (Auerbach and Robson 1947; Auerbach 1978), and this made the investigation of mutagenesis more widely available. A great stimulus was provided by Gustafsson (1947), who was able to induce several very useful mutants in barley, especially yielding greater stiffness of straw. Table 12.1 shows how the world-wide releases of cultivars dependent on

Table 12.1

Cultivars of non-ornamental species developed either by mutant propagation or by using mutants in crossing programmes (modified from Micke 1976 and Gottschalk and Wolff 1983)

	pre-1950	1950–59	1960–69	1970–79
Direct utilization				
Barley		2	12	17
Other cereals		3	13	17
Other crop plants	1	7	20	19
Incorporation by crossing				
Barley			5	20
Other cereals			3	13
Other crop plants		2	4	10
Total	1	14	57	96

mutation breeding have increased over a 30 year period. They form a tiny proportion of all releases, even now.

Much of the more recent work has been funded and developed in association with the nuclear energy industry, and artificial mutagenesis, radiation genetics, and mutation breeding have developed a technology of their own which this chapter is not designed to cover. (It has been successful, in that about 100 varieties resulting from artificial mutagenesis have radiation rather than chemical treatment as their source of mutation (Gottschalk and Wolff 1983).) Our concern here is with the genetical aspects of mutation breeding, rather than the physiological, biochemical or engineering aspects. Auerbach (1976) and the publications of the International Atomic Energy Authority (e.g. 1970, 1974, 1976) should be consulted for details.

It is not difficult to see why artificial mutagenesis should have had great attractions. First of all, it offered the possibility that totally novel properties might be induced in existing crop plants. Secondly, mutations at specific gene loci within established varieties might be induced, for example, for disease resistance, so that a simply modified variety (produced by say one or two generations of inbreeding) might thereby be very rapidly released. Thirdly, it provided the prospect, through the soon recognized fact that mutations of all kinds, from point mutations to alterations in ploidy, could be induced, that genetical material might be moved between species, and inter-specific crosses or even inter-generic crosses might be facilitated. Some of these hopes have been realized, but the field has perhaps not been as rewarding as it might have seemed originally.

12.2 Inbreeding species

Given the lower level of variability for any particular trait that is to be found in an inbreeding species as against an outbreeding species, mutation breeding may be expected to be more successful in inbred species than in outbred species, regarding artificial mutagenesis merely as a source of increased variability. Furthermore, there is the fact that in the generations of breeding after the mutagen has been applied (called M_1, M_2, etc.) it will be much easier to recognize mutants as they segregate out. (The generations of selfing will also be much easier to arrange.) This has in general been the case.

12.2.1 Single gene traits

The approximate rates of induction of different types of mutation are shown in Table 12.2, these results coming from Lyon *et al.* (1954). It is therefore to be expected, as Brock (1970) has pointed out, that since many, though by no means all, recessive mutants are indicative of a lost

Table 12.2

Relation between mutation rate and numbers of plants which must be grown to be 95 per cent confident of obtaining mutants of a particular type

Type of mutation	Rate of mutation u	Number of cell lineages to be examined N
Chromosomal	10^{-2}	300
Quantitatively inherited trait	10^{-2}	300
Recessive	10^{-4}	30 000
Dominant	10^{-5}	300 000

or damaged function, artificial mutagenesis may be particularly valuable in the domestication stage of development of a crop, where lowering of the concentration of toxic alkaloids, removal of spines or other breaking down of the plant's defence mechanisms may be desirable. As noted in Chapter 11, directed DNA modification will, where possible, be a more desirable way to achieve this end.

Whatever the trait sought, for single gene mutations, if the mutation rate is u and the probability of occurrence of a desired mutant is required to be p_1, then the number of cells to be examined is given by $N = \ln(1 - p_1)/\ln(1 - u)$ (Brock 1970) and the number of cell progenies needed to be examined may easily be calculated for any desired value of p_1. Table 12.2 shows the number of cell progenies to be examined to be 95 per cent confident of obtaining an appropriate mutation. Allowing for the segregation of recessives, these M_1 numbers need to be multiplied by 10 to 100 to obtain the size M_2 necessary. (For a single recessive, ten plants give a probability of $(3/4)^{10}$, i.e. 0.06, of no recessive homozygotes occurring. For two independent recessives, the probabilities are combined multiplicatively.)

In developing a high-yielding, high-protein cultivar of chickpea (*Cicer arieum* L.), Shaikh *et al.* (1982) irradiated seed of the best current commercial variety, grew 3500 M_1 plants, grew 75000 M_2 plants from the M_1 generation, and seleted 752 promising M_2 plants on the basis of morphological or yield variation. Of these, one was sufficiently promising to be multiplied and yield-tested at two sites over two seasons, before release as a variety. A mutant with both improved yield and improved protein content is expected *a priori* to be extremely rare, yet Hyprosola, as it is called, arose in only 75000 plants. As an example of the more usual very large numbers, consider the work of McKenzie and Martens (1974), on stem rust resistance in oats. No satisfactory resistance was known to exist to the prevailing major race of stem rust, since genes with resistance to many races produced other agronomic faults, while most resistance genes did not give a satisfactory cover of widely occurring rust races. Accordingly, five oat cultivars were irradiated with fast neutrons or gamma rays, or treated with diethyl sulphate, with the specific aim of finding resistance to the prevalent and virulent stem rust race, C10. In approximately two million M_2 plants, thirteen were found to possess moderate rust resistance, of which only four were resistant to race C10. Of the thirteen resistant plants, some were chlorophyll mutants, and others possessed other disadvantages. Thus, no more than five out of two million plants were likely to provide some additional resistance to prevalent rust races. In two million plants, natural mutation might be expected to produce variation in resistance of this order of magnitude. To test whether mutants detected are induced or naturally occurring, which may of course not be necessary, is a complex process (Jørgensen 1974). It is of interest in this context that in ten years of breeding work for high lysine content in barley and maize, the two most promising mutants were still naturally occurring ones from world collections (Munck 1976).

12.2.2 Polygenic traits

In any population of plants which have been strongly selected in one direction, e.g. upwards for yield, it is to be expected that alleles favouring the direction of selection will be at high frequencies (or fixed in self-fertilizing species). Accordingly, random mutation would be expected to reduce or diminish the effect of selection. This has usually been found to be the case (Brock 1965, 1967; Gregory 1966). The general result appears to be that mutation depresses the mean of a trait such as yield, but may increase the variance in the trait, thereby allowing the opportunity for selection. This effect has been obtained with yield in peanuts, rice, soyabeans, barley, oats, and wheat.

In one substantial experiment on spring wheat, Singh and Sharma (1976) conducted a 9 × 9 diallel of EMS-treated and untreated cultivars, and assessed the effect of mutation on components of variance of grain yield/plant, 1000-grain weight, and grains/spike in the F_1 and F_2. In no case was V_A or V_D significantly increased by the treatment, but V_D for grain yield/plant was significantly lower in the treated group, a result not at all easy to interpret. The range of grain yield/plant was increased in almost all treated F_2 progenies, though not significantly, and this was not all achieved by an increase in very low yielding progenies, for the number of F_3 progenies exceeding the better parent in the best ten crosses was greater for the treated group than the untreated group in eight of these ten crosses. Thus, the results, while inconclusive, were promising.

In a careful comparison of hybridization, hybridization of irradiated parents and irradiation of the F_1 from non-irradiated parents, Emery and

Wynne (1976) found no evidence that irradiation increased the response to selection for yield practised over the F_2 to F_6 generations in peanuts. However, as they used only 55 F_1 plants in deriving the best five F_5 lines each with three F_6 sublines, these results are also inconclusive.

Gregory (1966) has suggested that small mutations are as likely to be positive as negative in their effects on traits such as yield, but that the magnitude of effects is asymmetrical, large downward effects being more frequent than large upward effects, the probability of improvement declining approximately exponentially with the magnitude of the effect (cf. the analysis of natural mutants by Fisher (1930), and many later workers).

Accordingly, experiments for using mutations in breeding for quantitative traits must be very large both to maximize the chance of producing big improvements, and to maximize the ability to detect small improvements. Thus, as Brock (1970) has emphasized, the fact that useful variation may be available from artificial mutagenesis need not mean that this will be the most efficient method for increasing a trait such as yield. Two points need to be considered: first, is more variability available from crossing, and secondly what happens to the rest of the genome? If selection is easy, as for earliness, for example, induced mutation may be advantageous, but if not, as with yield say, hybridization may well be better in most cases. In most experiments designed to compare the variability elicited by artificial mutagenesis with that from crossing, the mutationally induced variation is the smaller. This is illustrated by Kassem *et al.* (1976) and Salem *et al.* (1976), and the results of Emery and Wynne mentioned above are consistent with this finding also. Furthermore, even if the genetical variance for the trait of interest is increased, selection may do no more than restore the mean to its original value. After this, hybridization will be necessary to take advantage of the hypothetical different genes involved in attaining that particular mean value. The gain in time to homozygosity which is possible for single mutants is therefore not obtainable readily for mutant polygenes.

12.3 Outbreeding species

The general principles of mutation breeding with outbreds are the same as for inbreds, except that everything may be expected to be more difficult. Because selfing may be more difficult or impossible, other patterns of inbreeding may have to be used to obtain mutants in homozygous state, and larger numbers and special techniques may be necessary to distinguish segregants arising from existing variability from those arising from the artificial mutagenesis process. Furthermore, since outbreeding species are more variable for a given population size, the process may be inherently less rewarding.

12.3.1 *Single gene traits*

Breeding to homozygosity and failure to detect useful recessives will be problems here, as already noted. However, there will be particular areas, such as the breakdown of self-incompatibility, where artificial mutagenesis may be the most rapid effective technique. In the case of the single locus gametophytic system, where all of the many self-incompatible alleles found naturally have different specificities (i.e. all crosses are possible, but no selfings), Lewis (1951) and many other workers since have shown that changes in *S* gene specificity cannot be induced by radiation, but that self-compatibility may readily be induced. (The generation of new specificities may be possible through induced or spontaneous inbreeding, sometimes following mutagenic treatment, but that is a different problem; see de Nettancourt (1977).)

12.3.2 *Polygenic traits*

That mutation depresses the level of a trait and selection on the greater revealed genetical variation usually does not allow selection past the previous best value seems to be well established (Brock 1970). Thus, as noted for autogamous species, a generation of crossing or back-crossing of the best mutants with the best of the original line may be necessary, followed by several generations of truncation selection, which would mean that the mutation breeding process was no more rapid than the ordinary breeding programme.

12.3.3 *Heterosis*

Artificial mutagenesis has two applications in hybrid breeding. First, the induction of mutants in inbred lines may be advantageous on occasion. Secondly, and more importantly, the induction of male steriles to facilitate hybrid production may be very important. Male sterility is not always easy to obtain, even by artificial mutagenesis; the experiment of van der Veen and Wirtz (1968) reflects a peculiarly favourable case. Seeds of the experimental organism *Arabidopsis thaliana* were treated with EMS to carry out a selection programme for nuclear genic male sterility. In a little over a hundred M_2 lines, six mutations arose at five different loci, while no cytoplasmic male steriles were obtained.

Driscoll and Barlow (1976) noted that in 34 different species, 25 male steriles were chromosomal mutations, four cytoplasmic mutations and five nuclear or cytoplasmic. Accordingly, chromosomal mutations would seem to be the most promising source of male sterility, especially for crops where cytogenetical manipulations are workable technology (Driscoll 1972, 1978). Widening the sources of male sterility has been recognized as a highly desirable aim since the epiphytotic of southern corn leaf blight in 1970 in the United States, which arose because a single source of cytoplasm (Texas cytoplasm) had been used in a very high proportion of all maize hybrids, and the strain of the pathogen which arose was virulent on hybrids containing this cytoplasm (National Academy of Sciences 1972).

12.4 Vegetatively propagated species

Vegetatively propagated crops include potatoes, sweet potatoes, sugar-cane, and cassava, with all of which sexual reproduction is the normal path to the incorporation of new variants obtained through mutation breeding, whereas for most tree crops and most ornamental woody plants, the generation time is so great that mutation breeding may offer special advantages. This will be particularly the case when the plant in question is self-incompatible, so that it will be both highly heterozygous and difficult to inbreed. As noted by Nybom (1970), since most point mutations are recessive, without the use of highly homozygous material mutation breeding in vegetatively propagated plants will result in very few mutations being detected. A special advantage of mutation breeding for vegetatively propagated plants is that somatic mutations which are unstable or not reliably transmitted through the sexual process may be incorporated into existing varieties or indeed, especially with ornamentals, form new varieties or cultivars.

Sexuality may be induced in apomictic plants, but it is very rare for a species to be completely apomictic, and it may be that a search for sexually reproducing forms will be more rewarding than mutation breeding, given the likely heritable reduction in vigour which may accompany the mutagenic process.

12.5 Chromosomal mutations

As noted already, most mutagenic agents cause far more chromosomal damage, and hence rearrangement, than point mutations. Indeed, as Auerbach (1976) has noted, an important topic in applied mutation research is the increase in efficiency of mutagens, i.e. gaining an increased yield of point mutations per dose of mutagen, or per number of chromosomal mutations (see also Kaul and Bhan 1977). However, advantage may also be taken of chromosomal mutations in several different ways.

12.5.1 Chromosomal modification

For many cytogenetical mapping techniques, such as those discussed in Section 4.6.2, rearranged chromosomes will be necessary. Mutation breeding is the simplest way to obtain these. For example, if some seeds are treated with a mutagen, then M_2 plants may frequently be found to be partially sterile, which indicates the presence of a reciprocal translocation, which should give progenies in M_3 which segregate as 50 per cent fertile and 50 per cent partially sterile. Among the fertile individuals, half will be homozygous for the original pair of chromosomes, the other half homozygous for the translocated chromosomes. Such translocated chromosomes make very good marker chromosomes for mapping.

Mutation may also be used for elimination of unwanted genetical material after crosses between species or genera. In 1956, Sears transferred a gene for resistance to leaf rust from *Aegilops umbellulata* to common wheat. *A. umbellulata* was first crossed with *Triticum dicoccoides* to form an amphiploid ($2n = 42$), this hybrid was then backcrossed to wheat twice and selection was made for leaf rust resistant plants. This yielded a resistant plant with 43 chromosomes, the 43rd being from *Aegilops* and carrying the resistance (*R*) gene. This chromosome had deleterious effects on fertility and vigour. A plant bearing an isochromosome with the *R*-gene duplicated was irradiated and used as a male parent for a cross to obtain wheat with a small translocation carrying the *R*-gene, with the other disadvantageous factors lost. Here, the mutation breeding was far simpler than the cytogenetical manipulations. Since then, many similar transfers have been made, and for stem rust in wheat, two genes have been transferred from einkorn wheat, three from Agropyron species and one from rye (Knott and Srivastava 1977). As noted by Knott and Dvorák (1976), the elucidation of the genetics of homoeologous pairing in wheat has relegated the use of irradiation to those crosses where the species are so dissimilar that homoeologous pairing is impossible.

12.5.2 Alteration of ploidy

Polyploidy is a frequent feature of crop plants, and while most of these are natural, one potentially important one, Triticale, has been artificially produced by crossing wheat and rye. Autopolyploids are readily formed by treatment with chemicals such as colchicine, and also occur spontaneously, though rarely. Similarly, spontaneous haploidy occurs infrequently, though it can also be induced by treatment with particular chemicals. Haploid individuals will be particularly useful, as discussed in Chapter 14, for the production of pure lines, but they have other uses as well.

13
Disease resistance

13.1 Introduction

As was noted in Chapter 1, breeding for resistance to disease is one of the major activities of plant breeders in all countries working on almost all crops. Before discussing breeding methods as such, one should see why this should be the case (Abdallah and Hermsen 1971).

13.1.1 Balance in natural populations

In principle, there is no reason why a host attacked by a pathogen should evolve a balance; after all, extinction has been found to be a very frequent phenomenon throughout the evolutionary record. Nonetheless, there is very good evidence that joint evolution of plants and their pathogens has been the rule rather than the exception, from fungal diseases (see Shepherd and Mayo (1972) for an account of the complexities of the gene-for-gene system described in Chapter 6), through insect herbivores (see for example Feeny 1975), to marsupial herbivores (see Oliver *et al.* 1977). Evidence of a different kind for joint evolution of host and pathogen may be found in local differentiation of similar species or populations of a single species in their response to particular pathogens. Pielou (1973), for example, has demonstrated very substantial differentiation in the genus *Solidago* and its aphid parasites.

A theoretical understanding of the dynamics of genetical host–pathogen interaction will be very valuable, but it is so complex that only the simplest models have been treated. Jayakar (1970), for example, has considered the simplest possible model of one gene determining susceptibility in a host and one gene determining avirulence in the pathogen, with interaction between genotypes. He has shown that equilibria are possible with all hosts the same and susceptible to all pathogen genotypes; with the host polymorphic, the parasite monomorphic; and with both host and pathogen monomorphic (the pathogen having the most virulent genotype, of course). This simple general model therefore allows no broad predictions, but instead suggests that many different host–pathogen interactions will be found, and that their explication will come through an understanding of their individual ecology, rather than from population genetics. This view is reinforced by the work of Mode (1958), who considered the gene-for-gene system first elucidated by Flor (1956) for flax and its rust. Mode pointed out that host and pathogen exert selective pressures on each other, so that the tendency must be either for extinction or for an intermediate equilibrium, extremely resistant hosts and extremely virulent pathogens both increasing the selection pressures for virulence and resistance respectively.

Following Flor, Mode considered a diallelic locus in the host and two diallelic loci in the pathogen, thus:

		Host genotype		
		R^1R^1	R^1R^2	R^2R^2
	$A_1 - A_2 -$	R	R	R
Pathogen	$A_1 - a_2a_2$	R	R	S
genotype	$a_1a_1A_2 -$	S	R	R
	$a_1a_1a_2a_2$	S	S	S

Mode was able to derive conditions for a stable genetical equilibrium, but because fitness depends both on gene frequency and on the reduction in viability of infected hosts, as well as on the presence or absence of pathogenicity, the general conditions may only be stated explicitly for specific cases, and are not very illuminating even there. They agree with the intuitively appealing idea that stable equilibria will develop with a moderate average degree of virulence in the pathogen and a moderate average degree of resistance in the host. It is not clear whether selection would favour tighter or looser linkage between resistance genes, since pressure to do this to preserve favourable gene combinations would lessen adaptibility in the face of new mutation. Levin (1975b) has claimed that there will be in general a relationship between breeding system, and the predatory and pathogenic pressure a population is likely to experience, and that the amount of recombination within the range of a species is related to the geography of pest pressure within that range, such that 'in both breeding system and chromosome system, central populations have greater recombination potential than marginal populations'. This could merely reflect the sum of all selection pressures on marginal populations, should it prove to be the norm.

13.1.2 Monoculture

The common potato is attacked by at least 18 different viruses, six different bacteria, 46 different fungi, and five different nematodes, and suffers at least 40 other diseases, according to Johnson (1972). Rice may be affected by even more, and more serious, pathogens (Khush 1977). Elliott (1958) pointed out that economic plants in the United States of America suffer from over 2500 different recorded diseases, not all important or controllable by genetical means. Evidently, any plant will suffer from disease, but the opportunities for disease to develop are greatly enhanced by monoculture, especially when, in addition to large areas being under one particular crop, that crop is a highly selected variety consisting of one or at most a very few genotypes. Under such conditions, previously tolerable diseases may increase to epidemic proportions, while introduction of monoculture to new areas may elicit the development of new diseases entirely.

Given that monoculture enhances the probability of severe disease or infestation, and given that a substantial use of monoculture appears to be necessary to mankind in its present state of development, there are four strategies for the alleviation of these problems: management practices, such as altered rotation to prevent the build-up of soil-borne pathogens, or changed planting dates to minimize the likelihood of infestation (as with the hessian fly in the United States); the use of chemical pesticides; the use of biological control (so far mainly successful against insect pests); and resistance breeding. Although all these approaches obviously have their place, and indeed should be complementary rather than mutually exclusive, we are concerned here only with the fourth. It should be noted, however, that unexpected interactions between genotypes for disease resistance and pesticides are not unlikely (see for example, Heyland and Fröhling 1977). Furthermore, there are two distinct approaches to biological control (Murdoch *et al.* 1985), which may not be equally appropriate for all crops. In the first, the aim is to use rapidly multiplying, host-specific parasites or pathogens so as to achieve a low, stable pest population. This may be appropriate for pests of perennial crops. In the second, the aim is to achieve local extinction, in which case parasites or pathogens with more than one possible host may be desirable, so that the parasite or pathogen is not also locally extinguished. This may be appropriate for high value crops, where even a low level of infestation lowers the value of the product greatly.

13.2 Genetics of resistance

Resistance may be classified functionally or genetically (Day 1974). The former suggests general or broad resistance (horizontal in van der Plank's (1963) terminology), specific (vertical) resistance, or tolerance. From the genetical point of view, one can have oligogenic, polygenic, or cytoplasmic resistance. The strategies adopted in breeding will depend upon what types of resistance are available, and also upon the type of disease.

The classification of McNew cited in Elliott (1958) is illuminating in this regard. He classified diseases according to their degree of specialization. This is shown in Table 13.1. It is not completely precise; for

Table 13.1
Specialization of disease organisms (from Elliott 1958)

Type of disease	Mode of action
Soft rots, seed-decay organisms	Enter through wounds, destroy stored products
Damping-off diseases	Same organisms as soft rots, etc., but attack plants, usually juvenile tissues, through wounds
Root rots	Cause cortical necrosis, i.e. prevent normal uptake of water and minerals. Unspecialized leaf blights are aerial analogues
Wilt diseases	Adapted for extensive spread through cortical tissue and along vascular bundles
Cell-stimulating and gall diseases	Modify regulation of host plant's cells, and so must be specialized, yet can be free-living
Foliage-destroying diseases	Include leaf blights, downy and powdery mildews pathogens and rusts. More specialized, highly polymorphic
Viral and other totally specialized diseases	Can only multiply in the presence of functional host cells

example, foliage-destroying pathogens may range from unspecialized to highly specialized. Nonetheless, from this classification, it can be seen that specific resistance would usually not be sought for the first two or three categories, but breeding might be directed towards, for example, thicker skins in products which were to be stored for a considerable time, or transported. (This has happened in tomato breeding.) Past breeding work may well have been directed, especially at the stage of domestication, towards the elimination of toxic products, rough skins, or thick integuments, all of which might in the ancestral plant have protected against disease. In many cases, particularly with toxic products, a reversal of the breeding direction is impossible. Where the disease organism has a specialized interaction with the host organism, then specific disease resistance may be sought, generally with some success, and incorporated into productive varieties.

Given that disease resistance has a substantial genetical component, which may initially be detected by screening existing varieties and progeny-testing them, it is of some importance to determine the mode of inheritance, as this decides the breeding strategy.

13.2.1 Single gene resistance

In Chapter 6 it was pointed out that major genes control resistance of plants to a wide variety of parasites, and that these genes interact with major genes for avirulence in the pathogens themselves. Resistance genes are frequently linked, whether closely (Shepherd and Mayo 1972) or loosely (Patterson and Gallun 1977). In order to detect major gene resistance precisely, the plant breeder may choose plants which have minimal general resistance, in order to maximize the efficiency of his breeding programme. While this is undesirable, it is not clear how it is to be avoided while the emphasis in breeding programmes is on single gene resistance. Nevertheless, given that disease resistance is frequently an objective sought in an emergency in a breeding programme, it is to be expected that single gene resistance will usually be sought first.

One important consequence of breeding for specific resistance has been the ever-widening search for new sources of resistance from mutation breeding (Chapter 12), from wild, ancestral, or relic populations, and from related species. As noted by Knott and Dvořák (1976), transfer of genes from alien species becomes harder the greater the evolutionary divergence between the species, and of course the closer the species the more likely it is that the new mechanism of resistance will be the same as in the crop plant, leading to more rapid breakdown. Thus, Hooker (1977) has suggested that much of this effort has been misplaced, as a result of the tactical rather than strategic approach of most resistance breeding.

Shepherd and Mayo (1972) have shown that the gene-for-gene hypothesis described in Chapter 6 is a simplification of the true situation, which (on the host side at least) involves complex loci, whose evolutionary origin is unknown and whose gene products are unknown, so that not much may be said about the nature of resistance and pathogenicity from a genetical point of view. The genetics of pathogens is even less well-defined (Person and Mayo 1974). A proper understanding of the physiological basis of resistance is long overdue (Callow 1977).

13.2.2 Other genetically determined resistance

Although Ellingboe (1982) has suggested that virtually all the thoroughly analysed cases of heritable disease resistance are of the gene-for-gene kind, there has been considerable investigation of and theorizing about general or durable resistance, the terms being used almost interchangeably (Johnson 1984).

General resistance has been defined by some through it being 'enduring and stable', as shown by 'experience and adequate testing' (Caldwell 1968). It is therefore to be recognized retrospectively. It is not always polygenic, but this is usually the case. In principle, specific resistance, i.e. that associated with a specific virulence-avirulence locus in the pathogen, could also be polygenically determined, as a threshold phenomenon (Falconer 1965), but it is not clear that any such case has been thoroughly elucidated though several may have been recognized (e.g. Caten 1974). The interaction of selection between host and pathogen will probably render such a form of resistance unstable; its analysis will certainly be very difficult (Jenns and Leonard 1985).

General resistance slows the disease's spread rather than preventing the development of the disease through local necrosis. From the theoretical considerations outlines above, it is to be expected that stabilizing selection might be found in natural populations, since a pathogen which kills everything will die out, while a host that is resistant to every strain of the pathogen is a very powerful selective factor. Thus, in established situations generalized resistance may be the rule, but this will not apply, as noted, in monoculture. The long-lasting resistance of potato cultivars to potato late blight following the 1845 epidemic is perhaps the classic case, though stem rust in wheat may also be one (Farrer 1898). (Specific resistance may also be very durable, as in the *Agropyron* rust resistance transferred to wheat.)

Resistance to attack by insects has been described by Painter (1951) as arising from non-preference, antibiosis, and tolerance. It seems likely that both tolerance, which is measured through yield depression being minor, and non-preference are polygenically determined (Ellis and Hardman 1975).

As noted by Simons (1972), breeding for generalized resistance is more difficult than breeding for specific resistance. First, there is the general problem of selecting for a polygenic trait, which as will have been seen from the earlier chapters is a slower and less precise problem than selecting for single gene traits. Secondly, because general resistance is inherently more variable, it may be easy to detect in the field in a real-world epiphytotic, but hard to investigate in the greenhouse. The technical problems involved in such work are outside the scope of this book, however. Thirdly, back-crossing techniques which are most appropriate for incorporation of specific resistance genes are not appropriate for generalized resistance, since enormous numbers of plants will need to be grown to have even a modest probability of incorporating both the desirable traits of the susceptible parent and the individually undetectable genes for resistance from the other parent. (For example, Heijbroek (1977) found that partial resistance of sugarbeet to beet cyst eelworm was lost after only two back-crosses.) Fourthly, genotype by environment interactions, especially those involving day length and temperature, will make the assessment of generalized resistance under experimental conditions hazardous in the extreme. Fifthly, the one specific method which has been suggested for breeding for general resistance, that is, selection among the progeny of crosses between susceptible genotypes, may in fact yield genotypes which have epistatically determined gene-for-gene resistance (Johnson 1984). Finally, it may be difficult to distinguish between tolerance and generalized resistance.

13.3 Breeding methods

The first step in any breeding programme for resistance will be the detection of resistance. In the case of a new or unexpected epiphytotic, field populations of crop plants may be surveyed for unattacked plants, on the reasonable assumption that these must have some resistance. Once some degree of resistance has been found, it must be thoroughly investigated, in order that its genetics be understood, as well as its physiological basis. Its likely stability may then also be assessed. Antibiosis, which includes the gene-for-gene systems, is a specific resistance, though not always monogenic, and therefore may be rapidly selected for. Non-preference resistance, which is largely a phenomenon of the pathogen's biology, may be related to plant defence mechanisms (Maxwell 1972) such as secondary plant substances like nicotine in tobacco or gossypol in cotton or to pubescence, as with several insect pests of cotton. Problems may arise with such resistance through conflicting aims, e.g. breeders have previously aimed to reduce gossypol content in cotton, in order to make cotton-seed oil more commercially attractive, but this increases the chance of insect infestation. Similarly, other cotton pests attack pubescent rather than glabrous cotton plants.

If the resistance is merely tolerance, i.e. yields are not reduced to a commercially damaging extent, but a high level of pathogen infestation is maintained, then this clearly raises problems. Nearby stands of the same crop, not possessing the same tolerance, may be infected, since the tolerant stand constitutes a reservoir of the pathogen. Equally, the presence of a high level of infestation is likely to facilitate the spread of virulent mutants which overcome the tolerance.

For all these reasons, though breeders recognize that incorporation of the single gene resistance is generally a series of tactical steps rather than a grand strategic plan, nonetheless, monogenic resistance is normally what is bred for. Incorporation of individual resistance in genes may be carried out by crossing and selection in F_2 and subsequent generations, or by back-crossing after the initial cross.

It is now widely recognized that incorporation of a single gene for resistance to the prevailing strain of a pathogen will usually very rapidly bring about the evolution of a new virulent strain of pathogen. For this reason, it is now normal practice to incorporate a substantial number of resistance genes into a new variety or cultivar. The earliest suggestion for the systematic exploitation of multigene cultivars seems to have been that of Watson and Singh (1952). These have been developed in flax and oats (Knott 1972b).

Multiple resistance can be incorporated into crops in four ways: multilines (i.e. mixtures of different resistance genotypes) as suggested by Browning and Frey (1969); gene deployment (i.e. successive incorporation of new resistance genes into existing cultivars); pyramiding of genes i.e. the incorporation of several resistance genes into one genetical background); and use of anticipatory selection, sometimes confusingly called stabilizing selection, than is, incorporation of resistance genes for virulence genes which have not yet been detected in populations of pathogens found in the region for which the crop in question is being bred, on the assumption that the more 'unnecessary' virulence genes a pathogen has, the less fit it is (van der Plank 1968). This last procedure requires further theoretical and experimental justification.

13.3.1 Multilines

Marshall and Pryor (1978, 1979) have considered multilines in some detail. They contrast 'dirty' multilines, i.e. mixtures of lines each carrying a unique resistance gene, with 'clean' multilines, i.e. mixtures of lines each resistant to all prevalent races of the disease. The formation of a multiline by incorporation of different resistance genes into the same genetical background, through a series of individual sets of back-crosses, is a much simpler procedure than the incorporation of multiple disease resistance into one genetical background. This is the main advantage of the approach, though in the development of the theory (Browning and Frey 1969), multilines have been supposed to stabilize the pathogen's race structure by minimizing the development of races with multiple virulence genes, to act as 'spore traps' because each plant would be attacked by only one genotype of the stabilized population, and hence to prolong the useful life of resistance genes. However, it is not clear that pathogens with multiple virulence genes must be less fit than those with a single virulence gene, so these advantages may not be real (see especially Watson 1970). (In one striking case, MacKey (1974) showed that for oat stem rust in Sweden rust strains carrying many virulence genes were far more common than those carrying few virulence genes, strong circumstantial evidence against van der Plank's hypothesis.) It is also not clear that stabilizing selection will be strong enough to prevent the development of multiply virulent pathogens.

Borlaug (1958) held the advantages of multilines to include, in addition to the prolongation of resistance genes' useful life, the lowered risk of a catastrophic epidemic. CIMMYT's work on multilines follows this approach.

Marshall and Pryor (1978) have built on the work of Groth (1976) to attept to determine the level of stabilizing selection needed to prevent the development of such multiple virulence and to assess how often this might occur in practice. They assumed that pathogens containing 'necessary' genes have equal fitness, and that each successive 'unnecessary' virulence gene would reduce fitness by an amount s. They considered the cases where virulence genes act additively and multiplicatively.

They showed that for n resistance genes, if $s > \frac{1}{2}$, 'simple' (in van der Plank's terminology) races will dominate, if $s < 1/\{2(n-1)\}$ 'super races' attaching everything will develop, and for $\frac{1}{2} \geqq s \geqq 1/\{2(n-1)\}$ intermediate numbers of genes will be found in the pathogen population. This is the additive case, while for multiplicative fitnesses, the lower bound is $1/n$ not $1/\{2(n-1)\}$. Leonard (1977) estimated that $0.42 > s > 0.12$. Thus, to prevent a super race developing, $n > (1+2s)/(2s)$ for the additive case or $n > 1/s$ for the multiplicative case. Accordingly, $n > 9$ is the most pessimistic requirement, and would be sufficient to meet the responses of virtually all pathogens examined in any detail. Breeding such a variety would require substantial effort, with nine back-crossing programmes conducted simultaneously, in effect.

To overcome the objection mentioned above that multilines are dirty, and act as sources of infestation for other non-resistant plantings, it is necessary to have a low proportion of susceptibles in a crop. If this is of the order of 5–10 per cent, then 20 to 200 different lines may be needed in the multiline. Combinations of resistance genes, either disjoint (e.g. phenotypes AB, CD, EF, . . .) or overlapping (e.g., phenotypes AB, BC, CD, . . .), provide in principle more variety from the limited numbers of resistance genes usually available (Marshall and Pryor 1979). However, mixtures of plants carrying such sets of genes will only be more effective than mixtures of plants each carrying a single resistance gene if the fitnesses of the unnecessary virulence genes in the pathogen are strictly additive. It is surprising how little experimental work has been carried out on the fitness of different pathogen genotypes, given its importance in determining the merit of different resistant genotypes.

Since selective forces will rarely be constant, since their nature has not been elucidated clearly, and since it is by no means clear that unnecessary virulence genes will always be disadvantageous, it seems clear from this analysis that multilines will rarely be useful purely for disease control. (As implied above, little is known of the pleiotropic fitness effects of virulence genes; Marshall and Burdon 1981.) Should mixtures of genotypes (cultivars) be grown commercially for other reasons (e.g. to increase yield), then the relative ease of production of multilines might increase their use. In Britain, the simple practice of growing a range of wheat or barley cultivars on each farm has been advocated for some time, and more recently a mixture of cultivars has been recommended (Wolfe 1978). This is only practicable if enough cultivars of similar agronomic and processing character but different disease resistance are available; in such a case, one effectively obtains a multiline by simple mixture of seed. There is evidence that growth of such mixtures can both increase yield and decrease the damage caused by epidemics of aerially spread disease. White (1982) grew pairwise mixtures of six spring barley cultivars in Northern Ireland over three seasons, during one of which there was a severe outbreak of powdery mildew. Both disjoint and overlapping resistance combinations were used, and also a combination of an extremely susceptible cultivar with a resistant one. In all cases, the level of infection in the mixture was lower than in the more susceptible cultivar grown as a pure stand. In most cases, even when disease incidence was negligible, mixtures yielded more than the mean of the two components grown as monocultures. The advantages of mixtures are rarely so unequivocally displayed, however (Sharma and Prasad 1978; Williams *et al.* 1978).

13.3.2 Pyramided lines

The principle of pyramiding genes in a line is simple. Suppose that the rate of mutation to new specificities at the virulence–avirulence locus in a pathogen is u, and that the population size of the pathogen is N. Then the chance of no mutation to virulence, for a rate of mutation of 10^{-5} and a population size of the pathogen 10^6, is less than 5×10^{-5}. However, for the same mutation rate and population size, the equivalent probability for no mutation to virulence to overcome two resistance genes is 0.9999. In general, it is $(1 - u^k)^N$, for k distinct mutations from avirulence, and this will be very high for modest values of k. Even with the enormous values of N which may be expected to arise in an epiphytotic of a major crop plant, this probability will still be substantial. Trenbath (1977) has presented results which bear on this problem, from simulation using a less simplified model. His results provide limited evidence for the concept that resistance will evolve more slowly in pyramided lines than in multiline or single-gene varieties, but he has suggested that the differences are not important.

There are more obvious problems. First, there is the very powerful selective force which such a highly protected crop generates, so that stability is an unlikely outcome of the use of such varieties. Secondly, if the concept of stabilizing selection in the host–pathogen interaction has any validity, long-term protection will require the cyclical use of many pyramided lines. This will require an immense breeding effort. Thirdly, the computations just outlined assume independence among virulence-avirulence loci, which may not be the case. They also assume that the resistance will be totally effective until it breaks down, whereas changes in the environment may render plants susceptible for other reasons.

No completely successful strategy for resistance breeding exists, and as has frequently been pointed out, breeding for disease resistance is in effect a zero sum game. It is not possible to answer the question, how much resistance is enough?

14

Cytogenetical manipulation

14.1 Introduction

At least one topic which might have been included in this chapter has already been discussed because of its greater appositeness elsewhere: aneuploidy in linkage analysis (Section 4.6). Here I consider in further detail a number of related cytogenetical techniques of potential or realized practical importance.

14.2 Haploids

Haploid plants may be produced by a wide range of techniques, not all of which come under the heading of cytogenetical manipulation (Table 14.1). Their natural occurrence is a widespread but low-frequency phenomenon (Kimber and Riley 1963). Of the artificial techniques, pseudogamy, anther culture, and pollen culture have been most widely used. Gametophytic culture is also possible in brown algae, as well as higher plants (Fang *et al.* 1978). Pollen culture (Nitsch 1974a,b) perhaps has many advantages over other methods, for example avoidance of competition between haploid plants, as occurs with callus culture from other tissues, simplification of mutagenesis and transformation experiments on account of the single cell origin of a culture, and simpler investigation of embryogenesis, though this is not actually a plant breeding aim as such.

The major advantage of haploids in plant breeding is that the haploid chromosome set may be doubled, allowing immediate achievement of complete homozygosity in the resulting dihaploids. Thus, crosses would be made in the normal way, and haploids cultured from the F_1 giving a completely homozygous F_2 after chromosome doubling. This would eliminate the early segregating generations of a selection programme for an autogamous species, with advantages and disadvantages as set out in Chapter 10.

Walsh (1974) has considered the efficiency of this haploid method of breeding in some detail. In general, he has concluded that given unlimited F_2 numbers, unselected haploids do not provide as much

Table 14.1
Origin of haploid plants (modified from Lacadena 1974)

Origin of haploid plants
Spontaneous
Monoembryony
Parthenogenesis
Androgenesis
Polyembryony
Parthenogenesis
Androgenesis
Induced
Emasculation and isolation
Delayed pollination
Pseudogamy
Abortive pollen
Distant hybridization
Semigamy
Cytoplasm–chromosome interactions
Chromosome elimination (following distant hybridization)
Alloplasmy
Genetical selection techniques
Culture methods
Anther culture
Pollen culture
Protoplast culture
Tissue culture
Mutagenic treatments

potential for obtaining valuable gene combinations as pedigree selection. With practical limits on the size of the F_2, haploids are only a little less satisfactory than pedigree selected plants, while for very low heritability, haploid results are very similar to pedigree results. He further suggested that if repulsion linkage of useful genes is common, as may well be expected in a breeding programme which is choosing parents with different attributes or different genes affecting particular traits of importance, doubling haploids will miss many of the advantageous recombinants possible. Jinks and Pooni (1981b) also emphasized the fact that dihaploids would not allow gametic association to break down, and suggested forming them from the F_2 or F_3 rather than the F_1. This would, however, either require larger numbers of dihaploids to be produced or risk the loss of favourable gene combinations through sampling. Walsh suggested that rigorous testing of F_7 or later pedigreed lines will be necessary to demonstrate differences between the methods. Since the haploids do not have to be taken this far, the advantage lies here, for, as has frequently been mentioned in this text, time is of the essence in a breeding programme, so that the disadvantage of the haploid method is more apparent than real. In addition, it allows early generation selection uncomplicated by transient heterosis and other problems of segregation. Thus, comparing unselected haploids with pedigreed lines, though technically appropriate, does not allow expression of the full advantage of the haploid method.

While the first workers to identify a haploid flowering plant and a doubled haploid (Blakeslee *et al.* 1922; Blakeslee and Belling 1924) recognized the potential of such doubled haploids in plant breeding, through the rapid production of inbred lines, there are other advantages which can accrue to the use of doubled haploids. Griffing (1975) has discussed these in detail, as applied to methods of recurrent selection. Chase (1952) had earlier made some of the same suggestions.

Griffing considered four diploid selection procedures: individual selection where the individual plant's phenotype is the criterion of selection; clonal selection, where the individual's genotype is evaluated on the basis of its mean clonal performance; general combining ability (gca) selection, where the individual's genotype is evaluated on the mean performance of its half-sib progeny; and reciprocal recurrent selection, where two populations are used as gca testers for each other, so that the criterion of selection is crossing performance. (These procedures have been discussed in Chapters 7 and 9.) Then a cycle of selection consists of the appraisal step, followed by random mating among the chosen individuals, to generate the next population to be appraised. Haploid procedures considered by Griffing were the same as diploid, except that doubled haploids were used throughout, i.e. the initial population consists of doubled haploids. Then Table 14.2 shows the response to one cycle of selection. In this table, the subscripts ind, c, and gca, serve notice that the intensity of selection i will not necessarily be the same for all methods. The variable n is the number of propagules per clone or the number of progeny for gca selection. The model used by Griffing is for two alleles at equal frequency. For more complex cases, the advantage of

Table 14.2

Response to one cycle of selection (from Griffing 1975)

Selection method	Diploid	Haploid
Individual	$i_{\text{ind}} V_A (V_A + V_D + V_E)^{-\frac{1}{2}}$	$2 i_{\text{ind}} V_A (2V_A + V_E)^{-\frac{1}{2}}$
Clonal	$i_c V_A \left(V_A + V_D + \frac{V_E}{n} \right)^{-\frac{1}{2}}$	$2 i_c V_A \left(2V_A + \frac{V_E}{n} \right)^{-\frac{1}{2}}$
GCA	$\frac{1}{2} i_{\text{gca}} V_A \left(\frac{n+3}{4n} V_A + \frac{V_D}{n} + \frac{V_E}{n} \right)^{-\frac{1}{2}}$	$i_{\text{gca}} V_A \left(\frac{n+1}{2n} V_A + \frac{V_D}{n} + \frac{V_E}{n} \right)^{-\frac{1}{2}}$

doubled haploids over diploids may in some cases be reduced. The key points to be noted in Table 14.2 are that the variance on which selection operates is doubled in the haploid relative to the diploid (cf. the effect of inbreeding set out in Section 5.2.2), and dominance is eliminated as a complicating factor from individual and plant selection. In all cases, haploid methods are better, the differences being greatest with large V_E or V_D.

For reciprocal recurrent selection, each doubled haploid cycle of selection produces a new set of single crosses which can be evaluated and possibly released, whereas this is not the case with ordinary reciprocal recurrent selection, approach to the 'best' cross being asymptotic.

Because of their many advantages, haploid-derived lines are in very widespread use in breeding programmes and commercial cultivation.

14.3 Homoeologous pairing

The chromosomes of the ancestral diploid parents of the polyploid wheats are sufficiently similar to be capable of allosyndetic conjugation in hybrids, and have not suffered major structural alterations since the polyploids were formed (Riley 1960). Furthermore, the similarity in genetical activity of homoeologous chromosomes is quite substantial (see for example, Hart and Langston 1977). However, meiotic pairing between homoeologues from different genomes does not occur. It has long been known that a locus on chromosome 5B suppresses homoeologous chromosome pairing (Riley and Chapman 1958; Okamoto 1957; Sears and Okamoto 1958). The mechanism of such suppression is as yet unknown, though it has also been demonstrated in other groups, e.g. *Lolium* species (Taylor and Evans 1976, 1977). In these ryegrasses, both suppression by B chromosomes and suppression by genes in the normal chromosome complement have been demonstrated; Riley (1960) referred to suggestive though not conclusive evidence for many other species.

The formal genetics of partially diploidized species is different from either diploid inheritance or polysomic inheritance (Sved 1965), and the evolutionary implications are of great interest, possibly affecting our understanding of the evolution of gene action (Hart and Langston 1977) as well as particular problems such as the origin of multi-locus gametophytic self-incompatibility (Mayo 1978). However, the main area of practical concern is that manipulation of homoeologous pairing suppression will allow much wider crosses and more ready incorporation of alien genetical material.

Riley *et al.* (1968) were able to incorporate resistance to yellow rust (*Puccinia striiformis*) from *Aegilops comosa* into the wheat variety Chinese Spring by modulation of homoeologous suppression. Initially, a back-crossing programme from progeny of Chinese Spring × *Aegilops comosa* back to Chinese Spring gave a line with one chromosome from *A. comosa* determining rust resistance added to the full complement of Chinese Spring. This 43rd chromosome was homoeologous to group 2 of wheat, and was designated 2M. To induce recombination between 2M and its wheat homoeologues, crosses were made with *Aegilops speltoides*, which suppresses the gene on chromosome 5B which prevents homoeologous pairing. Further back-crossing to Chinese Spring was then carried out with the 29 resulting chromosome hybrids carrying haploid complements of Chinese Spring and *A. speltoides* together with chromosome 2M. Selecting for rust resistance eventually yielded a resistant plant with 42 chromosomes forming 21 bivalents at meiosis. The plant was heterozygous for dominant rust resistance from *A. comosa.* Homozygotes derived from it yielded a rust resistant breeder's variety, Compair.

Evidently, such a procedure is very lengthy; in this case it took over ten generations, several of which were in the crosses with *A. speltoides* and the subsequent back-crosses. Accordingly, mutants of the gene on chromosome 5B permitting homoeologous meiotic pairing may be regarded as having many advantages, theoretical and practical. Such mutants have been isolated (Wall *et al.* 1971a,b; Sears 1977). Sears (1977) identified an apparent deletion of *Ph*, the chromosome 5B suppressor, by X-irradiating normal pollen and testing it on plants monosomic for a marked chromosome 5B, then testing the M_1 plants which lacked the maternal chromosome 5B, by crossing to *Triticum kotschyi* or rye and then screening for increased pairing. Further development of these methods will be possible in all species which allow a high degree of aneuploidy, and will allow a much more extensive 're-engineering' of appropriate genomes.

14.4 Hybrid breeding

The genetics of hybrid vigour and breeding for hybrid performance have already been discussed, but not the methods of achieving complete cross-fertilization, which are especially important in breeding autogamous species. The essential requirement is male sterility, so that all plants of a desired type will be outcrossed.

Male sterility may be genetical, i.e. determined by chromosomal genes, when it is usually recessive, thus:

$$\left.\begin{array}{l} MsMs \\ Msms \end{array}\right\} \text{ male fertile}$$
$$msms \quad \text{male sterile}$$

It may be cytoplasmic, in which case it is maternally inherited, thus:

Female		Male
S	×	F
	S	

Male sterility may also combine both genetical and cytoplasmic control, with both the *msms* genotype and the S cytoplasm needed for male sterility, and the *Ms* genes epistatic to the gene of the S cytoplasm. This allows simple restoration of male fertility, as may be seen from the pattern of segregation in the following crosses:

Female	Male	Progeny	
msms S	*MsMs* F	*MsMs* S	male fertile
msms S	*Msms* F	$\frac{1}{2}$ *Msms* S	male fertile
		$\frac{1}{2}$ *msms* S	male sterile
msms S	*msms* F	*msms* S	male sterile

Inbred lines may therefore readily be maintained and crossed, with the appropriate manipulation of genotypes and cytoplasm.

Although many sources of male sterility are known in maize, where the trait was first discovered (Rhoades 1931), commercial utilization has rested on but a few of these, as noted in the previous chapter, and it is only now that intensive genetical investigation of mitochondrial and chloroplast DNA is under way to engineer less vulnerable sterility–fertility relationships (Pring and Levings 1978). Many complexities have been revealed. For example, there appear to be at least two genes involved in the recovery of fertility in perennial ryegrass (Connolly and Wright Turner 1984). Such complexity does not alter the restoration method in principle, however, since the restorer plants will supply the necessary alleles for both genes.

An alternative method of sterility–fertility determination, appropriate for wheat and other autogamous crops where satisfactory restoration is difficult (Hughes and Bodden 1977; Johnson and Patterson 1977), has been proposed by Driscoll (1972), and illustrates further the kind of manipulation possible where aneuploidy is tolerable. The basis of the method is a line homozygous for a gene for recessive male sterility carried on a wheat chromosome, with a dominant homoeoallele which restores fertility borne on a single alien homoeologous chromosome which also carries a dominant marker *D*. Then the lines X and Y are set up and may be maintained as shown below, and the line Z, which is the basis of hybrid seed, may be obtained by selfing Y, thus:

		X	Y	Z
Genome	wheat	42	42	42
	alien	2	1	0
Genotype		*ms ms MsD MsD*	*ms ms MsD*	*ms ms*
Gametes	female	*ms MsD*	$\frac{3}{4}$ *ms*	*ms*
			$\frac{1}{4}$ *ms MsD*	
	male	*ms MsD*	*ms*	—
Selfed progeny		*ms ms MsD MsD*	$\frac{3}{4}$ *ms ms*	—
			$\frac{1}{4}$ *ms ms MsD*	

In fact, most hybrid wheats, including commercially released products, have used the nuclear-cytoplasmic method, the source of cytoplasmic male sterility being *Triticum timopheevi.* Gains of 5–10 per cent over established cultivars have yet to outweigh problems of variability and the extra cost of seed (Wilson and Driscoll 1983; Boland and Walcott 1985).

14.5 Polyploidy

The consequences of autopolyploidy for progress under selection are described where appropriate in this book. The main characteristic is the slower time to achieve a given level of genetical change, which applies also to cytogenetic manipulation (Dewey 1977). If all the genetical material comes from one species, then for the most part the consequences of autopolyploidy must be changes in emphasis of existing tendencies. While these have long been recognized to include substantial and even unexpected changes in morphology and size (Blakeslee 1941), application of such results is more limited. Prediction of the performance of different levels of ploidy is still very difficult, the optimum level of ploidy varying substantially from species to species, for example being triploid for sugarbeet (Burnham 1966) and tetraploid or higher for asparagus (Dorè 1975). Because of problems of sterility in higher polyploids, their application is of special advantage in clonally propagated species such as the last mentioned. In addition, as in the case of bananas where natural triploids were domesticated, being seedless, and watermelons, where artificial triploids have been used for the same purpose, advantage may specifically be taken of disturbance to meiosis.

Allopolyploidy's potential is qualitatively different. As was recognized very early by Haldane (1932), Muller (1935), and Bridges (1935), the existence of duplicate genes allows the evolution of a new function without impairing the existing function. Evidence for this hypothesis is now convincing (Hart and Langston 1977; Ferris and Whitt 1977). The tolerance of aneuploidy by tetraploid and hexaploid wheats shows, of course, that in a relatively young allopolyploid genomic differentiation is but modest.

Apart from the manipulative advantages which allopolyploidy offers, its greatest prospect lies in the engineering of new crops. Triticale (*X Triticosecale* Wittmack) is the prime example of such a crop. As noted by Rajhathy (1977), Triticale is unknown in nature. This would tend to imply that it has not formed or has not survived, the latter being more likely since the parent species are found together, hybrids are possible (and have long been made artificially, e.g. Kihara 1919), and spontaneous chromosome doubling occurs. Presumably, the natural disadvantages of the hybrid which prevented its spread were among the factors which had to be overcome by early Triticale breeders. The aim in creating this artificial cereal was to produce a breadmaking cereal of equal quality to wheat with the winter hardiness and adaptation to light marginal soils of rye. Initially, as in the Canadian programme, the problem was to develop satisfactory amphiploids, and most crosses were made between hexaploid wheat and rye, yielding, after doubling, an octoploid. These Triticales have largely been superseded by hexaploid Triticales, where the A and B genomes of wheat are combined with the R genome of rye. This naturally led to many problems, such as lowered vigour with wheat as the female parent in manipulative crosses (Gour and Singh 1977). As noted by Larter (1973), 10 years of development were needed for Triticale to yield as much as a control wheat variety, and during this time wheat breeding continued to advance. Since then, however, progress has been more rapid, as a result of both national breeding programmes and international ones under the auspices of CIMMYT, and there are many areas where Triticale has advantages over wheat or rye. Comparing the results of 40 years' careful plant breeding with thousands of years of domestication followed by a century of plant breeding, one must conclude that Triticale, which for many years looked like the answer to a question which had not been asked, is now a worthwhile minor cereal crop (Gupta and Priyadarshan 1982), and a modest breeding programme can provide a succession of commercially acceptable varieties (Driscoll *et al.* 1983). Problems occupying Triticale breeders have become agronomic rather than fundamentally genetical (e.g. Brouwer 1977). No doubt many other artificially engineered crop species will follow.

15
Perennial crops

15.1 Introduction

In most of the discussion so far, I have assumed that generation length for a crop is no more than a year, that all plants are equivalent in this fashion, and that generations are discontinuous, where this has been relevant. Some specific exceptions have been noted, such as the special problems and advantages of clonally propagated species, many of which are important perennial crops, or the detection of competition in forest trees, but in the main we have been concerned with annual crops, and with others, such as forage grasses, which may be reproduced sexually at least once a year.

While the formal genetics of many economically important perennial species has been intensively investigated (from Crane and Lawrence (1947) for garden plants to Pryor (1976) for *Eucalyptus*), the application of this work to plant breeding has in many cases been very modest. Meanwhile, tree crops, fruits, timber, and others, have special problems for the breeder. In this chapter, we are concerned to discuss some of them briefly, together with a few more general matters relating to perennial plants.

15.2 Environmental factors

15.2.1 Natural populations

Population size is a vital factor in determining the rate of genetical change under natural and artificial selection. For evolutionary purposes, the definition of a plant population is the same as for an animal population, but as emphasized by Harper (Kays and Harper 1974; Harper and White 1974), for many matters important in plant breeding and agronomy, the number of plants may be less relevant than the number of plant organs or parts (e.g. tillers in a cereal crop or a grass sward). For example, it is well known that log mean plant weight depends negatively on log plant density (de Liocourt 1898). When this is applied to the assessment of the interaction between genotypes of forage species, the interaction pattern may be quite different depending on how freely the different genotypes tiller. In forest trees, Goff and West (1975) have shown that the development of stand structure differs between small and large stands, whereby in small stands, interaction between canopy and understorey is important, while in a large, more uniform stand, the de Liocourt relationship is observed. Since data for six or more years may be needed for successful prediction to maturity 10 or 20 years later in forest species (Moser 1972), it is evidently important to have a proper understanding of plant competition, and this in general is not available (Hühn 1975b).

Many studies have been conducted of geographical variation in both single gene traits and morphological characters. For example, Hayashi *et al.* (1976) showed that there was significant regional variation in both isozyme patterns and certain morphological characters in Japanese populations of *Pinus thunbergii*, and that the morphological characters also varied with latitude. The problem with such studies is to know which variations are adaptive, and which are associated with timber productivity and quality.

There is some evidence that natural stands of outbreeding perennials are slightly inbred, so that initial gain on crossing selected naturally occurring parent plants, i.e. elimination of neighbourhood inbreeding (Burdon 1982a,b), can be considerable. Inbreeding depression is as important in artificial selection of perennials as annuals. Wilcox (1983), for example, reported inbreeding depression of between 10 and 20 per cent for several economic traits in selfed *Pinus radiata*.

15.2.2 Experimental design

In choosing from natural stands of trees to propagate or utilize in breeding programmes, experimenters have often used either individual selection, on the tree's own phenotype, or a comparison tree method where a candidate tree is scored on important traits with respect to the trees around it. Ledig (1974) has pointed out that this second selection method, which was introduced to overcome problems of common environment, is in fact a form of within family selection if neighbours are related. The relative efficiency of the two methods is then the ratio of the responses to selection.

$$\frac{R_{\mathrm{ct}}}{R_{\mathrm{i}}} = (1-r)\frac{(n-1)^{\frac{1}{2}}}{n(1-t)}$$

where n is the number of trees used for comparison, r the degree of relationship between the candidate tree and those around it, and t the intra-class correlation coefficient measuring the similarity within groups. Then the comparison tree method is more efficient than individual selection if r is very small and t is large, which will occur with either a large environmental variance or a large heritability. Selection of stands will probably increase efficiency, since the common environmental variance may thereby be increased. However, if the trees are closely related, individual selection will be more effective. Combined selection would also be possible; it should often be preferable. Owino *et al.* (1977) found a very marked upward bias (60 per cent or more over-estimation) in predicted response in loblolly pine based on $R = h^2S$, when additive by environmental variance was included in the numerator of h^2, as would tend to occur when estimation was conducted at only one location. Since the bias was least for simple truncation selection and increased as family information was included, in such cases data from more than one site are vitally necessary if combined selection is to be used. Burdon (e.g. 1979, 1982a) has argued very strongly for the use of combined family and individual selection because family information is usually available and because breeding population size may be a limiting factor.

In breeding *Pinus radiata*, Shelbourne and Low (1980) used an index calculated within sites and combined over sites to overcome such problems. They had up to ten traits in their index and combined ranks for up to five sites, thereby ensuring overall reliability at the expense of a small loss of site-specific adaptation. At 7 years, the expected gains were hardly less than if selection had been laboriously carried out separately in each locality.

Seven years is a short time in timber-tree breeding. However, the long replacement cycle of forest, plantation, or orchard species has the advantage that very long-running experiments can be carried out. For example, one experiment carried out on citrus under irrigation over the years 1941 to 1976 allowed the use of six factors, five with four levels and one with two levels applied to 128 main plots, together with four combinations of two rootstocks and two scions within the main plots (cf. Wood 1977). This experimental design was varied after 14 years, and then again after another 16 years, and valuable results about fertilizer response and orchard management were obtained (see e.g. Cary 1972). However, no trees died over the course of the experiment, and one statistician was associated with the experiment almost throughout. These conditions may not be widely met.

A great deal of effort has to be expended in breeding tree crops to develop seedling predictors of adult performance. For example, Tan (Tan *et al.* 1975; Tan and Subramaniam 1976; Tan 1977) has carried out a number of diallel crosses in *Hevea* to assess general combining ability, and has been able to show that most variation in yield and its components is additive. Furthermore, yield predicted from components of yield in seedlings is a satisfactory predictor of mature yield, and a seedling's general combining ability is a good predictor of clonal performance. Similarly, very high genetical correlations, likely to be useful in early selection, have been reported in the timber tree *Pinus patula* (Barnes and Schweppenhauser 1978).

15.3 Application of standard methods

Perennial species present particular opportunities for different methods of improvement, such as clonal propagation of novel or successful mature plants, responsible for most long-established varieties of fruit tree and vine, as well as for the recent great expansion of palm-oil production (James 1984). However, they may also be improved by the methods developed mainly for annual plants, though there will be differences related to the generation time. For example, selection will often be carried out before reproduction in perennials but after reproduction in annuals. For selection after reproduction, response will differ according to the breeding system, being that obtaining before reproduction multiplied by a factor $(1 + f)/2$, where f is the degree of selfing (Wright and Cockerham 1985).

Shelbourne (1969) reviewed the application of the standard methods for annual plants, showing that most of them had their place in tree improvement. Table 15.1 briefly summarizes the use of these methods in forest tree breeding practice. The seed orchards mentioned in the table may be of two kinds: clonal seed orchards, established with cuttings or grafts of selected trees, and seedling seed orchards, composed of seedlings from selected trees, which may have been self-pollinated, open-pollinated or control-pollinated, i.e. crossed with another selected tree. In the early stages of a tree improvement programme, this last category will rarely be available. Despite the fact that seedling orchards provide greater diversity, the constraint of time usually means that seed orchards are clonal in origin (Burdon 1982b). In fact, seedling orchards are very frequently a by-product of progeny trials.

The widespread availability of clonal orchards should allow the testing of ideas such as those of Donald (1968), mentioned in Section 2.4.2. This is, is a plant which performs well in mixed stands likely to perform poorly in a uniform stand of its own genotype, and vice versa? Again, if it is desirable to mix genotypes for disease control (Section 13.3), the availability of clones permits precise investigation of such mixtures.

As an example of the methods in Table 15.1, consider the response to a combination of mass selection and progeny testing. Here, open-pollinated seed from selected trees is planted as sets of half-sib progenies. These trees are then thinned to a stand of the best trees from the best families. The response after one cycle of selection is given by

$$R = i_1 \tfrac{1}{2} V_A/\sqrt{V_P} + i_2 \tfrac{1}{4} V_A'/\sqrt{V_P'} + i_3 \tfrac{3}{4} V_A/\sqrt{V_P''}$$

Table 15.1

Application of methods of selection for quantitative traits in forest tree breeding (modified from Shelbourne 1969)

Method		Application
1 Basis of selection	1.1 Individual	Universal
	1.2 Within-family	Widespread
	1.3 Index of individual values	Large industrial programmes
	1.4 Index of family values	Limited, e.g. *Pinus radiata* in New Zealand
2 Breeding system	2.1 Mass selection (collection of open-pollinated seed)	Universal
	2.2 Mass selection plus progeny testing (seedling tree seed orchard)	Widespread
	2.3 Simple recurrent selection (clonal seed orchard)	Widespread
	2.4 Recurrent selection for gca (clonal seed orchard plus progeny testing; crossed tree seed orchard)	Widespread
	2.5 Reciprocal recurrent selection (2.4 plus testing of crosses)	Limited, e.g. *P. pinaster* in France
	2.6 Heterosis breeding (crossing inbred lines for F_1 production; interspecific crosses)	Interspecific hybrids widespread
	2.7 Vegetative propagation (propagation of phenotypically superior trees, or of selected tested trees)	Almost universal

where i_1 is the intensity of selection of individuals from the whole population, i_2 is that among families grown and i_3 that within families retained. V_P includes V_G and variance due to small plot effects, family × replication interaction, and individual genotype–environment interaction. V_P' is the variance of half-sib family means. V_P'' is the variance within half-sib families. V_A' is less than V_A by an amount dependent on i_1.

Shelbourne considered the application of this method to *Pinus radiata* in New Zealand, where V_A, V_P, etc., had been estimated precisely. He suggested that with values of $i_1 = 0.001$, $i_2 = 0.25$ and $i_3 = 0.2$, genetical gains of about 50 and 13 per cent might be obtained for straightness and diameter, respectively, in 15 years. This could be contrasted with improvements after 1 year of about 20 per cent and about 5 per cent for simple collection of open-pollinated seed from the best trees. The number of years represents the time before planting improved trees; as the harvest is many years later, the extreme time disparity is not always as disadvantageous as it might appear. However, clonal orchards will in most cases allow greater progress because better genotypes can be chosen and propagated without loss through segregation and recombination of advantageous dominant and epistatic phenotypes.

15.4 Genotype–environment interaction

Existing experimental designs, then, are adequate for many purposes in investigating interaction between genotype and environment in perennial crop species, but the kinds of interaction which actually occur are remarkably diverse. For example, Summerfield *et al.* (1977) have considered the cropping of the 'serendipity berry' *Dioscoreophyllum cumminsii* as a producer of a sweetening agent. It is a tropical herbaceous dioecious perennial which grows in the heavily shaded understorey of broken tracts of forest, and therefore has special problems. First of all, since it normally grows in heavy shade, it will need to be cropped in heavy shade, secondly, spacing (assessed by the designs mentioned in Chapter 3) must be an unknown quantity under artificial cultivation, and thirdly, female plants grown in the glasshouse give only male flowers. Thus, there are many curious problems to be overcome before any yield is obtained at all, problems which would be much simpler if the crop were an annual. As a second example of quite different genotype–environment interactions, consider the effects of constant and variable cutting of Italian ryegrass (*Lolium multiflorum*) and perennial ryegrass (*Lolium perenne*) (Ollerenshaw and Hodgson 1977). Italian ryegrass dies on too frequent, too close cutting and cannot in general be cut as frequently as perennial rye. On the other hand, perennial rye responds better to variable cutting than to constant height cutting. Breeding therefore must be aimed at the particular agronomic practices likely to be in use, or alternatively, these may need to be changed to suit new cultivars (see also Lazenby and Rogers 1965).

15.4.1 Grafting

Grafting is a very old practice in orchard crops. It introduces problems of genotype–environment interaction at many different levels. Lefort and Legisle (1977) have suggested analysing scion by rootstock grapevine interactions with a diallel, using the model:

$y_{ij} = \mu + a_i + \beta_j + \gamma_{ij}$
α_i = effect of genotype i as rootstock
β_i = effect of genotype j as scion
γ_{ij} = interaction between ith rootstock and jth scion

The general associating ability of the *i*th genotype

$$= \frac{\alpha_i + \beta_i}{2} = gaa_i$$

specific associating ability of the *i*th and *j*th genotype

$$= \frac{\gamma_{ij} - \gamma_{ji}}{2} = saa_{ij}$$

general reciprocal effect

$$= \frac{\alpha_i - \beta_i}{2} = gre_i$$

specific reciprocal effect

$$= \frac{\gamma_{ij} + \gamma_{ij}}{2} = sre_{ij}$$

Using this model, and applying the methods described in Section 4.4.2, Lefort and Legisle were able to show that grafting as such has a strong initial effect on plant growth, but that this effect disappears in time. The rootstock by scion effect is similar, while the scion becomes more important with time.

An important aspect of rootstock–scion interaction is the uptake and use of nutrients and water from the soil. While the diallel analysis sketched above is useful detecting general effects of grafting, all reciprocal crosses where customary rootstocks (*Vitis riparia*, *V. rupestris*, etc.) are used as scions are irrelevant to normal vineyard practice. [Grape quality is so important in wine grape breeding that very few new scions, i.e. varieties, have been introduced in the past century, though clonal selection has allowed great increases in yield; and of course rootstocks do not produce satisfactory grapes. See Wagner (1975) for a general discussion of wine grape breeding.]

For many purposes, the critical variables will be those related to the uptake and content of nutrients and other elements from the soil. Sarič *et al.* (1977) showed that the scion influenced the uptake and content of nitrogen, phosphorus, and magnesium more than the root-stock, while the reverse was true for potassium and calcium. Downton (1977) showed that in general vines on rootstocks took up less chlorine, less sodium, and more potassium than vines on their own roots, but that the effect depended very much on scion variety; for example, Cabernet Sauvignon did not show the same sodium effect as other varieties. It is of particular interest that these gross genotypic effects are also to some extent related to nematode resistance, since rootstocks selected for nematode resistance may also be chloride excluders.

Related effects exist in other species. In citrus, Embleton *et al.* (1973)

found that rootstocks were most important for variation in sodium and chloride response, using existing commercial cultivars as scions. Smith (1975) showed in mandarin-type citrus that scion and rootstock had similar large effects, and that the scion most affected sodium, aluminium, copper, and magnesium uptake and content, and least affected phosphorus uptake, while the rootstock affected boron, potassium, copper, phosphorus, and sodium, with no effect on nitrogen. In mandarins, rootstocks can also affect juice volume, ascorbic acid and carotenoid content, and pH (El-Zeftawi and Thornton 1978).

These extremely complex results do not exhaust the peculiarities of grafting. For some centuries, interstocks have been used to a limited extent in apple orchards. These are dwarf cultivars which are grafted in between rootstock and scion, to limit vigour and induce early fruiting. Parry and Rogers (1968, 1972) discuss their use. In general, it seems that for apples their use offers no advantage over straightforward scion-rootstock trees, but where good dwarfing rootstocks are not available, as with cherries, dwarf interstocks may be helpful. The interactions between interstocks and plant growth are such that the greater the length of an interstock, regardless of its genotype, the more effect it has in retarding growth of the tree.

15.5 Economic factors

Given the normal production life cycle of a tree crop, replacement with improved varieties will generally only be of economic significance towards the end of the existing plant's reproductive span. Although there is some planting every year, so that continuous improvement is possible, this must be carried out without interfering with agronomic practices or changing product quality. For example, in Burgundy, given the 30-year lifetime of a vine, a grower will hesitate before changing from his current Pinot Noir clone to another of the 300 distinct clones of that variety (Hanson 1982). Thus, introduction of improved varieties or hybrids will be a slow process. Mukherjee (1976) has considered the optimal replacement strategy for coconut trees, which can be replaced by hybrid trees yielding, at maturity, up to twice as much as those trees replaced. Since the old trees have an economic life of 70 years, and new trees take several years to bear, it would appear to be critical to optimize the time of replanting. However, an analysis using ordinary discounted cash flow techniques showed the optimum economic life to be insensitive to fluctuations in product price of $12\frac{1}{2}$ per cent up or down, changes in interest rates in the range 5–10 per cent and shortfalls in yield of 0–20 per cent. This is an area where much work needs to be done.

There is as yet little theoretical integration of economics into tree-breeding strategy, though practical economic decisions are taken every

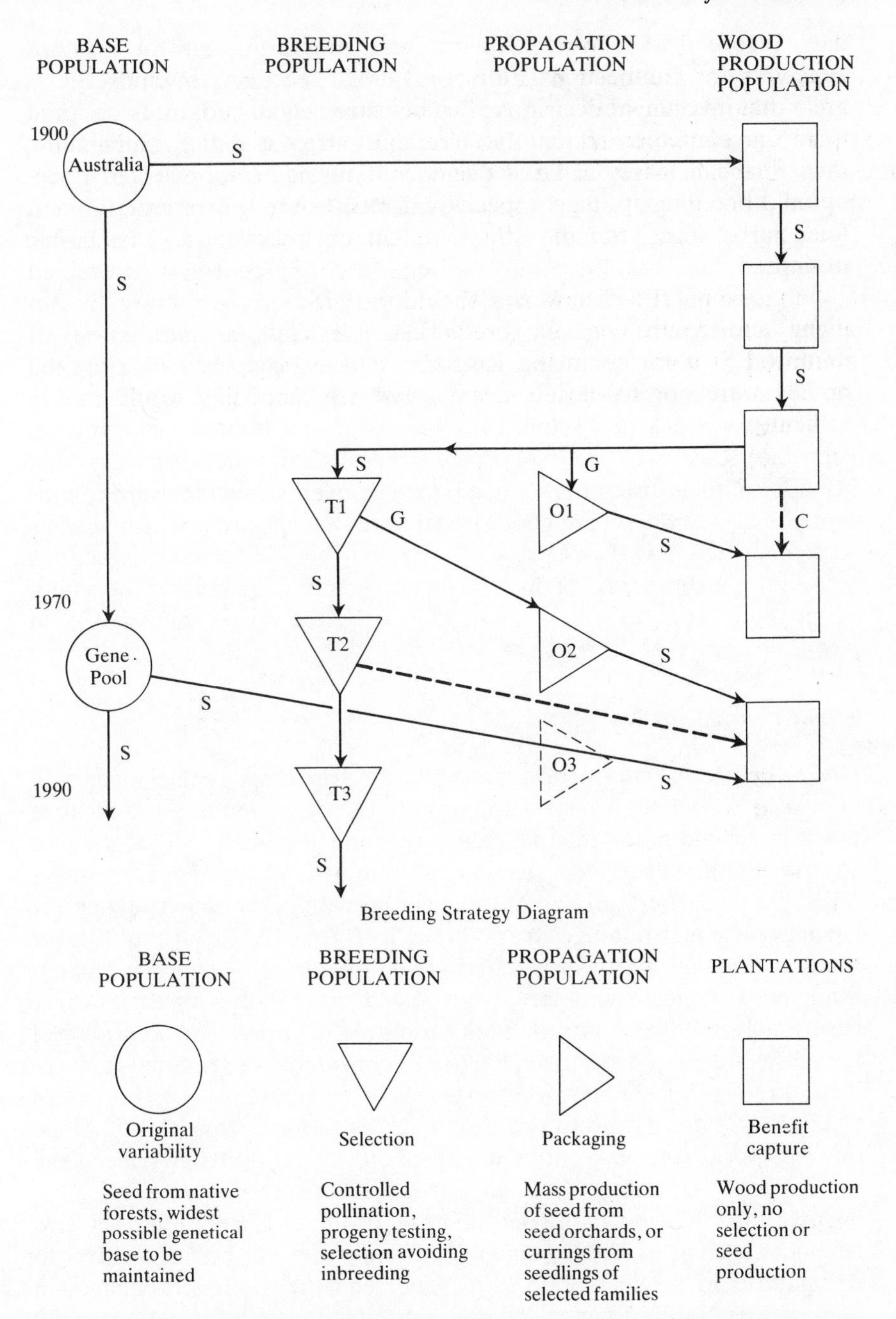

Fig. 15.1. Flow chart of domestication and breeding strategy for eucalypts, (Eldridge 1983; based on Libby 1973). S, seed; C, cuttings; G, grafts; T, progeny test; O, seed orchard.

day. Figure 15.1 shows a logical and practical plan for breeding eucalypts for combustion (Eldridge 1984). The times involved are so great that investment decisions must be extremely uncertain. If the future is an era of timber scarcity, as frequently suggested (e.g. Hora 1981), then financial losses will not occur as a result of adoption of a sub-optimal breeding strategy, especially if returns are indeed insensitive to financial changes, but the effects of current mistakes will be greatly magnified.

'What wonderful discoveries should we make in astronomy, by out-living and confirming our predictions,' as Gulliver thought of the immortal Struldbrugs in the Kingdom of Luggnagg. Even if time and money were not so closely coextensive, tree breeding would still be difficult.

16
Conservation of germplasm

16.1 Introduction

It is only relatively recently that the importance of the conservation of variability has been widely realized (Simmonds 1962). The long-term effect of modern plant breeding methods has been to enlarge the acreages under a very few crop species, and to make uniform the plants within those acreages, as landraces were replaced by pure lines, open-pollinated varieties of outbreeders replaced by hybrid varieties, and seedlings replaced by clones. Although it had been frequently noted that particular varieties were outstandingly successful over long periods of time, e.g. the Majestic potato which, bred in 1911, still represented 50 per cent of the British crop acreage 50 years later, this was generally regarded as being related to such varieties' reliability. As Simmonds pointed out, a more likely reason was the very small gene pool of the cultivated potato. Thus, the view was very widespread that variability sufficient for any need existed, and it was only a matter of uncovering it. The words of MacKenzie (1832, cited by Hunter and Leake 1933) would have been agreed with by many during that century: 'If a farmer cannot find a potato adapted for the soil of his farm, he has nothing to do but to raise new varieties from seed, and to go on until he obtains a variety possessing the desired qualities'. As a clonally propagated outbreeding tetraploid, the cultivated potato would be expected to be highly heterozygous, so that seedling potatoes might be expected to be very variable, despite the narrow genetical base of the crop. The utility of the variability, however, might be more apparent than real.

In the 1920s and 30s, Vavilov was active in investigating the origin and diversity of major crop plants (see Vavilov 1951). As the first major explorer, collector, and evaluator of world crop genetical resources, Vavilov established one of the finest germplasm banks in the world, and set the pattern for the systematic introduction of new genetical material which has become more widespread in recent years. (Plant introduction as such has of course been widespread throughout history, and crop plants are usually most successful outside their centres of origin,

Jennings and Cock 1977.) Vavilov regarded it as absolutely vital to explore and classify world plant genetical resources, and while conservation was implicit in what he did and wrote, especially in his advocacy of the maximum use of local material in breeding programmes, the loss of locally adapted variable material was not then as catastrophically rapid as it has been in recent years. For example, Ochoa (1975) has shown how, in potato-collecting expeditions in centres of diversity of the potato in Latin America, over a period of 40 years more than half of the important native varieties have vanished. Certainly, centres of variability are not a permanent phenomenon (Kupzow 1976), but the rapidity of change is frightening.

This loss of variability is almost universal (see Frankel and Hawkes 1975), and today there is widespread recognition of the need for action, which to some extent has begun. The problem is that what is happening is not just the loss of local varieties, together with the opportunity for valuable hybridization (see e.g. Wilkes 1977, for the relationship of the natural hybridization of maize and teosinte to hybrid vigour in maize), but the introduction of very uniform germplasm which is, as noted in Chapter 12, highly susceptible to mass epiphytotics, and which also, while agronomically superior in some ways, e.g. high yield under favourable conditions, does not incorporate all desirable attributes of adaptation.

Quite apart from the need to maintain variability at as high a level as possible, there is the fact that use of wild germplasm may provide significant short-term economic benefits. For example, incorporation of genes from wild populations has produced yield increases over the cultivated parent of well over 10 per cent in crops as diverse as oats (Frey 1976) and grain sorghum (*Sorghum bicolor* (L.) Moench) (Cox *et al.* 1984). In the latter case, improved cultivars were chosen from the U.S.A. and India, wild races from Egypt, the Ivory Coast and South Africa, so that great geographical diversity was achieved. After crossing, back-crosses were made for four generations, after which four generations of selfing were carried out. Between 2 and 43 per cent of the BC_4-derived lines exceeded the grain yield of the recurrent parent, i.e. the improved cultivar, and mean yield increases of the order of 15 per cent could confidently be expected from such a program. Back-cross reduction of the wild genome's contribution to no more than one eighth was recommended because of the agronomic defects of the wild genotypes, but this represents no real problem to the breeder, given the potential gains.

Genetical conservation on a large scale is more a matter for the agreement of governments and international agencies than for the individual plant geneticist, so that the rest of this chapter deals with the theoretical underpinnings of rational plant conservation.

16.2 Population sampling

Given that a decision has been made to sample primitive landraces, weedy relatives, or wild ancestors of some particular crop plant, a general strategy is needed for such surveys which would ensure that an adequate sample of variation is obtained as efficiently as possible. What form the sampling survey takes depends very much on the population structure of the species of interest (Jain 1975a). Population structure is determined by the plant's reproductive system, its distribution and abundance, hybridization between and within species, karyotypic variation, recombination system, demography, and patterns of dispersion and gene flow. We have seen in Chapters 7 to 10 how these factors influenced observed variability and the strategies to be taken for plant breeding. Under domestication, population structure changes. Initially, there will be many small crop populations together with wild and weedy relatives, while later much larger crop populations together with the wild and weedy relatives will still coexist. Under these conditions, many favourable chance hybridizations will occur, and the alert plant breeders and domesticators of the past have taken great advantage of these. Even in relatively recent times, there have been many very large populations of crop plants with fringe populations of wild and weedy relatives. Today, this is changing, with the elimination of the fringe populations, and many problems remain with the definition of ancestral populations and their relationships (Pickersgill 1971). The sampling strategy will be very different for these different stages of development.

Marshall and Brown (1975) have suggested that exploration for genetical conservation consists of two stages, the location of gene pools and assessment of threats to their continued existence, and the sampling from such gene pools once they have been assessed in this manner. Important gene pools may exist even where particular crops have never been important, e.g. oats in Ethiopia (Ladizinsky 1975), so that surveys need to be detailed as well as extensive. Even if the population structure of a species in the locality is known, it may be necessary to assess its diversity.

16.2.1 Surveys of variability

Marshall and Brown (1975) have suggested the use of heterozygosity

$$H_e = 1 - \sum p_i^2 = \sum_{i \neq j} p_i p_j$$

to measure genetical diversity or variability in a population, as was done in Chapter 5. Useful variability is much harder to define. As the work of Slatkin (1978) shows, spatial variation in selective forces is potentially the strongest factor in increasing heterozygosity, but such increases will not normally be a result of gene flow between differentiated populations. This would happen only when the populations with different mean values for a trait were mixed substantially every generation or every few generations. Heterozygosity at individual loci will therefore not necessarily be informative with respect to additive variance for a given trait of interest. Genetical variability *per se*, as has been noted, is not the same as additive variance. Furthermore, phenotypic variance and heterozygosity may not change with change in the mean of the trait. Nonetheless, heterozygosity is certainly associated with variability, though not necessarily in the desired fashion as noted in Section 9.2.1.

Explanation of the extraordinary level of polymorphism in most populations of outbred organisms is one of the greatest current problems in theoretical population genetics. Theoretical allele distributions for neutral or selective models are very hard to distinguish on the basis of observational evidence, but some of the inferences from neutral theory may be used in designing sample surveys.

As pointed out by Marshall and Brown, the design of such a survey consists of setting aims for three elements: the number of plants or seeds to be sampled per site, the number of sites to be sampled per region, and the distribution of sites within the region. For the case of an annual crop, the choice of site is relatively simple, being the individual field or farm. For wild or weedy species this is not so simple. However, given the target population, the aim is to select a certain number of plants, n say, such that the probability of the presence of all alleles with a frequency greater than some minimum value is greater than a particular chosen probability. If the minimum frequency for an allele of interest is taken as 1 per cent, then in most cases random samples of about 100 plants will generally allow the collector to be about 90 per cent confident that he has all alleles of interest. Ewens (1972) has shown that in a sample of n gametes, the expected value for the number K of neutral alleles is approximately

$$E[K] = 1 + \frac{\psi}{1+\psi} + \ldots + \frac{\psi}{n-1+\psi}$$

with variance

$$\mathrm{Var}[K] = E[K] - 1 + \frac{\psi^2}{(1+\psi)^2} + \ldots + \frac{\psi^2}{(n-1+\psi)^2}$$

where

$$\psi = 4N_e u$$

(As before, N_e is effective population number and u mutation rate.) Then, if three common alleles is the general aim for any particular locus in the given sample, if $\psi > 1$ (i.e. N_e very large for $u \leqq 10^{-5}$), about 100 gametes will yield this target in most cases. Heterozygous advantage will usually increase the number of alleles obtained in a sample.

The optimum number of sites in a locality will be the maximum possible, while the distribution of sites will depend on distribution and gene flow of the local population. Oka (1969) has suggested a way of estimating what proportion of variation has been obtained as $1 - \{(1 - P) + P(1 - p)^n\}^N$ where P equals the proportion of available variability found per site, p the proportion per plant or minimal sample unit per site, N the number of sites sampled, and n the number of minimal sampling units sampled per site. However, in many cases P and p will be unknown. Where they are known, effort in sampling may be taken as being proportional to $N(a + bn)$ where a represents effort expended per site and b that expended per plant. On the additional assumptions that specific variability is uniformly distributed over sites, that equal numbers of plants should be taken per site, and that distribution of samples is not a problem (i.e. small samples are satisfactory), then a maximum number of sites with a minimum number of plants per site will be best. This is not appropriate for perennial crops, or vegetatively propagated crops. However, it is of interest that optimal sample sizes need, on these assumptions, simply knowledge of the number of alleles per locus for a number of populations, as against the optimal distribution of sites which requires detailed ecological and demographic data.

Brown (1978b) has extended many of these ideas, and in particular has emphasized the importance of locally very frequent alleles, which may be 20–30 per cent of variants for a range of species. The question is whether these alleles are important in adaptation, and much research is needed here. Brown contends that isozymes are the best method of assessing population structure, because different variants are usually unambiguously distinguishable, loci are generally distinguishable, a high proportion of all variants are detectable, and the loci assayed may be a random sample of all loci. This last assumption is probably not correct (Johnson 1977; Mayo and Brock 1978). Nonetheless, the limitations are at least readily apparent, and no better method exists as yet.

16.2.2 Collection of variants

As summarized by Bradshaw (1975), very rapid genetical change occurs in samples grown at a particular site for conservation purposes. This means that conservation is not simple; it is not sufficient to obtain seeds or plants from a wide variety of localities and grow them in a number of large collections at specific sites. Local adaptation will occur, and much valuable variation will be lost. Brown (1978a,b) has emphasized that what the conservationist seeks to preserve are co-adapted gene complexes, and it is self-evident that removal of adapted varieties from their region of local adaptation will tend rapidly to the breakdown of such co-adapted gene complexes. Brown *et al.* (1977) have provided evidence at the level of isozyme polymorphism in *Hordeum spontaneum* that such complexes are widespread in natural populations. Certainly, they are not to be expected in even the widest composite crosses (Suneson 1956). Bradshaw has therefore recommended that collection of plants should not be by simple stratified random sampling as recommended by Marshall and Brown, but rather by very wide stratified systematic sampling, followed by further systematic collection in areas of particular potential interest.

16.3 Maintenance of variability

The key need is two-fold: to maximize variability maintained for the future, and to make the variability available for current use. There are two possible approaches to these aims.

16.3.1 Gene pools

Simmonds (1962) and others have urged the maintenance of mass collections, using sexual reproduction. However, as already noted, such mass reservoirs lose variability very rapidly, so that bulking different varieties and accessions together will accelerate the process of loss of variability. A second reason for not using mass reservoirs is that the definition of what is available within a gene pool becomes much harder. Even if plants are maintained by regular growing in appropriate nurseries, it may well be better for small plots of well-defined types to be grown, rather than massive populations. Smith (1984) has emphasized the importance in conservation of maintaining diverse stocks in each storage centre and of identifying stocks with special traits, such as dwarfing, restricted branching, storage protein variants, or disease resistance, so that these can be drawn on as necessary. It will, however, be rare for this knowledge to allow one to quantify the expected annual benefit of conservation, given by Smith as $(B - B_o)Pr - nC$, where Pr is the probability that one of the n stocks conserved at a cost C has an economic return B greater than that of the original stocks, B_o. In the case of tree-breeding, where the long generation interval makes a change of objective both costly and slow, Burdon and Namkoong (1983) have therefore recommended subdivision of breeding populations to provide subpopulations which will certainly differ to some extent and so should respond differently to altered selection.

Jain (1975b) and others have for the reasons in the above argument recommended conservation *in situ* for locally adapted populations, whether of wild or weedy relatives of crop plants, or of primitive races.

While recognizing that this is very difficult, Jain has emphasized that local populations have both local adaptation and a population structure which may well maintain a slow rate of genetical change and a satisfactory level of variability without outside intervention. This is provided that such local populations are allowed to persist in relative isolation, which is unfortunately unlikely. A knowledge of plant demography will be necessary in most cases to assure the success of such a 'genetical reserve', even assuming that agreement to establish it is possible. (The importance of the potential genetical contribution of wild and weedy relatives of crop plants is not yet widely agreed; Hawkes 1977.) There is also likely to be merit in maintaining stocks of imported species as well, so that the genetical base of such plants is not dangerously narrow (Eldridge 1978b).

16.3.2 Seed and pollen storage

It is now possible to store seeds for a very long period by lowering temperature and moisture content, without significantly affecting viability (Roberts 1975; Roberts and Ellis 1977). Alternatively, they may be stored by holding them fully imbibed at low temperatures. There are many recalcitrant species, from important tropical tree crops such as cocoa (*Theobroma cacao* L.) and rubber (*Hevea brasiliensis*) to relatively minor ones like durian (*Durio zibenthimus* Murr.) and chestnut (*Castanea* spp.) (Roberts and Ellis 1982). However, since most major cereals and many other crop species may be held in this way, it is the most promising method for long-term storage. It is undemanding in terms of both space and money, for while the numbers of seeds to be stored are prodigious, it appears likely that at −20°C and 5 per cent moisture, seeds of common wheat, *Triticum aestivum*, may have a regeneration period, i.e. a time after which they need to be regrown to maintain viability, of as much as 400 years. Viability has been determined to be a function of time of storage, t, storage temperature, c, and moisture content, m, of the form:

$$\text{viability} = k_1 - k_2 mte^{(-k_3 + k_4 c + k_5 c^2)}$$

(Ellis and Roberts 1981). Regrowing is regarded as necessary after viability has fallen by 5 per cent. The work of estimating the parameters of the equation would be worthwhile only for a large-scale storage program. Thus, once the storage facilities are set up, the more demanding aspects of conservation, viz. the regrowing at intervals, are minimized. Even at 0°C, and a higher moisture content, several decades at least would pass before regrowth became necessary. The genetical stability of stored seeds is unfortunately poor, so that shorter times may be necessary to avoid an excessive build-up of mutants. Pollen storage at the moment is less successful, but when it becomes a practical proposition, space requirements will become even more modest.

Techniques of storage of plant parts for vegetatively propagated species are as yet not so well developed, so that in the case, for example, of the potato, enormous fields are necessary for the annual regrowth of potatoes at the International Potato Centre (CIP). However, for a clonally propagated species, the loss of variability due to natural selection is quite slow. The problem for CIMMYT, say, with over 11 000 entries for maize and related species (Hanson 1977), is actually more difficult.

16.4 Large-scale co-operation

It will be very evident from the foregoing discussion that conservation of variability is a large-scale activity. As emphasized by Simmonds (1962) and others, museum or type collections are generally assets of diminishing value, on account of the loss of genetical variability consequent on local adaptation. Similarly, mass reservoirs, while they are important for making adult plants available in the case of perennial crops, and other related purposes, also have the disadvantage of loss of genetical variability. On the other hand, mass storage at low temperatures of enormous numbers of seeds requires computer technology for cataloguing and making available seed to interested investigators, as well as for the maintenance of the germplasm banks themselves. Thus, while important national institutes such as the N. I. Vavilov Institute in Leningrad exist, and serve a very important function, a real need for international co-operation remains. Frankel (1975) has summarized the needs for worldwide genetical resource conservation. Of primary importance are the linkage of existing institutions and the designation of certain places as base collections. (In some cases these exist *de facto*; IRRI, for example has over 40 000 different accessions in its collection; Khush and Coffman (1977).) This would allow the development of a co-operative network of resource banks, which in turn would allow the making of emergency collections where these were seen to be necessary. An extensive training programme is required, both for people to work in such centres, and also to make plant breeders and other agricultural workers aware of the problems and needs in genetical conservation. Overall, such a worldwide network must be co-ordinated by an acceptable neutral organization. FAO has been advanced as such an organization, but 'few *Politicians* with all their Schemes are half so useful Members of a Commonwealth, as an *honest Farmer*', as Swift put it in *The Drapier's Letters*.

Much work needs to be done on assessing the costs and benefits of large-scale co-operative genetical conservation. First, it should be possible to demonstrate the benefit of past conservation measures. Secondly, the merits of existing and planned conservation schemes should be assessed. Thirdly, the possibility of new kinds of problems arising should be examined. For example, losses of germplasm during disasters, natural or man-made, are perhaps more likely with concentrated mass storage.

17
Plant breeding strategy

17.1 Introduction

Having now considered most of the components of theoretical plant breeding, we can begin to formulate a strategy for large-scale breeding programmes. Evidently, two separate strategies will be needed, one where a decision is to be made about the crop to be grown, the other how most effectively to breed the chosen crop. In the former case, a range of different crops will need to be evaluated, and much of this can be done theoretically, from the known properties of potential candidates. This is not always the case, however. For example, for many years there was only one leguminous seed crop grown on a large scale in the United Kingdom, the broad bean *Vicia faba*, with peas grown for seed on a smaller scale, so that it has frequently been suggested that others should be grown. Froussios (1970) considered the potential of Navy beans, *Phaseolus vulgaris* L., as a seed crop (rather than as a green vegetable). A substantial collection of at least 1200 types had been developed at Cambridge, representing the five or more major races (Rathjen 1965), but while the properties of many of these were well-established in the environments from which they were taken, it was not clear which if any would be suitable for growing in the United Kingdom. Accordingly, about 100 were grown under a number of different spacing regimes, so that a choice could be made of likely breeding candidates. Later, other promising varieties were imported, and breeding programmes are now well established. However, despite the crop's importance in other countries and the consequent existence of much improved germplasm, it is not yet a major crop in Britain, requiring much more breeding work to develop early maturity and cold spring tolerance (Innes and Hardwick 1975; Scarisbrick *et al.* 1977; Davis and Evans 1977).

For an existing crop to be successfully bred on a large scale, it is initially necessary to determine the reproductive system, any special agronomic requirement, what germplasm is available, and in what environments the crop is to be evaluated and grown. Generally, the reproductive system will be well understood for a major crop plant, while special agronomic or processing demands will usually have been met. However, the germplasm available will frequently be problematical in extent and nature, and the environments in which the crop is to be grown will not be well understood at all, insofar as they relate to the new crop. This will be particularly the case with introduced perennial crops. For example, when the wine grape, *Vitis vinifera*, was introduced to Australia, settlers generally took what cuttings they could obtain in South Africa on their way to Australia, and attempted to grow these wherever they settled. Thus, quite inappropriate varieties were grown in many regions, in some of which they performed unexpectedly well (Mayo 1986), and it is only today that any kind of systematic investigation of appropriate varieties for particular localities, or even the selection of the highest yielding clones of particular varieties is being carried out.

Once a breeding programme's objectives have been set, and as noted earlier this is not an easy matter, this programme will consist of three stages: evaluation of existing varieties; design and conduct of the breeding programme itself, i.e. what crosses are to be made, what cytogenetical manipulations are needed, what physiological investigations need to be made (for example, with respect to disease resistance), and so on; and evaluation of potential varieties resulting from the breeding programme.

17.2 Evaluation of existing varieties

When a very large number of varieties must be evaluated, which should be carried out over a large number of sites, the problem becomes comparison of lines across environments, for which there are essentially three methods. First, there are standard analysis of variance techniques, as described in Chapter 3. These are subject to the restrictions generally applicable to linear models, such as homogeneity of error variance and additivity of effects. Thus, the larger the trial the less useful such methods will be.

Secondly, there are regression techniques such as those developed by Yates and Cochran (1938), and Finlay and Wilkinson (1963). In these methods, as described in Chapter 6, environmental (site) means are used as reference points, and the regressions of the varieties on the site means are used to indicate differential response to environment, and hence to evaluate important aspects of response. In such analyses, values obtained at particular sites in different years may be used as different environments, and in this way more than one aspect of environment may be assessed, without attempting to specify the important aspects of environment which contribute to the observed patterns of response. These models, as noted in Chapter 6, and also emphasized by Mungomery *et al.* (1974), have the disadvantage that all effects are expected to be linear, or at least linear after transformation (i.e. of other known form such as quadratic), so that if several different response patterns are in fact being assessed, important interactions will be neglected or misinterpreted.

For these reasons, many workers have turned to multivariate methods, so that the number of comparisons which must be made is reduced by grouping lines or varieties which respond similarly, and then comparing such groups. These techniques may also be applied to environments rather than genotypes, so that suitable environments for testing may be delineated. Thus, Abou-El-Fittouh *et al.* (1969) used cluster analysis to classify locations to minimize the within-cluster genotype by environment interaction, and thereby maximize the difference between clusters, so as to obtain a number of different regions which could be evaluated separately in breeding upland cotton. Goodchild and Boyd (1975) used principal component analysis to assess regional and temporal variation in wheat yield in Western Australia, using data for more than 40 years' production, and were able to show that in Western Australia most evaluation of genotypes was conducted in highly variable localities, which might be expected to maximize the genotype–environment component in yield, whereas concentration on breeding for less variable localities where the bulk of the crop was grown might have been far more profitable. Fox and Rosielle (1982), who assessed 99 wheat lines in most of the Western Australian trial environments over two years, were unable to derive useful subsets of lines for prediction or even description of genotype–environment interactions (see also Ramey and Rosielle 1983). In this context, it is of great interest that it is rare in Australia in large-scale trials for the best genotypes in any particular year at any particular site to have been bred near or for that site. In contrast, Antonovics and Wu (1978) noted that Japonica rice varieties grown in the localities of Taiwan in which they were bred consistently yielded 4–5 per cent more than varieties bred elsewhere. Gains from breeding for specific sites were therefore a practicable objective.

The relationship between experimental yields and farm yields, whereby in poor years there is not much difference but in good years experimental yield are far greater than farm yields (Davidson and Martin 1968), further reinforces the need to achieve, perhaps breed for, stability of yield.

Although the international nurseries conducted by the International Centre for Maize and Wheat Improvement in Mexico (CIMMYT) and other such organizations are essentially designed for the evaluation of new varieties, they constitute one particular model for large-scale evaluation of existing genotypes, since genotypes are compared without regard to the breeding programmes from which they have arisen. The Australian Interstate Wheat Variety Trials constitute another, smaller, model. We shall examine these in slightly more detail.

The interstate trials were set up in 1968 to assess, over the long term, the agronomic and yield characteristics of advanced hybrid material from breeders throughout Australia. This would allow testing for yield in all states before possible introduction of new varieties, give opportunity for breeders to study and obtain potentially useful breeding lines, allow a wide range of environments to be used for particular varieties, permit the assessment of adaptability, and possibly accelerate breeding programmes. The pattern of the trial has been constant and is conducted in two stages. First is the series A first year trial (A^1), where a single trial is conducted in each state or region, with a maximum of 30 entries contributed by breeders throughout Australia by agreement. In the series B or second year trial (B^1), the same trial and entries from series A are grown on as many as three sites in each region. In this second year another series A trial with new entries (A^2) is initiated in each state or region; this is essential as breeding programmes constantly generate new material of high promise. Thus, over 1982–1984, the trial was as follows:

1982	1983	1984
Series 13 Year 1 24 genotypes at 6 sites	Series 14 Year 1 22 genotypes at 6 sites	Series 15 Year 1 22 genotypes at 6 sites
Series 12 Year 2 28 genotypes at 17 sites	↘ Series 13 Year 2 24 genotypes at 19 sites	↘ Series 14 Year 2 22 genotypes at 22 sites

When these trials were started, five check varieties, Eagle, Heron, Gamenya, Halberd, and Pinnacle were included. Since then, some of these varieties have been replaced and others have been included as check varieties, but the procedure continues as outlined.

Fox (1977) analysed the results of the 1975 trial by several different methods, and was able to show first and most remarkably that two sites, one in Western Australia and one in Queensland, more than 2000 km apart, showed extremely similar responses for varieties. Should such a relationship have persisted, this would have had both the practical implication that one or other site could be used for some varieties with prediction to the other and also the theoretical implication that, in a sense, Australia, despite the apparent disparity of its wheat growing environments, could in fact be regarded as being far more similar than might have been suspected. Unfortunately, analyses of a number of years did not confirm this result. Secondly, it was shown that principal component analysis was very effective in grouping sites. Thirdly, simple ranking of varieties at sites was a very useful technique in assessing relative merit. Fourthly, and related to the previous point, the differences in relative performance of varieties between sites were not determined by site mean yield differences. This would have the implication that the adaptation analysis of Finlay and Wilkinson should be a useful preliminary technique.

Analyses of later seasons' results, including graphical extensions to adaptation analysis (Hancock *et al.* 1983), have tended to confirm the latter two results. Principal component analysis has been found to add rather little to the Finlay–Wilkinson analysis; this will be discussed further below in a broader context. Non-linear genotype–environment interaction has been shown to be important. The analyses of 1981–1983 trials have shown that genotypes from a given breeding centre usually perform similarly over a range of environments and that their fertility response is predictable, at least to some extent, from the relative fertility of the region in which they were selected. Thus, genotypes selected under conditions of low fertility tend to yield well under such conditions but not to respond strongly to a better environment.

Turning now to a much larger scale of experiment, consider the international spring wheat nurseries conducted by CIMMYT. The results of the seventh Nursery 1970–1 (CIMMYT 1973) constitute a suitable set of data for the illustration of different methods for detecting major patterns of genotype–environment interaction. In this nursery, 50 spring wheat varieties from 18 different countries were tested in 42 countries at 67 sites. This yielded usable results for 49 varieties at 66 sites in 41 countries, representing 16 of the 18 contributing countries plus 25 others. Evidently, given this distribution of sources and test sites, no pattern of local adaptation could be present in the data.

Initially, a principal component analysis (Banfield 1978; Holland 1969; Pearce 1969) was conducted on varieties over sites as replicates, so that metric values for the same trait of different varieties constituted different variables. Analyses of both yield and plant height data showed that over 80 per cent of variance was associated with the first latent root, and the loadings of all 49 varieties in the first latent vector were of the same sign and of similar magnitude. Thus, with respect to both yield and plant height, sites were well characterized by means, the essentially one-dimensional relationship suggesting that there was limited genotype–environment interaction only. To evaluate this further, regression of varieties on site means was carried out for all 49 varieties, in the manner of Finlay and Wilkinson (1963) described in Chapter 6. In all cases, simple linear regression was highly significant, the lowest proportion of variance absorbed by the linear fit being 49 per cent. This regression analysis is summarized in Table 17.1; significant departures from collinearity allow preliminary assessment of environmental response of different varieties. Plant height showed a similar pattern to yield.

Although simple linear regression was in most cases adequate as a

Table 17.1
Regression analysis for yield and plant height, using data from Seventh International Spring Wheat Nursery (CIMMYT 1973)

	Yield				Plant height		
	Degrees of freedom	Sums of squares ($\times 10^7$)	Mean squares ($\times 10^7$)	Variance ratio	Sums of squares ($\times 10^6$)	Mean squares ($\times 10^6$)	Variance ratio
Regression on site mean, ignoring variety	1	642.2	642.2		1.578	1.578	
Deviations from collinearity	48	16.3	0.34	8.16	0.048	0.0001	9.96
Residual	3184	132.6	0.04		0.318	0.00001	
Total	3233	843.2			2.576		

preliminary description of the pattern of variation of varieties over sites, it could be argued that genotype–environment interaction would be observable in the residuals remaining after this fit. Accordingly, a further principal component analysis was conducted on these residuals. Fig. 17.1 shows the pattern of sites revealed by the first two components for the yield data. No systematic effects are obvious, in contrast to the results mentioned above for trials within Australia. For plant height, outlier sites were revealed, but no clustering or trends were apparent. Thus, for many practical purposes, especially in assessing varieties' responses to different sites in terms of high upward response to favourable conditions and low downward response to unfavourable conditions, the linear univariate methods may be expected to be helpful, given that several conditions hold. First, that test sites are representative of commercial growing conditions in their localities; this is often known not to be the case. Secondly, that the genotypes under test are in fact widely disparate in genotype. Since many, perhaps most, carried Norin 10 dwarfing genes, many others carried similar genes for *Septoria* resistance, and very many used CIMMYT material in other ways, this condition is probably not met in this case.

Byth (Mungomery *et al.* 1974; Byth *et al.* 1976; Shorter *et al.* 1977) has by contrast emphasized the limitations of simple linear methods, and has suggested that ordination followed by some form of pattern analysis will allow detection of either groups of similar genotypes or groups of similar environments; the analysis outlined above should indicate that in many cases linear, univariate analyses will allow rapid and reliable assessment of large numbers of genotypes and environments. The problem of assessing a number of genotypes over a smaller range of

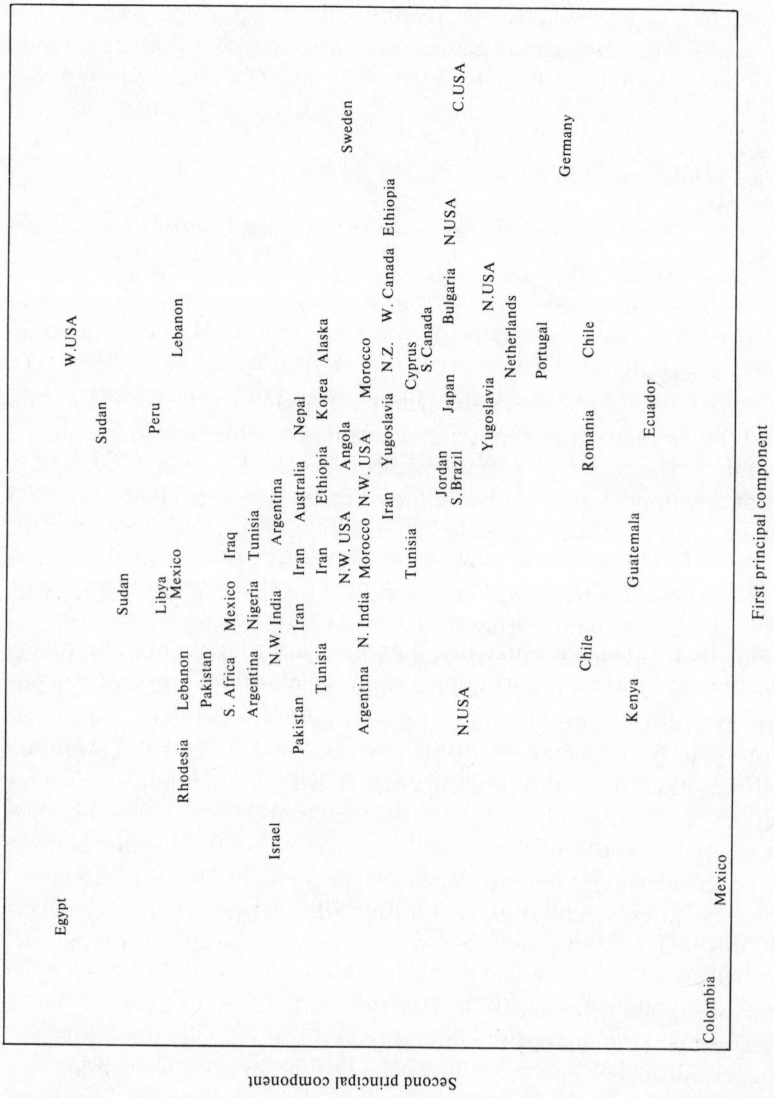

Fig. 17.1. Plot of sites against the first two principal components, the analysis having been carried out on residuals from a regression analysis. (Yield data from Seventh International Spring Wheat Nursery, CIMMYT 1973.)

environments, once major yield increases are not easily won, is a more difficult and as yet unsolved problem.

Should selection be carried out on some index of several traits (Sections 7.4.2.2–7.4.2.4), the problems of choosing sets of genotypes and environments in which genotype–environment interaction is negligible become much greater (Namkoong 1985). The linear combination of traits into an index does not guarantee that genotype–environment interaction in index levels will not be present, despite the preliminary clustering on individual traits. Similarly, if the clustering into interaction-free groups is on the basis of the index values, interactions may arise for the individual traits.

17.3 Design of a breeding programme

At present, the two most important determinants of the form of a breeding programme are the crop plant's reproductive system, and whether the crop is produced by an annual or by a perennial plant.

The influence of the reproductive system, as has already been made clear, is twofold: first, for autogamous species, the problem has frequently been reduced to that of obtaining as many homozygous descendants from crosses between good genotypes as possible, and assessing these in as many environments as possible, while for allogamous species, it is certain that at some stage hybrid breeding techniques will be necessary to utilize heterosis or recover from inbreeding depression. If a plant may be propagated asexually, then both approaches are potentially practicable, and the choice will depend on which offers more advantage. In the case, however, of perennial species, evaluation of many genotypes in many environments may be impractical, so that breeding has generally conformed far more to the pattern of a skill or craft than to that of a science, since what is needed is experience in assessing which individual plants are likely to perform well and to cross well. (With annual plants, since particular genotypes are in general not maintainable, this experience is diluted, and in addition the opportunity to demonstrate its error or to improve it is much greater.)

It is highly probable, and indeed desirable, that in the future totally different fundamental considerations will apply. For example, somatic cell genetics may allow one to design a plant from the cell upwards, so to speak, thereby bypassing generations of inbreeding to ensure genetical homogeneity, or indeed back-crossing, where, as in the case of disease resistance, it will be possible to use selection techniques on cell cultures and reconstitute plants which are resistant to particular pathogens. However, these considerations are all very general; we should consider in more detail some specific attributes of a breeding programme.

17.3.1 Autogamous crops

The main problem in breeding selfed crops is selection in the segregating generations following crossing. This is because the stage of selection among homozygotes amounts to large-scale testing of individual genotypes followed by yield trials, which is essentially the same problem as selection among existing genotypes.

For many years, most breeders used a plan of selecting only for single gene traits, or traits which were at least discrete in manifestation, in the early generations, until the plants were largely homozygous. This was done for two reasons: first, it was simpler; and secondly, it was claimed that time was needed for balanced linkage combinations to break up, with the formation of new, possibly advantageous gene combinations. Evidence on this latter point is tenuous, to say the least.

Accordingly, many methods have been devised to accelerate the early generation examination, with a view to obtaining homozygosity as rapidly as possible, and allowing more intense selection in the generations from say F_5 onwards. Two different approaches have been used, first, the single seed descent approach, where individual plants are taken from the F_2, and their descendants are taken through under largely artificial conditions to (at the latest) the F_8, with up to four generations a year, thereby assuring rapid approach to homozygosity. Secondly, early generation yield-testing has been used, whereby pedigreed plants or lines are yield-tested in each generation and selections are made in the early generations on this basis. Both methods have great disadvantages. Early generation yield testing is far more costly in terms of space and far slower and may be vitiated by heterosis and competition between plants of different genotype. Single seed descent, on the other hand, allows the effective testing of very few of the lines potentially available in the very early generations. If the number of genes affecting the trait is small, this will matter only if linkage is tight. It is clear that the lower the additive genetical variance in yield, or the genotypic variance in yield, the less useful early generation selection for yield will be, and the more valuable the single seed descent and related methods will be.

The conflict between these methods has not been resolved, but possibly the very large-scale use of doubled haploids will do so, since this way homozygosity may be achieved at a very early stage, e.g. by doubling gametes from F_1 individuals. Given this very rapid attainment of homozygosity, early generation yield testing on a large scale could be carried out in what would effectively be a multiplied-up but homozygous 'F_2'. If evidence were obtained that substantial blocks of linked genetical material existed contributing positively to traits such as yield, which does not seem to be the case, a large-scale segregating generation could be allowed before the doubling of haploids was carried out, thereby combining the best of both methods. Given reasonable progress in this field, such techniques should soon be in use where appropriate, and it is to be hoped that many of the approximate genetical methods which have been developed to overcome our ignorance of the genetical architecture of yield will thereby be rendered obsolete.

17.3.2 Allogamous crops

The classical theory of selection response in panmictic populations would be applicable to outbreeding crops, especially those with self-incompatibility or other forced outbreeding systems, but for historical reasons the use of hybrid breeding to take account of the still largely unpredictable phenomenon of heterosis has been critically important. It is possible that truncation selection given suitable initial populations of large enough size, would achieve similar results to the hybrid breeding programmes, if both started from the same point, but this is not the case; the hybrid breeding programmes have been under way for many years. Nonetheless, the success of the Illinois selection programme for increasing protein and oil content in maize over 76 generations (Dudley 1977) does suggest that simple truncation selection, with suitable precautions against inbreeding depression and other problems, still has much to offer in sophisticated breeding programmes for important, highly developed crop plants.

This said, it is obvious that heterosis is ideal for perennial or vegetatively propagated crops, and the problem here will be to find the best heterotic combinations. Suitable designs for obtaining this information exist, and have been described earlier. The important consideration from a strategic point of view is that testing for heterotic value must be carried out continuously, both among material incorporated into a breeding programme and with other material which may become available as time passes.

The results cited earlier make it clear that unless heterosis is minimal in extent, reciprocal recurrent selection is better than other methods such as recurrent selection to a tester or pure line selection. As reciprocal recurrent selection has not been as widely used as might have been expected, presumably because the returns in practice have been inadequate, the improvements or modifications based on half-sib progenies or crossing to lines selected downwards may be advantageous, especially in the presence of strong dominance effects. However, this area needs much further investigation.

Overall, selection for combining ability is very difficult, and it is of great interest that methods used have been successful despite the fact that they essentially use only additive genetical variance and additive by additive interaction variance.

For crops which cannot be inbred, and for crops which perform better in mixed stand, synthetic varieties have long been bred. Forage crops are the most important examples. Synthetic varieties overcome self-incompatibility problems, may be expected to be advantageous in relation to disease resistance, since it is easy to incorporate diverse genes into the variety, and allow rapid production of new varieties. Thus, both for their economy and for the relatively large gene pool incorporated, they may be expected to be far more widely used in future. For this reason, more work should be put into developing the theory.

17.3.3 Plant design

This is the breeder's objective most likely to influence his strategy. In the case of major crops of low unit price, such as cereals, plant design criteria will be closest to physiological optima, since it is with such crops that real biological constraints are likely to be important. For example, it is important to know whether grain yield is more limited post-anthesis by the supply of assimilate to the growing grain (known as the source) or by the capacity of the grain to accumulate assimilate (the sink). (In the case of modern wheats and Triticales, source limitation appears more important, whereas for the old, lower yielding, tall cultivars this is not the case; Fischer and Hille Ris Lambers 1978.) Thus, the idea of an ideotype is likely to increase in importance for such crops. Davis and Evans (1977) have suggested the use of selection indexes for what is essentially breeding towards an ideotype in Navy beans, where the relevant traits all vary quantitatively, but this will not always be possible.

For horticultural crops, on the other hand, opportunistic or tactical behaviour may well be more rewarding. This may be direct, as with the development of a novelty or a convenience, e.g. seedless watermelon through triploidy, or indirect, as with the construction of a new tomato variety from various 'pieces'. For example, the tomato cultivar, Flora-Dade, was developed as a verticillium wilt-resistant variety for the fresh market, and therefore required resistance to this disease, as well as to some races of fusarium wilt. In addition, the cultivar required the j_2 recessive gene for jointlessness, to allow the fruit to be mechanically harvested free of stem. Finally, it had to have satisfactory appearance and storing quality after being picked green. Thus, several breeders' materials were combined and crossed with a commercial cultivar, Walter, and carried through to the F_8 with single plant selections in the F_2 and F_4. The resultant variety was satisfactory in all respects but quality, where it appeared to be inferior to Walter in all attributes examined. Despite this, it is a highly successful variety (Volin and Bryan 1976).

It is very difficult to breed for product quality, except where this can be quantified, as with certain aspects of malting quality (Sparrow 1970). Flavour is the most recalcitrant aspect of quality. A number of studies have shown that flavour can to a certain extent be associated with measurable physicochemical traits. For example, tomato flavour depends on attributes such as soluble solids, pH, sugar/acid ratio and locular/pericarp tissue ratio (Stevens *et al.* 1977), and there are also genotypic differences in minor volatile compounds important to flavour (Stevens *et al.* 1977). This is also true in coffee (*Coffea arabica* L.) (Romero Lopes *et al.* 1984). In carrots, Simon and Lindsay (1983) assessed the relationship between subjectively assessed flavour and certain chemical constituents for several processing treatments and several genotypes, and found that variation in volatile terpenoids and reducing sugar content accounted for much of the variation in preference. When methods such as these have been developed and proven for a number of major fruit and vegetable crops, it may be possible to draw general conclusions about selection for improved flavour, but until then the complexity of joint selection for yield, shape, size, colour, elasticity, and durability will make selection for flavour a minor part of most breeding programs.

17.4 The breeder's market

Little has been said in this book about the market which plant breeders serve. This has been deliberate, since by and large considerations of genetical theory are irrelevant to those of microeconomics. However, it should be noted that the strategy of plant breeding is affected dramatically by the goals of the breeding organization.

Organizations primarily engaged in research as such will tend to have objectives related to what is scientifically interesting, to what may be conducted within the span of a research project or grant, or to very long-term considerations having little to do with commercially useful products. Thus, one may expect breeding effort in such organizations, in so far as this is directed towards the release of commercial varieties, to be aimed at crops where the varieties do not provide property rights, partly at least to avoid competition with commercial interests. For example, in Australia, the main academic breeding thrust has been towards inbreeding cereals, where a variety, once released, is available to all. In part, this reflects funding by levies on all sales, so that breeders in research institutes and other publicly supported bodies could not in the ordinary run of things be expected to benefit commercially from their work.

Where applied research institutes or departments of agriculture are directly funded by government, they may be expected to conduct similar research to academic institutions, but with a more short-term orientation and a concentration on commercial release of usable varieties. Here, also, commercial considerations of property rights in new varieties or cultivars will be less important. It is, in this context, of great interest that the change in breeding allogamous crops in Japan about 50 years ago from a diffuse pattern of local breeding organizations to a centralized system had led to an increase in internal rates of return from about 25 per cent to about 75 per cent (Hayami and Akino 1977). In the centralized system, national breeding stations are responsible for hybridization programmes and the first few segregating generations, after which regional stations conduct adaptation trials, following which prefectural (local) stations test material for commercial release.

Profit-making organizations will evidently have rather different criteria; research may be expected to be much more directly related to specific objectives. It is of some interest that many regard such research as far more difficult than pure research, since the latter may be justified by appeal to the results, should these be interesting, regardless of the original objectives, whereas in research such as that aimed at releasing new varieties, a failure to release new varieties may be regarded as failure overall. The international institutes whose work is directed specifically towards the improvement of particular crops, such as IRRI, CIMMYT, and CIP, in many ways lie closer to the commercial organizations than to the national research organizations or academic institutions, for these crop-related institutions must justify their existence by a continual supply of either commercially useful material, or breeder's varieties. Given their international non-profit funding, they have the opportunity, which has been taken, to maintain a long-term view, but they also have to justify their results. Possibly for these reasons, they have been very successful, showing internal rates of return higher than the 20–50 per cent common for national breeding programmes (Arndt and Ruttan 1977; Dalrymple 1977; Evenson 1977b).

The choice of crop and breeding techniques for a commercial organization will usually depend on the ability to acquire property rights in material which has commercial potential. Possibly this is why hybrid breeding has been so enthusiastically and successfully pursued by commercial organizations. I do not intend to imply in any way that the hybrid varieties so produced are not of great benefit, but the breeder embarking on a major programme for a particular crop would need to evaluate his breeding objectives in the light of his institution's revenue objectives, which will be quite different in the different types of institution discussed above. The introduction of Plant Breeder's Rights in many countries allows commercial organizations to pursue other breeding methods than hybrid breeding because their proprietary rights are protected (see for example, Sehgal 1977), but this is a relatively recent innovation, and it is not yet clear what the long-term results will be. Not all crops will benefit or suffer equally. This is because the protection offered cannot be identical for all crops and may indeed be varied legislatively with the mode of propagation of a crop or in other ways (Bagwill 1981).

17.5 A prospect

I began this book by suggesting that the next major advances in crop production are unlikely to come from quantitative genetics, but rather from molecular biology. The passage of seven years since I made this suggestion in the first edition has not made it any less true, but the contribution of genetical engineering to crop production remains modest. This is not because of legislative and other restrictions on recombinant DNA research (Watson and Tooze 1981), but rather because the problems are inherently very difficult.

Many gaps remain in the theoretical understanding of the molecular genetics of plants. However, present techniques of genetical engineering and somatic cell genetics should be sufficient to underpin the massive investment of human ingenuity necessary to breed plants and plan cropping systems which will approach more nearly the theoretical limits of crop production. Applied research is both more difficult and more expensive than pure research, but the likely prizes are immense. Meanwhile, Mendelian and quantitative genetics provide the theory and the methods for steady progress in yield, in product quality, and in resistance to disease for all existing crops.

The last major cycle of plant breeding progress may have ended for the most important cereals (Evenson 1977a), but advances have been so great that they could not in any case have continued at the same rate. Internal rates of return of 40 per cent or more have been common in the major breeding programmes (Arndt and Ruttan 1977); slower progress will be worthwhile but less astonishing. Timber production, on the other hand, should still yield very high rates of return, since most cultivated timber species have been domesticated rather than highly selected (Eldridge 1975).

New crops for both new and old purposes should be sought, to widen the range of production and guard against ecological disaster, and with such crops the methods described here will all be used again, the processes being speeded by the newer techniques. The purely genetical approach to plant improvement may perhaps be regarded as an intermediate technology, but it is not a mature technology. The bridge between molecular biology and quantitative genetics is still under construction. 'The solution of which Difficulty, I shall leave among Naturalists' (Swift 1724).

Glossary

Active site. A specialized part of an enzyme directly involved in the reaction of substrates.

Adaptation. The process by which individuals, populations or species change in form or function in such a way as to survive better under given environmental conditions. Also the result of this process.

Alien chromosome. A chromosome from a related species transferred to a crop plant.

Allelle or allelomorph. One of a pair or series of genes which behave as alternatives in inheritance (because they are situated at the same locus in homologous chromosomes).

Allopolypoid. A polyploid containing sets of chromosomes of genetically different origin, for example, sets from two or more different species.

Allosteric. A term which describes alteration in the behaviour of a protein as the result of change in its conformation induced by the binding of a small molecule at a site other than its active site.

Allosyndesis. Meiotic chromosome pairing, in aneuploids and polyploids, of completely or partially homologous (homoeologous) chromosomes which have been introduced into the zygote by the different gametes.

Amino acid. An organic compound containing both carboxyl (—COOH) and amino groups ($—NH_2$).

Amphidiploid. A polyploid whose chromosome complement is made up of the entire somatic complements of two species (or an allopolyploid which behaves as a diploid at meiosis).

Analysis of variance. A statistical technique for partitioning the total variance in a trait into portions arising from different sources.

Anaphase. The stage in mitosis (or meiosis II) at which daughter centromeres separate and migrate to opposite poles. That stage in the first meiotic division when co-ordinated centromeres move apart.

Aneuploid. Containing a chromosome number other than the haploid number of chromosomes or a multiple of it (e.g. $2n$-1 or $2n + 1$) where n is the haploid number.

Apomixis. Reproduction in which sexual organs or related structures take part but fertilization does not occur, so that the resulting seed is vegetatively produced.

Asexual reproduction. Reproduction which does not involve the union of gametes.

Asynapsis. Failure of pairing of homologous chromosomes during meiosis.

Autogamy. Self-fertilization.

Autopolyploid. A polyploid arising through multiplication of the complete haploid or diploid set of a species.

Autoradiograph. Pattern produced on photographic film after exposure to a radioactive compound, e.g. to tritium incorporated into chromosomes in thymidine, which gives information regarding the incorporation of the radioactive precursor.

Auxin. A plant hormone that promotes such plant functions as cell elongation and callus production.

Auxotroph. A strain which fails to grow on a medium containing the minimum nutrients essential for the growth of the wild type.

Avirulent. A strain of a parasite unable to infect and cause disease in a host plant.

Back-cross. A cross of a hybrid to either of its parents, usually a cross of a heterozygote to a homozygous recessive (*See* **Test cross.**)

Backcross breeding. A system of breeding whereby recurrent backcrosses are made to one of the parents of a hybrid, accompanied by selection for a specific character or characters.

Bacteriophage. A virus which infects bacteria.

Balanced polymorphism. A polymorphism at a locus maintained by selective advantage of the heterozygote over both homozygotes.

Basic number. The haploid number of chromosomes in diploid ancestors of polyploids, represented by x.

Bias. A consistent departure of the expected value of a statistic from its parameter.

Biotype. A group of individuals with the same genotype. Biotypes may be homozygous or heterozygous.

Bivalent. A pair of homologous chromosomes associated in the first meiotic division.

Bulk breeding. The growing of genetically diverse populations of self-pollinated crops in a bulk plot with or without mass selection, generally followed by a single-plant selection.

Carrier. Heterozygote for a particular, usually recessive, trait.

Cell cycle. The sequence of cellular events between two mitoses. It is customarily divided into four phases: mitosis; synthetic or S phase, when the DNA is being replicated; G_1, the phase between mitosis and the S phase; and G_2, the phase between S and the next mitosis.

Cellular transformation. A stable heritable alteration in the phenotype of a cell usually brought about by viral infection; may also be produced by a chemical carcinogen or may appear in an apparently spontaneous fashion.

Centromere. That region of a chromosome which interacts with the spindle and is associated with chromosome movement during nuclear and cell division. It is the last region of sister chromatids to separate. Sometimes called **kinetochore**.

Character. An attribute of an organism resulting from the interaction of a gene or genes with environment.

Chiasma. The (cross-shaped) configuration formed when non-sister chromatids of a bivalent or multivalent remain in contact during the first meiotic division.

Chimaera. An individual containing two or more genetically different cell lines.

Chromatid. The individual daughter strands resulting from the replication of a chromosome. When daughter centromeres separate, the strands are called daughter chromosomes.

Chromsomal aberration. An abnormality of chromosome number or structure.

Chromosomes. Thread-like, deeply-staining bodies found within the nucleus and visible during mitosis and meiosis. They are composed of DNA and protein and carry the genetical information. An **acrocentric** chromosome is one with the centromere located near the end of the chromosome, a **metacentric** chromosome is one with the centromere located near the middle of the chromosome.

Cistron. The functional unit of the hereditary materials. In eukaryotes, defined by the phenotype of a heterozygote carrying two recessive mutations from different parents; if the phenotype is mutant, the genes concerned belong to the same cistron; if the phenotype is normal, the genes belong to different cistrons.

Clone. In eukaryotes, all the cells derived from a single cell by repeated mitoses, all having the same genetical constitution.

Co-dominance. The expression of both alleles in the heterozygote.

Codon. A triplet of bases in the DNA or RNA molecule which codes for one amino acid.

Cohort. A group of individuals that enter a breeding program in the same season.

Colchicine. An alkaloid that arrests spindle fibre formation and disjunction of daughter chromosomes. In mitotic cells it may lead to the doubling of the chromosome number.

Combining ability. General, average performance of a strain in a series of crosses. **Specific**, deviation in a particular cross from performance predicted on the basis of general combining ability.

Complementation. The complementary action of gene products in a common cytoplasm. The support of a function by two homologous pieces of genetical material present in the same cytoplasm, each carrying a recessive mutation and unable by itself to support that function.

Constitutive heterochromatin. The material basis of chromosomes or chromosomal regions which exhibit heterochromatic properties under most conditions, e.g. centromeric heterochromatin. (*See also* **Facultative heterochromatin**.)

Correlation. The co-relationship of two variables. A correlation coefficient is designed to measure the degree of association of the two variables.

Cotyledon. The first leaf or one of the pair of first leaves developed by the embryo of a seed plant.

Coupling. The linkage phase in a double heterozygote for two linked loci which has received both dominant factors from one parent and both recessives from the other, to be contrasted with **repulsion** phase (q.v).

Covariance. The mean of the product of the deviation of two variates from their individual means. A statistical measure of the inter-relationship between variables.

Crossing-over. The reciprocal exchange of genetical material between homologous chromosomes at meiosis. Non-homologous crossing-over: exchange of non-homologous segments of chromosomes. This event results in reciprocal deletion and duplication of the chromosome segments involved.

Cultivar. An identifiable category within a crop species. Plants of a cultivar are related by descent.

Cytokinin. A plant hormone that stimulates mitosis.

Cytoplasm. All the contents of a cell except the nucleus and the cell wall. It consists of the cytoplasmic matrix in which are situated organelles, e.g. plastids, mitochondria, ribosomes, endoplasmic reticulum, etc.

Cytoplasmic inheritance. Transmission of hereditary determinants through genetic material in the cytoplasm as distinct from transmission by genes carried by chromosomes. Often detected by differing contribution of the two parents in reciprocal crosses.

Deficiency. The absence or deletion of a segment of genetical material.

Degenerate codons. Two or more codons that code for the same amino acid.

Degrees of freedom, number of. The number of independent comparisons that can be made in a set of data.

Deletion. The type of chromosomal aberration in which there is a loss of part of a chromosome (also called a **deficiency**).

Detassel. Removal of the tassel (male inflorescence) in maize.

Deviation. Departure of an observation from its expected value.

Diakinesis. The last stage in the prophase of meiosis I, when the paired homologous chromosomes are highly contracted but before they have moved onto the metaphase plate.

Diallel cross, complete. The crossing in all possible combinations of a series of genotypes.

Dihybrid. Heterozygous with respect to two genes.

Dioecious. Plant producing male and female gametes in different individuals, i.e. staminate and pistillate flowers occur on different individuals.

Diploid. The condition in which the cell contains two homologous sets of chromosomes.

Diplontic Selection. Competition betweeen different cell types in a chimaera.

Diplotene. The stage in the prophase of meiosis I, when the paired homologous chromosomes separate except where they are held together by chiasmata.

Disjunction. The separation of chromosomes at anaphase.

DNA (Deoxyribonucleic acid). A polymer of nucleotides where the pentose sugar (ribose) is in the deoxy form. DNA is found mainly in chromosomes. It is the genetical material.

Dominant. A trait which is expressed in individuals which are heterozygous for a particular pair of allelic genes. Partial, incomplete, or semi-dominants are traits which are expressed in a reduced, incomplete or intermediate form in individuals which are heterozygous for a particular pair of allelic genes.

Donor parent. In plant breeding, the parent from which one or a few genes are transferred to the recurrent parent in backcross breeding.

Dosage effect. The influence upon a phenotype of the number of times a genetical element is present.

Double cross. A cross between two F_1 hybrids.

Double fertilization. In flowering plants, the union of one sperm nucleus with the egg nucleus and one with polar nuclei to form the embryo and the endosperm.

Double reduction. The genetical outcome of chromatid segregation, as opposed to chromosome segregation, whereby two sister-chromatid segments are included in the same meiotic product.

Drift. Random fluctuation in gene frequency in a population of finite size.

Duplication. A type of chromosomal aberration in which part of a chromosome is duplicated.

Duplex. (*See* **Nulliplex**.)

Effective population size. For a given population, the effective size is the size of an equivalent ideal population which is expected to experience the same increase in homozygosity over time, i.e. drift, as the population in question.

The ideal population is one in which mating is at random in the absence of selection and in which all individuals make the same expected contribution to the next generation.

Electrophoresis. The separation of molecules differing in charge and size due to their different mobilities when subjected to an electric field in a liquid medium. (*See also* **Gel electrophoresis.**)

Emasculation. Removal of the anthers from a flower.

Endoplasmic reticulum. A system of minute tubules within the cytoplasm. Two types are recognized, rough and smooth. Respectively, these are particularly concerned with pathways of protein and steroid synthesis.

Enzyme. A protein which acts as a catalyst in biological systems.

Epigenotype. The hereditarily differentiated constitution of a cell.

Epiphytotic. An unarrested spread of a plant disease.

Epistasis. Gene interaction where one gene interferes with the phenotypic expression of another non-allelic gene so that the phenotype is determined effectively by the former. The latter is described as hypostatic. More generally, the term epistasis is used to describe all types of non-allelic interaction where manifestation at any locus is affected by genes at any other loci.

Error variance. Variance, arising from unrecognized or uncontrolled factors in an experiment, with which the variance of recognized factors is compared in tests of significance.

Euchromatin. The material basis of chromosomes and chromosomal regions which show the cycle of chromosome coiling and staining characteristic of the majority of the chromatin (in contrast to **heterochromatin**).

Eukaryote. The biological category which includes all organisms that have nuclei and membrane-bound cellular organelles.

Euploid. Containing either the haploid chromosome number or an exact multiple thereof.

Expressivity. The degree of manifestation of a genetically determined character in individuals in which it is detectable.

F_1. The first generation of a cross between two true-breeding parents.

F_2. The second filial generation obtained by self-fertilization or crossing *inter se* of F_1 individuals.

F_3. Progeny obtained by self-fertilizing F_2 individuals.

Factor. Synonymous with gene.

Facultative heterochromatin. The material basis of heterochromatin present in only one of a pair of homologues, e.g. one X-chromosome in female mammals. (See also **Constitutive heterochromatin**.)

Family. A group of individuals directly related by descent from a common ancestor.

Fertility. Ability to produce viable offspring.

Fertilization. Fusion of the nuclei of two gametes or their equivalent.

First-degree relatives. Parents, offspring, and sibs.

Fitness. The biological fitness of an individual is the relative contribution of that individual to the ancestry of future generations, usually measured by the number of offspring who reach reproductive age. Fitness is unity (or 100 per cent) if each individual has two such offspring, in a population of constant size.

Gamete. The mature functional reproductive cell whose nucleus fuses with that of another gamete with the resulting cell (zygote) developing into a new individual. In a diploid organism the gametes are haploid.

Gel electrophoresis. Electrophoresis performed under conditions where the liquid medium is held immobile in a gel. Certain gels (e.g. starch) separate molecules on the basis of size as well as charge, since the gel pore size is small enough to act as a molecular sieve.

Gene. The unit of inheritance, usually implying that section of a linkage group which behaves as a unit of function, i.e. the cistron, but also possibly the smallest unit defined by mutation or recombination. The commonest function is to code for a polypeptide chain.

Gene-for-gene relationship. The host–parasite interaction in which every gene determining resistance in a host plant is matched by a gene for pathogenicity in the pathogen.

Gene interaction. Modification of gene action by a non-allelic gene or genes, generally the interaction between products of non-allelic genes.

Genetical equilibrium. The condition in which successive generations of a population contain the same genotypes in the same proportions with respect to particular genes or combinations of genes.

Genetical transformation. The transfer of genetical information to one cell strain (usually of bacteria) by DNA extracted from another strain.

General resistance. Resistance against all biotypes of a pathogen.

Genome. All the genetical material present in a typical cell of an organism. Alternatively, a haploid chromosome complement.

Genotype. The genetical constitution of an individual.

Germinal cells. The gametes and their immediate precursors.

Germplasm. The sum total of the hereditary materials in a species.

Haploid. The condition in which the cells contain one set of chromosomes.

Haploidization. The process whereby diploid somatic cells become haploid during a **parasexual cycle** (q.v.). Although similar to meiosis in its ultimate effects on chromosome number, it differs from it by the gradual and disordered loss of chromosomes over a large number of cellular generations.

Hardy–Weinberg rule. An isolated population undergoing random mating in the absence of selection and mutation reaches genotypic frequency equilibrium at an autosomal locus after one generation.

Hemizygous. An individual generally diploid having a given gene or genes present once, e.g. the genotype of the sex chromosomes in the heterogametic sex of an organism having X-Y (or equivalent) chromosomal sex determination.

Heritability. The proportion of the total variance of a character attributable to genetical as opposed to environmental factors.

Hermaphroditism. Reproductive organs of both sexes present in the same individual or in the same flower in higher plants.

Heterobeltiosis. The departure of a character in an F_1 hybrid from the superior of the two parents.

Heterochromatin. Chromosomal material showing a different cycle of coiling or compression and hence stainability from **euchromatin** (q.v.). (*See also* **Constitutive, Facultative heterochromatin**.)

Heterokaryon. A cell containing two genetically dissimilar nuclei in a common cytoplasm.

Heterosis. The departure of a character in an F_1 hybrid from the mean (arithmetic or geometric) of the two parents.

Heterothally. Haploid incompatibility in fungi and other thallophytes (opposite of **homothally**).

Heterozygote. In a diploid, an individual with two different genes at a given locus. (*See also* **Homozygote.**)

Hexaploid. A polyploid with six sets of chromosomes.

Homoeologous chromosomes. Chromosomes which are partially homologous, having evolved from original complete homology.

Homologous chromosomes. Chromosomes, separately derived from two parental gametes, which pair during meiosis and contain identical loci.

Homonuclear cell line. A cultured cell line with uniform karyotype. A heteronuclear line has a non-uniform karyotype.

Homozygote. In a diploid, an individual with two identical genes at a given locus.

Hormone. A messenger chemical that controls cell functions in growth and development at a site different from the point of synthesis.

Hybrid. The progeny of a cross between two genetically different organisms.

Hybrid vigor. (*See* **Heterosis**).

I_1, I_2, I_3. . . Symbols used to designate first, second, third etc., inbred generations. (*See* **S_1, S_2. . .**)

Ideotype. Crop plant with model characteristics known to influence photosynthesis, growth and (in cereals) grain production.

Idiomorph. An allele (or allelomorph) which is rare. This term is usually used to indicate a contrast with common **polymorphic** alleles (q.v).

Inbred line. A line produced by continued inbreeding. In plant breeding a nearly homozygous line usually originating by continued self-fertilization, accompanied by selection.

Inbred-variety cross. (*See* **Topcross.**)

Inbreeding. The crossing of individuals who are related by ancestry.

Inbreeding coefficient. The probability that the two genes at any locus in a diploid individual are identical by descent, i.e. they originated from the replication of one gene in a previous generation.

Incidence. The frequency of occurrence of a trait or disease in a population, often over a specific interval of time.

Independence. The relationship between variables when the variation of each is not influenced by that of others, whence their correlation is zero.

Interference. The effect of recombination in one interval on the probability of recombination in an adjacent interval.

Internal rate of return. That interest rate r_i which makes $\sum_{t=0}^{T}(R_t - C_t)/(1 + r_i)^t = 0$. where R_t = return (social benefit) in year t, C_t = research cost in year t, and $R_T = 0$.

Interphase. The stage between two successive cell divisions during which DNA replication occurs.

Inversion. A type of chromosomal aberration in which the linear sequence of a chromosomal region is reversed.

Irradiation. Exposure of organisms or parts of organisms to X-rays or other radiations to increase mutation rates.

Isoalleles. Functionally indistinguishable alleles at a locus.

Isochromosomes. A type of chromosomal aberration in which one of the arms of a particular chromosome is duplicated and the other arm is deleted. The two arms of an isochromosome are therefore of equal length and contain the same genes.

Isogenic lines. In plant breeding, two or more lines differing from each other genetically at one locus only. Distinguished from clones, homozygous lines, identical twins, etc., which are identical at all loci.

Isolation. The separation of one group from another so that crossing between groups is prevented.

Karyotype. The number, size and shape (and bands) of the chromosomes of a somatic cell. A photomicrograph of the chromosomes in one cell arranged in a standard manner, intended to be representative for the individual as a whole.

Kinetochore. (*See* **Centromere**).

Land race. A cultivar domesticated, selected and used for cropping before records of crop breeding began to be kept. A primitive cultivar characteristic of a locality.

Liability. Tendency to manifest a trait or disease.

Line breeding. A system of breeding in which a number of genotypes which have been progeny tested in respect of some character or group of characters are composited to form a variety.

Linkage. The association of parental character combinations in offspring in excess of the non-parential combinations.

Linkage desequilibrium. Combinations of alleles of linked genes which have frequencies significantly different from those expected with random combination.

Linkage group. A set of genes each of which shows linkage with one or more other genes in the set.

Linkage map. A diagram showing the linear order and position of genes belonging to the same linkage group, derived from recombination frequencies.

Linkage value. Recombination fraction expressing the proportion of non-parental or recombinant versus parental types in a progeny. In diploids, the recombination fraction can vary from zero to one half.

Load. In a population, the difference between the mean fitness and the optimum fitness, relative to the optimum fitness, is defined as the genetical load.

Locus. The position in a linkage map recognized as the site at which allelic genes are situated.

M_1, M_2, M_3, . . . Symbols used to designate first, second, third etc. generations after treatment with a mutagenic agent.

Male sterility. Absence or non-function of pollen in plants.

Mass-pedigree method. A system of breeding in which a population is propagated in mass until conditions favourable for selection occur, after which pedigree selection is practised.

Mass selection. A form of selection in which individual plants are selected on their individual merits and the next generation propagated from the aggregate of their seeds.

Mating system. Any of a number of schemes by which individuals are assorted in pairs leading to sexual reproduction. **Random**, assortment of pairs is by chance. **Genetical assortative mating**, mating together of individuals more closely related than individuals mating at random. **Genetical disassortive mating**, mating of individuals less closely related than individuals mating at random.

Mean. A measure of the location of the distribution of the values of a trait in a population, usually the arithmetical average.

Median. A measure of the location of the distribution of the values of a trait in a population, namely, the value of the variate on each side of which there is an equal number of larger and smaller variates.

Meiosis. The type of cell division which occurs during gametogenesis and results in halving of the somatic number of chromosomes so that in a diploid each gamete is haploid.

Messenger RNA (mRNA). The RNA which is produced by transcription of DNA and which serves as the template for protein synthesis.

Metabolic co-operation. Restoration of normality in mixed culture of two metabolically deficient cell types.

Metaphase. The stage of meiosis or mitosis at which the individual centromeres of the chromosomes lie at or near the equator of the spindle.

Metaxenia. Influence of pollen on maternal tissues of the fruit (*See* **Xenia**).

Mimic genes. Different genes with similar effects.

Mitochondria. Organelles situated within the cytoplasm of eukaryotes which are concerned with cellular respiration and oxidative phosphorylation.

Mitosis. The type of nuclear division which leads to daughter nuclei with the same chromosome complement as the original nucleus.

Mode. The value of a variate in the class of greatest frequency in a frequency distribution.

Modifying gene. A gene that affects the expression of a non-allelic gene or genes.

Monoecious. In plants, describing the condition where male and female gametes are borne by different flowers of the same individual.

Monohybrid. Heterozygous with respect to two allelic genes.

Monolayer. In tissue culture, a confluent sheet of cells one or two cells deep.

Monoploid. An organism with the basic (x) chromosome number. (*See* **Haploid.**)

Monosomy. Loss of one member of a chromosome pair so that there is one less than the diploid number of chromosomes ($2n - 1$).

Mulitifactorial determination. Determination by many genes with small effects together with effects of the environment.

Multiline. A mixture of lines or cultivars bearing different genes for resistance to pathogens.

Multiple alleles. The existence of more than two alleles for a particular character.

Multivalent. (*See* **Univalent.**)

Mutagen. A physical or chemical agent capable of inducing mutations.

Mutation. An inherited change in the genetical material, either of a single gene (point mutation) or in the number or structure of the chromosomes. A mutation which occurs in the gametes may be inherited by the progeny, a mutation which occurs in the somatic cells (somatic mutation) may give rise to a mutant clone.

Mutation rates. The number of mutations of any one particular unit of inheritance (cistron, gene, etc.) which occur per gamete per generation, or per cell per generation or per nucleus per generation.

Non-disjunction. The failure of homologous chromosomes to separate during first division of meiosis or of daughter chromosomes at the second division of meiosis or at mitosis leading to progeny with more or less than the normal number of chromosomes.

Normal distribution (Gaussian distribution). A continuous symmetrical bell-shaped frequency distribution, given by $f_X(x) = 1/\surd(2\pi\sigma^2)e^{-(x-\mu)^2/2\sigma^2}$, where μ and σ^2 are, respectively, the mean and the variance of X.

Nucleolus. A spheroidal structure within the nucleus often associated with a particular region (the nucleolar organizer) on a specific chromosome. It is the site of ribosomal RNA synthesis.

Nucleotide. The structural unit of nucleic acid consisting of a nitrogenous base, a pentose sugar, and a phosphate group.

Nucleus. A structure within the cell which contains the chromosomes and nucleolus, bounded by a double membrane.

Null hypothesis. Hypothesis that there is no discrepancy between observation and expectation based on some set of postulates.

Nulliplex. The condition in which a polyploid carries a recessive gene at a particular locus in all homologues. Simplex denotes that the dominant gene is represented one, duplex two, triplex three, quadruplex four times, etc.

Nullisome. An otherwise $2n$ plant lacking both members of one specific pair of chromosomes, hence, with $2n - 2$ chromosomes.

Oligogenic. Inheritance due to a small number of genes with discernible effects.

Operator. The binding site for a repressor molecule, at one end of an operon.

Operon. A group of adjacent genes (originally defined in bacteria) which apparently affect different steps in a particular metabolic pathway and which are regulated as an integrated unit.

Out-cross. A cross, usually natural, to a plant of different genotype.

Over-dominance. Heterozygous advantage; at a locus with two alleles the heterozygote is fitter than either homozygote, giving rise to a **balanced polymorphism** (q.v.).

P_1, P_2, P_3, . . . First, second, third, etc., generations from a parent. Also used to designate different parents used in making a hybrid or series of hybrids.

Pachytene. The state in the prophase of meiosis I, when the homologous chromosomes are completely paired.

Panmixia. (*See* **Random mating.**)

Parameter. A numerical quantity which specifies a population in respect to some characteristic.

Parasexual cycle. A cycle of events which has the same outcome as the sexual cycle without the regular alternation of meiosis and fertilization. In fungi the cycle consists of the rare fusion of haploid nuclei to form a diploid. Genetical recombination may occur at this stage and is followed by a gradual loss of chromosomes until a haploid remains.

Parthenogenesis. Development of an organism from a sex cell without fertilization.

Pathogen. The causal agent of a disease.

Pedigree. A record of the ancestry of an individual, family or strain.

Pedigree breeding. A system of breeding in which individual plants are selected in the segregating generations from a cross on the basis of their desirability judged individually and on the basis of a pedigree record.

Penetrance. Proportion of a genotype manifesting a trait. If some with the genotype do not manifest the trait, there is reduced penetrance.

Phenocopy. A phenotype which is caused by environmental factors but resembles one which is genetically determined.

Phenotype. The appearance (with respect to some physical, biochemical, physiological, etc. character) of an individual which results from the interaction of genotype and environment.

Physiological races. Pathogens of the same species with similar or identical morphology but differing pathogenic capabilities.

Plasmid. A small circular DNA molecule, capable of self-replication, that can carry genes into a host organism.

Plasmagene. An extra-nuclear gene.

Pleiotropy. A gene with multiple effects is said to be pleiotropic.

Polycross. Open pollination of a group of genotypes (generally selected) in isolation from other compatible genotypes in such a way as to promote random mating *inter se.*

Polygenic. Pertaining to polygenes or systems of polygenes.

Polygenes. Genes whose differences or mutations lead to effects which are too small to identify individually.

Polyhaploid. Haploid plant derived from a polyploid individual.

Polymerases. A group of enzymes which catalyze the formation of DNA or RNA from precursor substances in the presence of DNA or RNA templates.

Polymorphism. The occurrence in a population of two or more genetically determined forms in appreciable frequencies. The alleles determining such forms are known as polymorphic alleles or polymorphs. (*See also* **Balanced polymorphism, Transient polymorphism.**)

Polypeptide. An organic compound consisting of three or more amino acids which are linked covalently through amide bonds; the functional product of the translation of messenger RNA.

Polyploid. Having any multiple greater than two of the haploid number of chromosomes, e.g. $3n$, $4n$, etc.

Polysome (or Polyribosome). A group of ribosomes associated with the same molecule of messenger RNA.

Polysomic. In an otherwise diploid individual, having three, (**trisomic**), four (**tetrasomic**) etc., of a particular chromosome.

Population. In genetics, a community of individuals which shares a common gene pool. In statistics, a hypothetical (often infinitely large) series of potential observations among which observations actually made constitute a sample.

Potence. The effect on a trait of the integrated dominance and interaction relations of all the polygenic alleles within the combinations possible.

Prepotency. The capacity of a parent to impress characteristics on its offspring so they are more alike than usual.

Prevalence. The observed frequency of a trait or disease in a population, often at a particular age or time.

Principal component analysis. A statistical method for the exhibition of a multivariate complex as dependent on a number of independent components, each being a linear function of the original variables, chosen so that the first component accounts for as much of the variation as possible, the second as much of the remainder, etc.

Probability. The proportion of times in which an event occurs in an infinitely large and hypothetical series of cases, each capable of producing the event.

Progeny test. A test of the value of a genotype based on the performance of its offspring produced in some definite system of mating.

Pro-metaphase. The stage in mitosis between the dissolution of the nuclear membrane and the organization of the chromosomes on the metaphase plate.

Promoter. The region of the DNA at which the RNA polymerase binds to initiate transcription.

Prophase. The first visible stage of cell division when the chromosomes are contracted and therefore thicker than previously; before metaphase.

Protandry. Maturation of anthers before pistils.

Protein. A macromolecule consisting of one or more of the same or different polypeptides with enzymatic, structural, or transport function.

Protogyny. Reverse of **Protandry** (q.v.).
Protoplast. The plant cell other than the cell wall.
Prototroph. A strain which has the same nutritional requirement as the wild type, either completely or with respect to a particular substance or nutrient. (*See* **Auxotroph.**)
Pure line. In plant breeding, a strain homozygous at all loci, ordinarily obtained by successive self-fertilization.
Quadrivalent. (*See* **Univalent**).
Quadruplex. (*See* **Nulliplex.**)
Qualitative character. A character in which variation is discontinuous.
Quantitative character. A character in which variation is continuous so that classification into discrete categories is arbitrary.
Rad. A measure of the amount of any ionizing radiation which is absorbed by the tissues. One rad is equivalent to 100 ergs of energy absorbed per gram of tissue.
Random. Arrived at by chance without discrimination.
Randomization. Process of making assignments at random.
Random genetical drift. (See **Drift.**)
Random mating. For a given population, where an individual of one sex has an equal probability of mating with any individual of the opposite sex or insofar as the genotypes with respect to given genes are concerned.
Recessive. A trait expressed in individuals who are homozygous for a particular gene but not in those which are heterozygous for this gene.
Reciprocal crosses. Crosses in which the sources of male and female gametes are reversed.
Recombination. The process whereby new combinations of parental characters may arise in the progeny.
Recombinant DNA. DNA molecules constructed by joining, outside the cell, natural or synthetic DNA segments to DNA molecules capable of replication in living cells.
Recurrent selection. A method of breeding designed to concentrate favourable genes scattered among a number of indivduals by selecting in each generation among the progeny produced by matings *inter se* of the selected individuals (or their selfed progeny) of the previous generation.
Regression. Originally, the tendency of offspring to revert in phenotype to a position between their parents' phenotypes and the population mean. Now, a technique for estimating the degree to which one variate is determined by one or more other variables so that, for example, the regression of offspring phenotype on parental phenotype gives a measure of genetical and familial influences on the character in question.
Regulator gene. A gene which synthesizes a repressor substance which alone or together with a co-repressor prevents transcription of a specific operon, according to the Jacob and Monod theory. In a more general way, a regulatory gene is any which influences the expression of a **structural gene** (q.v.), e.g. by controlling enzyme activity, enzyme synthesis, or enzyme breakdown.
Repressor. A protein which is coded for by a regulator gene and which interacts with inducer compounds and the operator to control the transcription of structural genes.
Repulsion. The linkage phase of a double heterozygote for two linked gene pairs which has received one dominant factor from each parent and the alternative recessive factor from each parent, e.g. for genes *A*, *a* and *B*, *b* the repulsion

heterozygote receives *Ab* from one parent and *aB* from the other where *A* and *B* are dominant, and *a* and *b* are recessive.

Restriction endonucleases. A class of enzymes which break both strands of a DNA molecule at specific points as a result of recognizing a given short sequence of bases.

Ribosomal RNA (rRNA). The ribonucleic acid component of the ribosomes.

Ribosomes. Organelles situated in the cytoplasm, which are associated with protein synthesis. They consist of RNA and protein and are the site of translation of messenger RNA to polypeptide.

RNA (Ribonucleic acid). Single-stranded nucleic acid where the pentose sugar is ribose. In eukaryotes, the nucleic acid which is found mainly in the nucleolus and ribosomes. Messenger-RNA transfers genetical information from the nucleus to the ribosomes in the cytoplasm and acts as a template for the synthesis of polypeptides. Transfer-RNA transfers activated amino acids from the cytoplasm to messenger-RNA.

Rogue. A variation from the standard type of a variety of strain. **Roguing**, removal of undesirable individuals to purify the stock.

S_1, S_2, S_3, . . . Symbols for designating first, second, third, etc., selfed generations from an ancestral plant (S_0).

Sample. A finite series of observations taken from a population.

Sampling error. Deviation of a sample value from the true value owing to the limited size of sample.

Segregation. The separation of alleles during meiosis so that each gamete contains only one member of each pair of alleles.

Selection. The forces which affect biological fitness and therefore the frequency or distribution of a particular trait within a given population.

Self-fertility. Capability of producing seed upon self-fertilization.

Self-fertilization. Fusion of male and female gametes from the same individual.

Self-incompatibility. Genetically controlled physiological hindrance to self-fertilization.

Sib (or **Sibling).** Brother or sister.

Sib mating. The mating of sibs.

Significance, test of. Statistical test designed to distinguish differences due to discrepancy between observation and hypothesis.

Single cross. A cross between two genotypes, usually two inbred lines, in plant breeding.

Somatoplastic sterility. Collapse of zygotes during embryonic stages due to disturbances in embryo–endosperm relationships.

Species. The unit of taxonomic classification into which genera are subdivided, usually determined by reproductive isolation.

Specific resistance. Resistance against certain biotypes of a parasitic organism.

Spindle. A spindle-shaped structure of fibrous protein which is organized at cell division and is concerned with the orientation of chromosomes at metaphase and their movement at anaphase.

Standard deviation. A measure of variability. The square root of the variance of a distribution. For the normal curve, the distance along the abscissa from the mean to the point of inflection.

Standard error. The square root of the variance of an estimate, used to indicate the reliability of the estimate.

Statistic. An estimate from a sample, of some population parameter.

Statistical test. A mathematical procedure carried out on a sample drawn from a population which enables one to test a hypothesis and so make inferences about the numerical properties of that population.

Strain. A group of similar individuals within a variety.

Structural gene. One whose base sequence determines the primary structure of its product. If the product is a polypeptide, the primary structure will be its amino acid sequence; if it is a ribosomal or a transfer RNA, its primary structure will be the sequence of its purine and pyrimidine bases.

Synapsis. Conjugation (pairing) at pachytene and zygotene of homologous chromosomes.

Syntenic. Genes which are known to be on the same chromosome, whether or not linkage can be detected.

Synthetic variety. A variety produced by crossing *inter se* a number of genotypes selected for good combining ability in all possible hybrid combinations, with subsequent maintenance of the variety by open pollination.

Telophase. The stage of cell division when the chromosomes have completely separated into two groups and each group has become enclosed in a nuclear membrane.

Test cross. A cross of a double or mutiple heterozygote to the corresponding multiple recessive to test for homozygosity or linkage.

Tetraploid. An organism with four haploid (x) sets of chromosomes.

Threshold. A critical value on an underlying scale of liability above which individuals manifest a trait or disease (*See also* **Liability**).

Tolerance. The ability of a plant to endure attack by a pathogen without severe loss of yield.

Top cross. A cross between a selection, line, clone, etc. and a common pollen parent which may be a variety, inbred line, single cross, etc. The common pollen parent is called the top cross or tester parent. In maize, a top cross is commonly called an inbred-variety cross.

Transcription. The process whereby genetical information is transmitted from the DNA in the chromosomes to messenger-RNA.

Transfer RNA (tRNA). Molecules of RNA able to combine with a specific amino acid and to pair with the appropriate codon for that amino acid in the mRNA.

Transformation. The transfer of genetical information to a recipient strain of bacteria by DNA extracted from a donor strain, and recombination of that DNA with the DNA of the recipient.

Transgressive segregation. Appearance in segregating generations of individuals falling outside the parental range with respect to some character.

Transient polymorphism. Polymorphism during the period in which an allele associated with greater fitness is replacing one associated with lower fitness.

Transition. A mutation caused by the substitution in DNA or RNA of one purine base for the other, or one pyrimidine base for the other. (*See* **Transversion**).

Translation. The process whereby genetical information from messenger-RNA is translated into a polypeptide.

Translocation. The transfer of part of one chromosome to another non-homologous chromosome. If there is an **exchange** of parts between two chromosomes, this is referred to as a **reciprocal translocation** or **interchange**. (The term translocation also refers to a detailed step in biosynthesis of polypeptide chains.)

Transversion. A mutation caused by the substitution of a purine for a pyrimidine, and vice versa, in DNA or RNA. (*See* **Transition.**)

Triplet. A series of three bases in the DNA or RNA molecule which codes for a specific amino acid.

Triplex. (*See* **Nulliplex.**)

Triploid. An organism with three haploid (x) sets of chromosomes.

Trisomy. Presence of one chromosome in triplicate instead of in duplicate ($2n + 1$).

Truncation selection. (*See* **Mass selection.**)

Unifactorial. Inheritance controlled by a single gene pair.

Univalent. An unpaired chromosome in meiosis. Bivalents, trivalents, quadrivalents, etc., are associations of 2, 3, 4, ... homologous chromosomes held together by chiasmata.

Variance. A measure of the dispersion of variability of a trait about its mean in a population. [The second moment about the mean, where the nth moment M_n of a distribution $f_X(x)$ is given by $M_n = \int x^n f_X(x) dx$.]

Variate. A single observation or measurement.

Variety. A subdivision of a species. A group of individuals within a species which are distinct in form or function from other similar arrays of individuals. Usage differs between countries.

Vector. A vehicle, such as a plasmid or virus, for carrying recombinant DNA into a living cell, or an organism, e.g. an insect, which transmits a pathogen, e.g. a virus.

Virulence. Capacity of a pathogen to induce a disease.

x. Basic number of chromosomes in a polyploid series.

Xenia. Effect of pollen on the embryo and endosperm. (*See* **Metaxenia.**)

X-linkage. Genes carried on the X-chromosome are said to be X-linked.

Zygote. The cell or organism resulting from the fusion of two gametes or their equivalents.

Zygotene. The stage of meiosis (part of prophase I) when the homologous chromosomes are associating side by side.

Bibliography

ABDALLA, M. M. F. and HERMSEN, J. G. T. (1971). The concept of breeding for uniform and differential resistance and their integration. *Euphytica* **21**, 351-61.

ABEL, W. O. (1972). Heterosis. *Z. PflZücht.* **67**, 45–52.

ABOU-EL-FITTOUH, H. A., RAWLINGS, J. O., and MILLER, P. A. (1969). Classification of environments to control genotype by environment interactions with an application to cotton. *Crop Sci.* **9**, 135–40.

ADAMS, S. E., JONES, R. A. C., and COUTTS, R. H. A. (1984). Occurrence of resistance-breaking strains of potato virus X in potato stocks in England and Wales. *Plant Path.* **33**, 435–7.

ALLARD, R. W. (1960). *Principles of plant breeding.* John Wiley, New York.

—— (1961). Relationship between genetic diversity and consistency of performance in different environments. *Crop Sci.* **1**, 127–33.

——, JAIN, S. K., and WORKMAN, P. L. (1968). The genetics of inbreeding plants. *Adv. Genet.* **14**, 55–131.

——, KAHLER, A. L., and WEIR, B. S. (1972). The effect of selection on esterase isozymes in a barley population. *Genetics, Princeton* **72**, 489–503.

ALTMAN, D. W. and BUSCH, R. H. (1984). Random intermating before selection in spring wheat. *Crop. Sci.* **24**, 1085–9.

ALVEY, N. G., BANFIELD, C. F., BAXTER, R. I., GOWER, J. C., KRZANOWSKI, W. J., LANE, P. W., LEECH, P. K., NELDER, J. A., PAYNE, R. W., PHELPS, K. M., ROGERS, C. E., ROSS, G. V. S., SIMPSON, H. R., TODD, A. D., WEDDERBURN, R. W. M., and WILKINSON, G. N. (1977). *GENSTAT: a general statistical program.* Rothamsted Experimental Station, Statistics Department.

AN, G., WATSON, B. D., STACHEL, S., GORDON, M. P., and NESTER, E. W. (1985). New cloning vehicles for transformation of higher plants. *EMBO J.* **4**, 277–84

ANDERSON, E. and BROWN, N. (1952). The history of the common maize varieties of the United States corn belt. *Agric. Hist.* **26**, 2–8.

ANDRUS, C. F. (1963). Plant breeding systems. *Euphytica* **12**, 205–28.

—— AND MCGILLIARD, L. D. (1975). Selection of dairy cattle for overall excellence. *J. Dairy Sci.* **58**, 1876–9.

ANTOINE, M. and NIESSING, J. (1984). Intron-less globin genes in the insect *Chironomus thummi thummi. Nature* **310**, 795–8.

ANTONOVICS, J. and WU, H. P. (1978). Twelve years of distinct yield trials of new Japonica rice varieties in Taiwan. I. Analysis, evaluation and recommendations. *Bot. Bull. Acad. sin., Shanghai* **19**, 145–70.

ARNDT, T. M. and RUTTAN, V. W. (1977). Valuing the productivity of agricultural research: problems and issues. In *Resource allocation and productivity in national and international agricultural research* (eds T. M. Arndt, D. G. Dalrymple, and V. W. Ruttan) pp. 3–25. University of Minnesota Press, Minneapolis.

ARNOLD, M. H. and BROWN, S. J. (1968). Variation in the host–parasite relationship of a crop disease. *J. agric. Sci., Camb.* **71**, 19–36.

——, INNES, N. L. and BROWN, S. J. (1976). Resistance breeding. In *Agricultural research for development: the Namulonge contribution* (ed. M. H. Arnold). Cambridge University Press.

ARUNACHALAM, V. (1977). Heterosis for characters governed by two genes. *J. Genet.* **63**, 15–24.

ATKINSON, A. C. (1969). The use of residuals as a concomitant variable. *Biometrika* **56**, 33–41.

AUERBACH, C. (1976). *Mutation research: problems, results and perspectives.* Chapman and Hall, London.

—— (1978). Forty years of mutation research: a pilgrim's progress. *Heredity, London.* **40**, 177–87.

—— and ROBSON, J. M. (1947). The production of mutations by chemical substances. *Proc. R. Soc. Edinb. B.* **62**, 271–83.

AYALA, F. and CAMPBELL, C. A. (1974). Frequency-dependent selection. *A. Rev. Ecol. Syst.* **5**, 115–38.

BABBEL, G. R. and SELANDER, R. K. (1974). Genetic variability in edaphically restricted and widespread plant species. *Evolution, Lancaster, Pa.* **28**, 619–30.

BAGWILL, R. E. (1981). The legal aspects of plant tissue culture and patents. *Env. Exp. Bot.* **21**, 383–7.

BAILEY, L. H. (1895). *Plant breeding.* Macmillan, New York.

BAILEY, N. T. J. (1961). *Introduction to the mathematical theory of genetic linkage.* Clarendon Press, Oxford.

BAKER, R. J. (1978). Issues in diallel analysis. *Crop Sci.* **18**, 533–6.

—— and MCKENZIE, R. I. H. (1967). Use of control plots in yield trials. *Crop Sci.* **7**, 335–7.

BANFIELD, C. F. (1978). Multivariate analysis in Genstat. *J. Statist. Comput. Simul.* **6**, 211–22.

BARNES, R. D. and SCHWEPPENHAUSER, M. A. (1978). *Pinus patula* (Schiede and Deppe) progeny tests in Rhodesia: genetic control of nursery traits. *Silvae Genetica* **27**, 173–216.

BARTLETT, M. S. (1937). Properties of sufficiency and statistical tests. *Proc. R. Soc. A.* **260**, 268–82.

—— (1978). Nearest neighbour models in the analysis of field experiments (with discussion). *J. R. Statist. Soc. B* **40**, 147–74.

—— and HALDANE, J. B. S. (1934). The theory of inbreeding in autotetraploids. *J. Genet.* **29**, 175–80.

—— and —— (1935). The theory of inbreeding with enforced heterozygosis. *J. Genet.* **31**, 327–40.

BARTON, K. A. and BRILL, W. J. (1983). Prospects in plant genetic engineering. *Science* **219**, 671–6.

BASSETT, M. J. and WOODS, F. E. (1978). A procedure for combining a quantitative inheritance study with the first cycle of a breeding program. *Euphytica* **27**, 295–303.

BECHOFER, R. E. (1970). An undesirable feature of a sequential multiple decision procedure for selecting the best one of several normal populations with a common unknown variance. *Biometrics* **26**, 347–50.

BECKER, W. A. (1984). *Manual of quantitative genetics* (4th edn). Pullman, Washington, Academic Enterprises.

BEHNKE, M. (1980). General resistance to late blight of *Solanum tuberosum* plants regenerated from callus resistant to culture filtrates of *Phytophthera infestans. Theor. Appl. Genet.* **58**, 19–26.

BELLIARD, G., PELLETIER, G. and FERRAULT, M. (1977). Fusion de protoplastes de *Nicotiana tabacum* à cytoplasmes différents: étude des hybrides cytoplasmiques néo-formés. *C. R. Acad. Sci. Ser. D.* **284**, 749–52.

BENNETT, J. H. (1976). Expectations for inbreeding depression on self-fertilization of tetraploids. *Biometrics* **32**, 449–52.

—— and BINET, F. E. (1956). Association between Mendelian factors with mixed selfing and random mating. *Heredity, Lond.* **10**, 51–5.

BERGER, E. (1976). Heterosis and the maintenance of enzyme polymorphism. *Am. Nat.* **110**, 823–39.

BHATT, G. M. (1977). Response to two-way selection for harvest index in two wheat (*Triticum aestivum* L.) crosses. *Aust. J. agric. Res.* **28**, 29–36.

BLACK, W., MASTENBROEK, C., MILLS, W. R. and PETERSON, L. C. (1953). A proposal for an international nomenclature of races of *Phytophthera infestans* and of genes controlling immunity in *Solanum demissum* derivatives. *Euphytica* **2**, 173–9.

BLAKESLEE, A. F. (1941). Effect of induced polyploidy in plants. *Am. Nat.* **75**, 117–35.

—— and BELLING, J. (1924). Chromosomal mutations in the Jimson weed, *Datura stramonium. J. Hered.* **15**, 195–206.

——, ——, FARNHAM, M. E., and BERGNER, A. D. (1922). A haploid mutant in the Jimson weed *Datura stramonium. Science, N. Y.* **55**, 646–7.

BLEASDALE, J. K. A. (1960). Studies in plant competition. *Symp. Br. Ecol. Soc.* **1**, 133–42.

—— (1967). Systematic designs for spacing experiments. *Exp. Agric.* **3**, 73–85.

—— (1982). The importance of crop establishment. *Ann. Appl. Biol.* **101**, 411–19.

BODMER, W. F. and FELSENSTEIN, J. (1967). Linkage and selection: theoretical analysis of the deterministic two locus random mating model. *Genetics, Princeton* **57**, 237–65.

BOERMA, H. R. and COOPER, R. L. (1975a). Comparison of three selection procedures for yield in soybeans. *Crop Sci.* **15**, 225–9.

—— and —— (1975b). Performance of pure lines obtained from superior-yielding heterogeneous lines in soybeans. *Crop Sci.* **15**, 300–2.

BOLAND, O. W. and WALCOTT, J. J. (1985). Levels of heterosis for yield and quality in an F_1 hybrid wheat. *Aust. J. agric. Res.* **36**, 545–52.

BOONE, D. M. and KEITT, G. W. (1975). *Venturia inaequalis* (Cke) Wint. XII. Genes controlling pathogenicity of wild-type lines. *Phytopathology* **47**, 403–9.

BORLAUG, N. D. (1958). The use of multilineal and composite varieties to control airborne epidemic disease of self-pollinated crop plants. In *Proc. First Int. Wheat Genet. Symp.*, pp. 12–29. University of Manitoba, Winnipeg.

—— (1983). Contribution of conventional plant breeding to food production. *Science, N. Y.* **219**, 689–93.

BOS, I. (1976). Selection for a qualitative trait in an autotetraploid crop. *Euphytica* **25**, 161–6.

—— (1977). Further arguments against intermating F_2 plants of a self-fertilized crop. *Euphytica* **26**, 33–46.

—— (1983a). Some remarks on honeycomb selection. *Euphytica* **32**, 329–35.

—— (1983b). About the efficiency of grid selection. *Euphytica* **32**, 885–93.

BOUKEMA, I. W. (1981). Races of *Cladosporium fulvum* Cke. (*Fulvia fulva*) and genes for resistance in the tomato (*Lycopersicon* Mill.) In *Genetics and breeding of tomato* (ed. J. Philouze) pp. 287–92, Versailles, INRA.

BOX, G. E. P. (1953). Non-normality and tests on variances. *Biometrika* **40**, 318–35.

—— (1954). The exploration and exploitation of response surfaces I. Some general considerations and examples. *Biometrics* **10**, 16–60.

—— and COX, D. R. (1964). An analysis of transformations. *J. R. statist. Soc. B.* **26**, 211–52.

—— and DRAPER, N. R. (1975). Robust designs. *Biometrika* **62**, 347–52.

—— and HUNTER, J. S. (1957). Multifactor experimental designs for exploring response surfaces. *Ann. math. Statist.* **28**, 195–241.

—— and Youle, P. V. (1955). Exploration and exploitation of response surfaces. *Biometrics* **11**, 287–332.

BOYD, W. J. R., GOODCHILD, N. A., WATERHOUSE, W. K., and SINGH, B. B. (1976). An analysis of climatic environments for plant-breeding purposes. *Aust. J. agric. Res.* **27**, 19–33.

—— and WADE, L. J. (1983). GA insensitivity as an aid to selection for high yield in wheat. In *Proc. Aust. Plant Breeding Conference*, Adelaide, South Australia, 14–18 February, 1983 (ed. C. J. Driscoll) pp. 323–4. Adelaide, Standing Committee in Agriculture.

BRADSHAW, A. D. (1975). Population structure and the effect of isolation and selection. In *Crop genetic resources for today and tomorrow* (eds. O. H. Frankel and J. G. Hawkes) pp. 37–51. Cambridge University Press.

BREAKWELL, E. J. and HUTTON, E. M. (1939). Cereal breeding and variety trials, at Roseworthy College, 1937–38. *J. Agric. S. A.* **42**, 632–41.

BREESE, E. L. (1969). The measurement and significance of genotype–environment interactions in grasses. *Heredity, Lond.* **24**, 27–44.

—— and MATHER, K. (1957). The organisation of polygenic activity within a chromosome in *Drosophila* I. Hair characters. *Heredity, Lond.* **11**, 373–98.

BRETTELL, R. I. S. and INGRAM, D. S. (1979). Tissue culture in the production of novel disease resistant crop plants. *Biol. Rev.* **54**, 329–45.

BREWIN, N. J. (1984). Hydrogenase and energy efficiency in nitrogen-fixing symbionts. In *Genes Involved in Microbe–Plant Interactions.* (eds. D. P. S. Verma and T. Hohn) pp. 179–203. Vienna, Springer-Verlag.

BRIDGES, C. B. (1935). Salivary chromosome maps with a key to the banding of the chromosomes of *Drosophila melanogaster. J. Hered.* **26**, 60–4.

BRIGGS, K. G. and SHEBESKI, L. H. (1968). Implications concerning the frequency of control plots in wheat-breeding nurseries. *Can. J. Plant Sci.* **48**, 149–53.

BRIGHT, S. W. J. and NORTHCOTT, D. H. (1975). A deficiency of hypoxanthine phosphoribosyl transferase in a sycamore callus resistant to azaguanine. *Planta* **128**, 79–89.

BRINKERHOFF, L. A. (1970). Variation in *Xanthomonas malvacearum* and its relation to control. *A. Rev. Phytopathol.* **8**, 85–110.

BROCK, R. D. (1965). Induced mutations affecting quantitative characters. In *The use of induced mutations in plant breeding.* [*Rad. Bot.* (suppl.) Vol. 5, pp. 505–13.] Pergamon Press, Oxford.

—— (1967). Quantitative variation in *Arabidopsis thaliana* induced by ionizing radiation. *Rad. Bot.* **7**, 193–203.

—— (1970). When to use mutations in plant breeding. In *Manual of mutation breeding.* Tech. Rep. Series No. 119, pp. 183–90. International Atomic Energy Agency, Vienna.

Brouwer, J. B. (1977). Developmental response of different hexaploid triticales to temperature and photoperiod. *Aust. J. exp. Agric. An. Husb.* **17**, 826–31.

Brown, A. H. D. (1971). Isozyme variation under selection in *Zea mays* L. *Nature, Lond.* **232**, 570–1.

—— (1975). Efficient experimental designs for the estimation of genetic parameters in plant populations. *Biometrics* **31**, 145–60.

—— (1978a). Enzyme polymorphisms in plant populations. *Theor. Pop. Biol.* **15**, 1–42.

—— (1978b). Isozymes, plant population genetic structure and genetic conservation. *Theor. appl. Genet.* **52**, 145–57.

—— and Allard, R. W. (1970). Estimation of the mating system in open-pollinated maize populations using isozyme polymorphisms. *Genetics, Princeton* **66**, 133–45.

——, Marshall, D. R., and Albrecht, L. (1974). The maintenance of alcohol dehydrogenase polymorphism in *Bromus mollis* L. *Aust. J. biol. Sci.* **27**, 545–59.

——, —— and Munday, J. (1976). Adaptedness of variants at an alcohol dehydrogenase locus in *Bromus mollis* L. (soft bromegrass). *Aust. J. biol. Sci.* **29**, 389–96.

——, Matheson, A. C., and Eldridge, K. G. (1975). Estimation of the mating system of *Eucalyptus obliqua* L'Hérit. by using allozyme polymorphisms. *Aust. J. Bot.* **23**, 931–49.

——, Nevo, E., and Zohary, D. (1977). Association of alleles at esterase loci in wild barley *Hordeum spontaneum* L. *Nature, Lond.* **268**, 430–1.

——, Zohary, D., and Nevo, E. (1978). Outcrossing rates and heterozygosity in natural populations of *Hordeum spontaneum* Koch in Israel. *Heredity, Lond.* **41**, 49–62.

Brown, W. V. (1972). *Textbook of cytogenetics.* C. V. Mosby, St. Louis.

Browning, J. A. and Frey, K. J. (1969). Multiline cultivars as a means of disease control. *A. Rev. Phytopathol.* **7**, 355–82.

Buiatti, M., Scala, A., Bettini, P., Nascari, G., Morpurgo, R., Bogani, P., Pellegrini, G., Gimelli, F., and Venturo, R. (1985). Correlations between *in vivo* resistance to *Fusarium* and *in vitro* response to fungal elicitors and toxic substances in Carnation. *Theor. Appl. Genet.* **70**, 42–7.

Bulmer, M. G. (1972). The genetic variability of polygenic characters under optimizing selection, mutation and drift. *Genet. Res.* **19**, 17–25.

—— (1980). *The theory of quantitative genetics.* Clarendon Press, Oxford.

Burdon, R. D. (1979). Generalization of multi-trait selection indices using information from several sites. *N. Z. For. J.* **9**, 145–52.

—— (1982a). Selection indices using information for the single-trait case. *Silvae Genetica* **31**, 81–5.

—— (1982b). The roles and optimal place of vegetative propagation in tree breeding strategies. In *Proc. IUFRO Joint Meeting of Working Parties on Genetics about Breeding Strategies including Multiclonal Varieties,* pp. 66–83. Staufenberg-Escherode, Lower Saxony For. Res. Inst.

—— and Namkoong, G. (1983). Short note: multiple populations and sublines.

Silvae Genetica **32**, 221–2.

BÜRGER, R. (1986). On the maintenance of genetic variation: global analysis of Kimura's continum-of-alleles model. *J. Math. Biol.* (in press).

BURNHAM, C. R. (1966). Cytogenetics in plant improvement. In *Plant breeding* (ed. K. J. Frey) pp. 139–87. Iowa State University Press, Ames.

BURR, B., EVOLA, S. V., BURR, F. A. and BECKMAN, J. S. (1983). The application of restriction fragment length polymorphism to plant breeding. In *Genetic engineering principles and methods,* Vol. 5 (eds J. K. Selow & A. Hollaender) pp. 45–59. New York, Plenum Press.

BURTON, J. W., STUBER, C. W. and MOLL, R. H. (1978). Variability of response to low levels of inbreeding in a population of maize. *Crop. Sci.* **18**, 65–8.

BUSBICE, T. G. (1969). Inbreeding in synthetic varieties. *Crop Sci.* **9**, 601–4.

—— (1970). Predicting yield of synthetic varieties. *Crop Sci.* **10**, 265–9.

——, GURGIG, R. Y., and COLLINS, H. B. (1975). Effect of selection for self-fertility and self-sterility in alfalfa and related characters. *Crop Sci.* **15**, 471–5.

BYTH, D. E. and CALDWELL, B. E. (1970). Effects of genetic heterogeneity within two soybean populations. II. Competitive responses and variance due to genetic heterogeneity of nine agronomic and chemical characters. *Crop Sci.* **10**, 216–20.

——, EISEMANN, R. C., and DE LACY, I. H. (1976). Two-way pattern analysis of a large data set to evaluate genotypic adaptation. *Heredity, Lond.,* **37**, 215–30.

CALDWELL, R. M. (1968). Breeding for general and for specific plant disease resistance. In *Proc. 3rd Int. Wheat Genet. Symp.,* pp. 263–72. Aust, Acad. Sci., Canberra.

CALLOW, J. A. (1977). Recognition, resistance and the role of plant lectins in host–parasite interactions. *Adv. Bot. Res.* **4**, 1–49.

CARLSON, P. S. (1973). The use of protoplasts for genetic research. *Proc. natn. Acad. Sci.,* U.S.A. **70**, 598–602.

——, SMITH, M. H., AND DEARING, D. (1972). Parasexual interspecific plant hybridization. *Proc. natn Acad. Sci., U.S.A.* **69**, 2292–4.

CARMER, S. G. (1977). Treatment designs to estimate optimum plant density for maximum corn grain yield. *Agron. J.* **69**, 803–7.

CARRASCO, A. E., MCGINNIS, W., GEHRING, W. J. and DE ROBERTIS, E. M. (1984). Cloning of an *X. laevis* gene expressed during early embryogenesis coding for a peptide region homologous to *Drosophila* homoeotic genes. *Cell* **37**, 409–14.

CARY, P. R. (1972). The residual effects of nitrogen, calcium and soil management on yield, fruit size and composition of citrus. *J. Hort. Sci.* **47**, 479–91.

CASALI, V. W. D. and TIGCHELAAR, E. C. (1975a). Breeding progress in tomato with pedigree selection and single seed descent. *J. Am. Soc. Hort. Sci.* **100**, 362–4.

—— and —— (1975b). Computer simulation studies comparing pedigree, bulk and single seed descent selection in self-pollinated populations. *J. Am. Soc. Hort. Sci.* **100**, 364–7.

CATEN, C. E. (1974). Intra-racial variation in *Phytophthora infestans* and adaptation to field resistance for potato blight. *Ann. Appl. Biol.* **77**, 259–70.

CERVANTES, T. S., GOODMAN, M., CASAS, E. D., and RAWLINGS, J. O. (1978). Use of genetic effects and genotype by environmental interactions for the classification of Mexican races of maize. *Genetics, Princeton* **90**, 339–48.

CHALEFF, R. S. (1980). Further characterization of picloram-tolerant mutants of

Nicotiana tabacum. *Theor. Appl. Genet.* **58**, 91–95.

—— (1983). Isolation of agronomically useful mutants from plant cell cultures. *Science, N.Y.* **219**, 676–82.

—— and Carlson, P. S. (1975). Higher plant cells as experimental organisms. In *Modification of the information content of plant cells* (eds R. Markham, D. R. Davies, D. A. Hopwood, and R. W. Horne) pp. 197–214. North-Holland, Amsterdam.

Chandraratna, M. F. and Sakai, K.-I. (1960). A biometrial analysis of matroclinous inheritance of grain weight in rice. *Heredity, Lond.* **14**, 365–73.

Chapman, V. and Riley, R. (1966). The allocation of the chromosomes of *Triticum aestivum* to the A and B genomes and evidence on genome structure. *Can. J. Genet. Cytol.* **8**, 57–63.

Chase, S. S. (1952). Production of homozygous diploids of maize from monoploids. *Agron. J.* **44**, 263–7.

Cheliak, W. M., Dancik, B. P., Morgan, K., Yeh, F. C. H., and Strobeck, C. (1985). Temporal variation of the mating system in a natural population of jack pine. *Genetics, Princeton* **109**, 569–84.

Chew, K. H. (1984). Alcohol dehydrogenase synthesis and waterlogging tolerance in maize. *Tropical Agriculture* **61**, 302–4.

Choo, T. M. and Kannenberg, L. W. (1979). Relative efficiencies of population improvement methods in corn: a simulation study. *Crop. Sci.* **19**, 179–85.

CIMMYT (1973). *Results of the Seventh International Spring Wheat Yield Nursery 1970–1971.* Mexico, D. F., Centro International de Mejoramiento de Maiz y Trigo.

Clarke, B. (1972). Density-dependent selection *Am. Nat.* **106,** 1–13.

Clarke, J. M. and McCaig, T. N. (1982). Evaluation of techniques for screening for drought resistance in wheat. *Crop Sci.* **22**, 503–6.

Cochran, W. G. and Cox, G. M (1950). *Experimental designs.* John Wiley, New York.

Cockerham, C. C. (1956). Effects of linkage on the covariance between relatives. *Genetics, Princeton* **41**, 138–41.

—— (1961). Implications of genetic variances in a hybrid breeding program. *Crop Sci.* **1**, 47–52.

—— (1963). Estimation of genetic variances. In *Statistical genetics and plant breeding* (eds W. D. Hanson and H. F. Robinson) pp. 53–94. NAS-NRC Publ. No. 982.

Coffman, C. B. and Gentner, W. A. (1977). Responses of greenhouse-grown *Cannabis sativa* L. to nitrogen, phosphorus, and potassium. *Agron. J.* **69**, 832–6.

Cohen, S. N. (1979). Experimental techniques and strategies for DNA cloning. In *Recombinant DNA and genetic experimentation* (eds J. Morgan, and W. J. Whelan) pp. 49–52. Oxford, Pergamon Press.

Compton, W. A. and Bahadur, K. (1977). Ten cycles of progress from modified ear-to-row selection in corn. *Crop Sci.* **17**, 378–80.

Comstock, R. E. and Moll, R. H. (1963). Genotype–environment interactions. In. *Statistical genetics and plant breeding* (eds W. D. Hanson and H. F. Robinson) pp. 164–96. NAS-NRS Publ. No. 982.

—— and Robinson, H. F. (1952). Estimation of average dominance of genes. In *Heterosis* (ed. J. W. Gowen) pp. 494–516. Iowa State University Press, Ames.

——, —— and Harvey, P. H. (1949). A breeding procedure designed to make maxium use of both general and specific combining ability. *Agron. J.* **41**, 360–7.

Connolly, V. and Wright-Turner, R. (1984). Induction of cytoplasmic male sterility into ryegrass *Lolium perenne. Theor. Appl. Genet.* **68**, 449–53.

Cornelius, P. L. and Dudley, J. W. (1974). Effects of inbreeding by selfing and full-sib mating in a maize population. *Crop Sci.* **14**, 815–9.

—— and —— (1976). Genetic variance components and predicted response to selection under selfing and full-sib mating in a maize population. *Crop Sci.* **16**, 333–9.

Cotterill, P. P. (1984). A plan for breeding radiata pine. *Silvae Genetica* **33**, 84–93.

—— and James, J. W. (1981). Optimising two stage independent culling selection in tree and animal breeding. *Theor. Appl. Genet.* **59**, 67–72.

Cox, T. S., House, L. R., and Frey, K. J. (1984). Potential of wild germplasm for increasing yield of grain sorghum. *Euphytica* **33**, 673–84.

Crane, M. B. and Lawrence, W. J. C. (1947). *The genetics of garden plants.* Macmillan, London.

Crawford, R. M. M. (1967). Alcohol dehydrogenase activity in relation to flooding tolerance in roots. *J. exp. Bot.* **18**, 458–64.

Cress, C. R. (1966). A comparison of recurrent selection systems. *Genetics, Princeton* **54**, 1371–9.

—— (1967). Reciprocal recurrent selection and modifications in simulated populations. *Crop Sci.* **7**, 561–7.

Crisp, P., Johnson, A. G., Ellis, P. R., and Hardman, J. A. (1977). Genetical and environmental interactions affecting resistance in radish to cabbage root fly. *Heredity, Lond.* **38**, 209–18.

Croughan, T. P., Stavarek, S. J., and Rains, D. W. (1981). *In vitro* development of salt-resistant plants. *Env. Exp. Bot.* **21**, 317–24.

Crow, J. F. (1952). Dominance and overdominance. In *Heterosis* (ed. J. W. Gowen) pp. 282–97. Iowa State University Press, Ames.

Cubillos, A. B. and Plaisted, R. L. (1976). Heterosis for yield in hybrids between *S. tuberosum* ssp. *tuberosum* and *S. tuberosum* ssp. *andigena. Am. Potato J.* **53**, 143–9.

Cunningham, E. P., Moen, R. A., and Gjedrem, T. (1970). Restriction of selection index. *Biometrics* **26**, 67–74.

Curnow, R. N. (1961). Optimal programmes for varietal selection (with discussion). *J. R. Statist. Soc. B* **23**, 282–318.

—— (1963). Sampling the diallel cross. *Biometrics* **19**, 287–306.

—— (1964). The effect of continued selection of phenotypic intermediates on gene frequency. *Genet. Res.* **5**, 341–53.

—— (1978). Selection within self-fertilizing populations. *Biometrics* **34**, 603–10.

Dalrymple, D. G. (1977). Evaluating the impact of international research on wheat and rice production in the developing nations. In *Resource allocation and productivity in national and international agricultural research* (eds T. M. Arndt, D. G. Dalrymple, and V. W. Ruttan) pp. 171–209. University of Minnesota Press, Minneapolis.

Daniel, C. (1976). *Application of statistics to industrial experimentation.* Wiley-Interscience, New York.

Darlington, C. D. (1971). *Evolution of man and society.* Allen and Unwin, London.

DARRAH, L. L., EBERHART, S. A., and PENNY, L. H. (1978). Six years of maize selection in 'Kitale Synthetic II', 'Ecuador 573', and 'Kitale Composite A' using methods of the comprehensive breeding system. *Euphytica* **27**, 191–204.

DARWIN, C. R. (1876). *The effects of cross and self fertilization in the vegetable kingdom.* John Murray, London.

DAVIDSON, B. R. and MARTIN, B. R. (1968). *Experimental research and farm production.* Agricultural Economics Research Report No. 7, University of Western Australia.

DAVIES, D. R. (1977). Creation of new models for crop plants and their use in plant breeding. *Appl. Biol.* **2**, 87–127.

DAVIS, J. H. C. and EVANS, A. M. (1977). Selection indices using plant type characteristics in Navy beans (*Phaseolus vulgaris* L.). *J. agric. Sci., Camb.* **89**, 341–8.

DAY, P. R. (1951). Mutation to virulence in *Cladosporium fulvum. Nature, Lond.* **179**, 1141–2.

—— (1956). Race names of *Cladosporium fulvum. Tomato Genet. Co-op. Rep.* **6**, 13–14.

—— (1960). Variation in phytopathogenic fungi. *A. Rev. Microbiol.* **14**, 1–16.

—— (1974). *The genetics of host–pathogen interaction.* W. H. Freeman, San Francisco.

—— (1977). Plant genetics: increasing crop yield. *Science* **197**, 1334–9.

DAY, R. H. (1965). Probability distributions of field crop yields. *J. Farm Econ.* **47**, 713–41.

DEKKER, J. H., MEGGITT, W. F., and PUTNAM, A. R. (1983). Experimental methodologies to evaluate allelopathic plant interactions: the *Abutilon theophrasti-Glycine max* model. *J. Chem. Ecol.* **9**, 945–81.

DE LIOCOURT, F. (1898). De l'amenagement des sapinieres. *Bull. Soc. for. Franche-Comté.* 396–409.

DE NETTANCOURT, D. (1969). Radiation effects on the one locus gametophytic system of self-incompatibility in higher plants (a review). *Theor. appl. Genet.* **39**, 187–96.

—— (1977). *Incompatibility in Angiosperms.* Springer-Verlag, Berlin.

DE PAUW, R. M. and SHEBESKI, L. H. (1973). An evaluation of an early generation yield-testing procedure in *Triticum aestivum. Can. J. Plant Sci.* **53**, 465–70.

DESSUREAUX, L. (1959). Introduction to the autotetraploid diallel. *Can. J. Genet. Cytol.* **1**, 94–101.

DEVINE, T. E. and WEBER, D. F. (1977). Genetic specificity of nodulation. *Euphytica* **26**, 527–35.

DEWEY, D. R. (1966). Inbreeding depression in diploid, tetraploid, and hexaploid crested wheatgrass. *Crop Sci.* **6**, 144–7.

—— (1969). Inbreeding depression in diploid and induced-autotetraploid crested wheatgrass. *Crop Sci.* **9**, 592–5.

—— (1977). A method of transferring genes from tetraploid to diploid crested wheatgrass. *Crop Sci.* **17**, 803–5.

—— and LU, K. H. (1959). A correlation and path-coefficient analysis of components of crested wheatgrass seed production. *Agron. J.* **51**, 515–18.

DHILLON, B. S. and SINGH, J. (1978). Evaluation of circulant partial diallel crosses in maize. *Theor. appl. Genet.* **52**, 29–37.

DOBZHANSKY, T. (1952). Nature and origin of heterosis. In *Heterosis* (ed. J. W.

Gowen) pp. 330–5. Iowa State University Press, Ames.
Donald, C. M. (1963). Competition among crop and pasture plants. *Adv. Agron.* **15**, 1–118.
—— (1968). The breeding of crop ideotypes. *Euphytica* **17**, 385–403.
—— (1978). The ideotype approach in plant breeding. In *Australian wheat breeding today*, a treatise of the 2nd assembly of The Wheat Breeding Society of Australia (ed. G. J. Hollamby) p. 41. Roseworthy Agricultural College.
—— and Hamblin, J. (1976). The biological yield and harvest index of cereals as agronomic and plant breeding criteria. *Adv. Agron.* **28**, 361–405.
Dooner, H. K. (1979). Identification of an R-locus region that controls the tissue specificity of anthocyanin formation of maize. *Genetics, Princeton* **93**, 703–10.
Dorè, C. (1975). La multiplication clonale de l'asperge (*Asparagus officinalis* L.) par culture *in vitro*: son utilisation en sélection. *Ann. Amél. Pl.* **25**, 201–24.
Dowker, B. D., Fennell, J. F. M., Jackson, J. C., and Arthey, V. D. (1975). Genotypic and environmental variation in some colour characteristics of carrots. *Ann. appl. Biol.* **81**, 377–83.
Downton, W. J. S. (1973). *Amaranthus edulis:* a high lysine grain amaranth. *Wld Crops* **25**, 20–1.
—— (1977). Influence of rootstocks on the accumulation of chloride, sodium and potassium in grapevines. *Aust. J. agric. Res.* **28**, 879–89.
Driscoll, C. J. (1966). Gene-centromere distance in wheat by aneuploid F_2 observations. *Genetics, Princeton* **54**, 131–5.
—— (1972). XYZ system of producing hybrid wheat. *Crop Sci,* **12**, 516–17.
—— (1978). *Mapping alien segments.* International Conference on Cytogenetics and Crop Improvement, Varanasi, India, Feb. 1978.
—— and Anderson, L. M. (1967). Cytogenic studies of Transec—a wheat–rye translocation line. *Can. J. Genet. Cytol.* **9**, 375–80.
—— and Barlow, K. K. (1976). Male sterility in plants: induction, isolation and utilization. In *Induced mutations in cross-breeding,* pp. 123–31. International Atomic Energy Agency, Vienna.
—— and Bielig, L. M. (1968). Mapping of the Transec wheat–rye translocation. *Can. J. Gent. Cytol.* **10**, 421–5.
——, McLean, M. A., Napier, K. V. and Johnson, R. J. (1983). Triticale breeding, interstate trials and quality evaluation. *Proc. Aust. Pl. Breed. Conf.*, Adelaide, Feb. 1983.
—— and Sears, E. R. (1965). Mapping of a wheat–rye translocation. *Genetics, Princeton.* **51**, 439–43.
Dubetz, S. and Bole, J. R. (1973). Effects of moisture stress at early heading and of nitrogen fertilizer on three spring wheat cultivars. *Can. J. Plant. Sci.* **53**, 1–5.
Dudley, J. W. (1977). 76 generations of selection for oil and protein percentage in maize. In *Proc. Int. Conf. Quant. Genet. 1976* (eds E. Pollack, O. Kempthorne, and T. B. Bailey) pp. 459–73. Iowa State University Press, Ames.
—— (1984). A method for identifying populations containing favourable alleles not present in elite germplasm. *Crop Sci.* **24**, 1053–4.
—— and Lambert, R. J. (1969). Genetic variability after 65 generations of selection in Illinois high oil, low oil, high protein, and low protein strains of *Zea mays* L. *Crop Sci.* **9**, 179–81.
——, Hill, R. R., and Hanson, C. H. (1963). Effects of seven cycles of recurrent phenotypic selection on means and genetic variances of several

characteristics in two pools of alfalfa germ plasm. *Crop Sci.* **3**, 543–6.

Dunbier, M. W. and Bingham, E. T. (1975). Maximum heterozygosity in alfalfa: results using haploid-derived autotetraploids. *Crop Sci.* **15**, 527–31.

Dyke, G. V., George, B. J., Johnston, A. E., Poulton, P. R. and Todd, A. D. (1982). The Broadbalk Wheat Experiment 1968–1978: yields and plant nutrients in crops grown continuously and in rotation. *Annual Report* Part 2, pp. 5–44. Rothamsted Experimental Station.

—— and Shelley, C. F. (1976). Serial designs balanced for effects of neighbours on both sides. *J. agric. Sci., Camb.* **87**, 303–5.

Eden, T. and Fisher, R. A. (1927). Studies in crop variation IV. The experimental determination of the value of top dressing with cereals. *J. agric. Sci., Camb.* **17**, 548–62. [Reprinted in *The collected papers of R. A. Fisher* (ed. J. H. Bennett). University of Adelaide.]

—— and —— (1929). Studies in crop variation VI. Experiments on the response of the potato to potash and nitrogen. *J. agric. Sci., Camb.* **19**, 201–13. [Reprinted in *The collected papers of R. A. Fisher* (ed. J. H. Bennett). University of Adelaide.]

e Gama, E. E. G. and Hallauer, A. R. (1977). Relation between inbred and hybrid traits in maize. *Crop Sci.* **17**, 703–6.

Eldridge, K. G. (1975). Eucalyptus species. In *Seed Orchards*, pp. 134–9. Forestry Commission Bulletin No. 54. London, HMSO.

—— (1978a). Genetic improvement of eucalypts. *Silvae Genetica* **27**, 205–9.

—— (1978b). Seed collections in California. *CSIRO Division of Forest Research Annual Report*, pp. 9–17.

—— (1984). Breeding trees for fuelwood. In *Procedings of the 15th International Congress of Genetics* (New Delhi 1983), Vol. 4 (eds. V. L. Chopra, B. C. Joshi, R. P. Sharma, and H. C. Bansal), pp. 339–49, Oxford & IBP, Bombay.

Ellerström, S. and Hagberg, A. (1967). Monofactorial heterosis in autotetraploid barley. *Hereditas* **57**, 319–26.

Ellingboe, A. H. (1982). Genetical aspects of active defence. In *Active defence mechanisms in plants* (ed. R. K. S. Wood) pp. 179–92. New York, Plenum.

Elliot, F. C. (1958). *Plant breeding and cytogenetics.* McGraw-Hill, New York.

Ellis, P. R. and Hardman, J. A. (1975). Laboratory methods for studying non-preference resistance to cabbage root fly in cruciferous crops. *Ann. appl. Biol.* **79**, 253–64.

Ellis, R. H. and Roberts, E. H. (1981). The quantification of ageing and survival in orthodox seeds. *Seed Sci. Technol.* **9**, 373–409.

El-Sayed Osman, H. and Robertson, A. (1968). The introduction of genetic material from inferior into superior strains. *Genet. Res.* **12**, 221–36.

El-Zeftawi, B. M. and Thornton, I. R. (1978). Varietal and rootstock effects on mandarin quality. *Aust. J. exp. Agric. Anim. Husb.* **18**, 597–602.

Embleton, T. W., Jones, W. W., Labanauskas, C. K., and Reuther, W. (1973). Leaf analysis as a diagnostic tool and guide to fertilization. In *The citrus industry* (ed. W. Reuther) Vol. 3, pp. 183–210. University of California Press, Berkeley.

Emery, D. A. and Wynne, J. C. (1976). Systematic selection for increased fruit yield in populations derived from hybridization only, F_1 irradiation, and hybridization following parental irradiation in peanuts (*Arachis hypogaea* L.). *Env. Exp. Bot.* **16**, 1–8.

England, F. (1977). Response to family selection based on replicated trials. *J.*

agric. Sci., Camb. **88**, 127–34.

Engledow, F. L. and Wadham, S. M. (1923, 1924a, 1924b, 1924c). Investigations on yield in the cereals. *J. Agric. Sci., Camb.* **13**, 390–439; **14**, 66–98; **14**, 287–324; **14**, 325–45.

Eskes, A. B. and Carvalho, A. (1983). 6. Variation for incomplete resistance to *Hemileia vastatrix* in *Coffea arabica. Euphytica* **32**, 625–37.

Evenson, R. E. (1977a). Cycles in research productivity in sugarcane, wheat, and rice. In *Resource allocation and productivity in national and international agricultural research* (eds T. M. Arndt, D. G. Dalrymple, and V. W. Ruttan) pp. 209–31. University of Minnesota Press, Minneapolis.

—— (1977b). Comparative evidence on returns to investment in national and international research institutions. In *Resource allocation and productivity in national and international agricultural research* (eds T. M. Arndt, D. G. Dalrymple, and V. W. Ruttan) pp. 237–64. University of Minnesota Press, Minneapolis.

Ewens, W. J. (1972). The sampling theory of selectively neutral alleles. *Theor. pop. Biol.* **3**, 87–112.

—— and Thomson, G. (1977). Properties of equilibria in multi-locus genetic systems. *Genetics, Princeton* **87**, 807–19.

Falconer, D. S. (1952). The problem of environment and selection. *Am. Nat.* **86**, 293–8.

—— (1965). The inheritance of liability to certain diseases, estimated from the incidence among relatives. *Ann. hum. Genet.* **29**, 51–71.

—— (1982). *Introduction to quantitative genetics* (2nd edn). Oliver and Boyd, Edinburgh. (1st edn 1960.)

Fang, T.-C., Tai, C.-H., Oü, Y. L., Tsuei, C.-C., and Chen, T.-C. (1978). Some genetic observations on the monoploid breeding of *Laminaria japonica. Scientia sin.* **21**, 400–7.

Farrer, W. (1898). The making and improvement of wheats for Australian conditions. *Agric. Gaz. N.S.W.* **9**, 131–68.

Fasoulas, A. C. (1973). *A new approach to breeding superior varieties.* Aristotelian University of Thessaloniki Publ. No. 3.

—— and Allard, R. W. (1962). Non-allelic gene interactions in the inheritance of quantitative characters in barley. *Genetics, Princeton* **47**, 899–907.

Federer, W. T. (1955). *Experimental design.* Macmillan, New York.

Feeny, P. (1975). Biochemical coevolution between plants and their insect herbivores. In *Coevolution of animals and plants.* Symposium V, First International Congress Systematic and Evolutionary Biology (eds L. E. Gilbert and P. H. Raven) pp. 3–19. University of Texas Press, Austin.

Feldman, M. W., Franklin, I. R., and Thomson, G. (1974). Selection in complex genetic systems. I. The symmetric equilibria of the three-locus deterministic viability model. *Genetics, Princeton* **76**, 135–62.

Ferguson, J. H. A. (1962). Random variability in horticultural experiments. *Euphytica* **11**, 213–20.

Ferris, S. D. and Whitt, G. S. (1977). Loss of duplicate gene expression after polyploidisation. *Nature, Lond.* **265**, 258–60.

Feyt, H. (1976). Étude critique de l'analyse des croisements diallèles au moyen de la simulation. *Ann. Amél. Pl.* **26**, 173–93.

Finlay, K. W. and Wilkinson, G. N. (1963). The analysis of adaptation in a plant breeding programme. *Aust. J. agric. Res.* **14**, 742–54.

Finney, D. J. (1956). The consequences of selection for a variate subject to errors of measurement. *Rev. Inst. Inter. Stat.* **24**, 1–10.
—— (1958). Statistical problems of plant selection. *Bull. Inst. Inter. Stat.* **36**, 246–68.
—— (1961). The transformation of a distribution under selection. *Sankhyā A* **23**, 309–24.
—— (1962). Genetic gains under three methods of selection. *Genet. Res.* **3**, 417–23.
—— (1966). An experimental study of certain screening procedures. *J. R. Statist. Soc. B* **28**, 88–109.
—— (1984). Improvement by planned multistage selection. *J. Am. Statist. Assoc.* **79**, 501–9.
Fischer, R. A. and Hille Ris Lambers, D. (1978). Effect of environment and cultivar on source limitation to grain weight in wheat. *Aust. J. agric. Res.* **29**, 443–58.
—— and Wood, J. T. (1979). Drought resistance in spring wheat cultivars. III. Yield association with morpho-physiological traits. *Aust. J. agric. Res.* **30**, 1001–20.
Fisher, R. A. (1918). On the correlation between relatives on the supposition of Mendelian inheritance. *Trans. R. Soc. Edinb.* **52**, 399–433. [Reprinted in *The collected papers of R. A. Fisher* (ed. J. H. Bennett). University of Adelaide.]
—— (1921a). Studies in crop variation I. An examination of the yield of dressed grain from Broadbalk. *J. agric. Sci., Camb.* **11**, 107–35. [Reprinted in *The collected papers of R. A. Fisher* (ed. J. H. Bennett). University of Adelaide.]
—— (1921b). On the 'probable error' of a coefficient of correlation deduced from a small sample. *Metron, Rovigo* **1**, 3–32. [Reprinted in *The collected papers of R. A. Fisher* (ed. J. H. Bennett) University of Adelaide.]
—— (1925). *Statistical methods for research workers.* Oliver and Boyd, Edinburgh.
—— (1930). *The genetical theory of natural selection.* Clarendon Press, Oxford.
—— (1936). The use of multiple measurements in taxonomic problems. *Ann. Eugen.* **7**, 179–89. [Reprinted in *The collected papers of R. A. Fisher* (ed. J. H. Bennett). University of Adelaide.]
—— (1941). Average excess and average effect of a gene substitution. *Ann. Eugen.* **11**, 53–63. [Reprinted in *The collected papers of R. A. Fisher* (ed. J. H. Bennett). University of Adelaide.]
—— (1942). The polygene concept. *Nature, Lond.* **150**, 154. [Reprinted in *The collected papers of R. A. Fisher* (ed. J. H. Bennet). University of Adelaide.]
—— (1947). The theory of linkage in polysomic inheritance. *Phil. Trans. R. Soc. B* **233**, 55–87. [Reprinted in *The collected papers of R. A. Fisher* (ed. J. H. Bennett) University of Adelaide.]
—— (1949). *The theory of inbreeding* (2nd edn. 1965). Oliver and Boyd, Edinburgh.
——, Immer, F. R. and Tedin, O. (1932). The genetical interpretation of statistics of the third degree in the study of quantitative inheritance. *Genetics, Princeton* **17**, 107–24. [Reprinted in *The collected papers of R. A. Fisher* (ed. J. H. Bennett). University of Adelaide.]
—— and Mackenzie, W. A. (1923). Studies in crop variation. II. The manurial response of different potato varieties. *J. agric. Sci., Camb.* **13**, 311–20. [Reprinted in *The collected papers of R. A. Fisher* (ed. J. H. Bennett).

University of Adelaide.]
—— and Yates, F. (1963). *Statistical tables for biological, agricultural and medical research* (6th edn). Oliver and Boyd, Edinburgh.
Flavell, R. B. (1981). The analysis of plant genes and chromosomes by using DNA cloned in bacteria. *Phil. Trans. R. Soc.* **B292**, 579–88.
—— (1984). DNA transposition—a major contributor to plant chromosome structure. *BioEssays* **1**, 21–2.
Fleischer, R. C., Johnston, R. F. and Klitz, W. J. (1983). Allozymic heterozygosity and morphological variation in house sparrows. *Nature, Lond.* **304**, 628–30.
Flor, H. (1942). Inheritance of pathogenicity in *Melampsora lini. Phytopathology* **32**, 653–69.
—— (1946). Genetics of pathogenicity in *Melampsora lini. J. agric. Res.* **73**, 335–57.
—— (1947). Inheritance of reaction to rust in flax. *J. agric. Res.* **74**, 241–62.
—— (1956). The complementary genic systems in flax and flax rust. *Adv. Genet.* **8**, 29–54.
Fonseca, S. and Patterson, F. L. (1968). Yield component heritabilities and inter-relationships in winter wheat (*Triticum aestivum* L.). *Crop Sci.* **8**, 614–17.
Ford, E. B. (1945). Polymorphism. *Biol. Rev.* **20**, 73–88.
Forkmann, G. (1977). Die Simulation quantitativer Merkmale durch Gene mit biochemisch definierbarer Wirkung VIII. Untersuchungen über das Absorptionsverhalten der Anthocyane. *Theor. appl. Genet.* **49**, 43–8.
—— and Seyffert, W. (1972). Die Simulation quantitativer Merkmale durch Gene mit biochemisch definierbarer Wirkung V. Untersuchungen zur Messmethodik. *Theor. appl. Genet.* **42**, 279–87.
—— and —— (1977). Simulation of quantitative characters by genes with biochemically definable action. VI. Modifications of a simple model. *Genetics, Princeton* **85**, 557–72.
Fox, P. N. (1977). Measurement of genotype x environment interaction in wheat yield. B. Agric. Sci. Thesis, Univesity of Adelaide.
—— and Rosielle, A. A. (1982). Reference sets of genotypes and selection for yield in unpredictable environments. *Crop Sci.* **22**, 1171–5.
Frakes, R. V., Davis, R. L., and Patterson, F. L. (1961). The breeding behaviour of yield and related variables in alfalfa. II. Association between characters. *Crop Sci.* **1**, 207–9.
Frankel, O. H. (1947). The theory of plant breeding for yield. *Heredity, Lond.* **1**, 109–20.
—— (1975). Genetic resources centres—a co-operative global network. In *Crop genetic resources for today and tomorrow* (eds O. H. Frankel and J. G. Hawkes) pp. 473–81. Cambridge University Press.
—— and Hawkes, J. G. (1975). *Crop genetic resources for today and tomorrow.* Cambridge University Press.
Frankham, R. (1980). Origin of genetic variation in selection lines. In *Selection experiments in laboratory and domestic animals* (ed. A. Robertson) pp. 56–80. Slough, Commonwealth Agricultural Bureaux.
Franklin, I. R. (1977). The distribution of the proportion of the genome which is homozygous by descent in inbred individuals. *Theor. Pop. Biol.* **11**, 60–80.
—— and Lewontin, R. C. (1970). Is the gene the unit of selection? *Genetics, Princeton* **65**, 707–34.

FREELING, M. (1973). Simultaneous induction by anaerobiosis or 2,4-D or multiple enzymes specified by two unlinked genes: differential Adh_1-Adh_2 expression in maize. *Mol. Gen. Genet.* **127**, 215–27.

—— (1974). Dimerization of multiple maize ADHs studied *in vivo* and *in vitro*. *Biochem. Genet.* **12**, 407–17.

—— and SCHWARTZ, D. (1973). Genetic relationships between the multiple alcohol dehydrogenases of maize. *Biochem. Genet* **8**, 27–36.

FREEMAN, G. H. and DOWKER, B. D. (1973). The analysis of variation within and between environments. *Heredity, Lond.* **30**, 97–109.

FREI, O. M., STUBER, C. W., and GOODMAN, M. M. (1986a). Isozyme markers as predictors of performance in single crosses. (Submitted to *Crop Sci.*)

——, —— and —— (1986b). Yield improvement in a corn population by selection for individual isozymes. (To be submitted).

FREY, K. J. (1976). Plant breeding in the seventies: useful genes from wild plant species. *Egypt. J. Genet. Cytol.* **5**, 460–82.

FROUSSIOS, G. (1970). Genetic diversity and agricultural potential in *Phaseolus vulgaris*. L. *Exp. Agric.* **6**, 129–41.

FUERST, E. P. and PUTNAM, A. R. (1983). Separating the competitive and allelopathic components of interference: theoretical principles. *J. Chem. Ecol.* **9**, 937–44.

GALE, M. D. (1977). Genetic analysis and the control of development of height. 184th Genet, Soc. Meeting, Cambridge, July.

—— and MARSHALL, G. A. (1975). The nature and genetic control of gibberellin insensitivity in dwarf wheat grain. *Heredity, Lond.* **35**, 55–65.

GALLAIS, A. (1967). Moyennes de populations tetraploïds. *Ann. Amél. Pl.* **17**, 215–27.

—— (1968). Etude théorique et comparée de la vigueur de differentes structures variétales chez les plantes allogames autotétraploïds. *Ann. Amél Pl.* **18**, 99–124.

—— (1973). Sélection pour plusiers caractères. Synthèse critique et généralisation. *Ann. Amél. Pl.* **23**, 183–208.

—— (1974). Covariances between arbitrary relatives with linkage and epistasis in the case of linkage disequilibrium. *Biometrics* 30, 429–46.

—— (1975). Prévision de la vigeur et sélection de parentes d'une variété synthétique. *Ann. Amél. Pl.* **25**, 233–64.

—— (1976a). Sur la signification de l'aptitude générale a la combinaison. *Ann. Amél. Pl.* **26**, 1–13.

—— (1976b). Development and application of prediction formulae for synthetics. *Ann. Amél. Pl.* **26**, 623–8.

—— (1976c). Effects of competition on means, variances and covariances in quantitative genetics with an application to general combining ability selection. *Theor. appl. Genet.* **47**, 189–95.

—— (1976d). The use of heterosis in autotetraploid cross-fertilized plants with some applications to lucerne and cocksfoot. *Ann. Amél. Pl.* **26**, 639–46.

—— (1977). Amélioration des populations, méthodes de sélection et création de variétés I. Synthèse sur les problèmes généraux et sur les bases théoriques pour la sélection récurrente intrapopulation. *Ann. Amél. Pl.* **27**, 281–329.

—— (1984). An analysis of heterosis vs inbreeding effects with an autotetraploid cross-fertilized plant: *Medicago sativa* L. *Genetics, Princeton* **106**, 123–37.

GALTON, F. (1874). *English men of science, their nature and nurture.* Macmillan,

London.

GARDNER, C. O. (1961). An evaluation of effects of mass selection and seed irradiation with thermal neutrons on yield of corn. *Crop Sci.* **1**, 241–5.

—— (1963). Estimates of genetic parameters in cross-fertilizing plants and their implication in plant breeding. In *Statistical genetics and plant breeding* (eds W. D. Hanson and H. F. Robinson) pp. 225–52. NAS-NRC Publ. No. 982.

GEDGE, D. L., FEHR, W. R., and COX, D. F. (1978). Influence of intergenotypic competition on seed yield of heterogeneous soybean lines. *Crop Sci.* **18**, 233–6.

GEIGER, H. H. and WAHLE, G. (1978). Struktur der Heterosis von Komplexmerkmalen bei Winterroggen-Einfachhybriden. *Z. PflZücht.* **80**, 198–210.

GENGENBACH, B. G., GREEN, C. E., and DONOVAN, C. M. (1977). Inheritance of selected pathotoxin resistance in maize plants regenerated from cell cultures. *Proc. natn Acad. Sci. U.S.A.* **74**, 5113–17.

GERLACH, W. L., LÖRZ, H., SACHS, M. M., LLEWELLYN, D., PRYOR, A. J., DENNIS, E. S. and PEACOCK, W. J. (1983). The alcohol dehydrogenase genes of maize: a potential gene transfer system in plants. In *Manipulation and expression of genes in eukaryotes* (eds. P. Nagley, A. W. Linnane, W. J. Peacock, and J. A. Pateman) pp. 213–20. Academic Press, Sydney.

GILBERT, N. E. (1958). Diallel cross in plant breeding. *Heredity, Lond.* **12**, 477–92.

—— (1961). Polygene analysis. *Genet Res.* **2**, 96–105.

—— (1967). Additive combining abilities fitted to plant breeding data. *Biometrics* **23**, 45–9.

GILLESPIE, J. H. (1984). Pleiotropic overdominance and the maintenance of genetic variation in a quantitative character. *Genetics, Princeton* **107**, 321–30.

GILLIS, P. R. and RATKOWSKY, D. A. (1978). The behaviour of estimators of the parameters of various yield-density relationships. *Biometrics* **34**, 191–8.

GOFF, F. G. and WEST, D. (1975). Canopy-understory interaction effects on forest population structure. *Forest Sci.* **21**, 98–108.

GOMEZ, K. A. and GOMEZ, A. A. (1976). *Statistical procedures for agricultural research with emphasis on rice.* International Rice Research Institute, Los Banos.

GOODCHILD, N. A. and BOYD, W. J. R. (1975). Regional and temporal variations in wheat yield in Western Australia and their implications in plant breeding. *Aust. J. agric. Res.* **26**, 209–17.

GOODMAN, M. M. and STUBER, C. W. (1983). Maize. In *Isozymes in plant genetics and breeding* Part B (eds S. D. Tanksley and T. J. Orton) pp. 1–33, Amsterdam, Elsevier.

GORDON, G. H. (1979). Computer applications in plant breeding and genetics. Ph.D. Thesis, Waite Agricultural Research Institite, University of Adelaide.

GOTTLIEB, L. D. (1975). Allelic diversity in the outcrossing annual plant *Stephanomesia exigua* spp. *Carotifera* (Compositae). *Evolution, Lancaster, Pa.* **29**, 213–25.

GOTTSCHALK, W. (1976). Monogenic heterosis. In *Induced Mutation in Cross-Breeding,* pp. 189–97. International Atomic Energy Agency, Vienna.

—— and WOLFF, G. (1983). *Induced mutations in plant breeding.* Berlin, Springer.

GOULDEN, C. H. (1939). Problems in plant selection. In *Proc. Seventh Int. Genet. Congr.* (ed. R. C. Punnett) pp. 132–3. Cambridge University Press.

GOUR, V. K. and SINGH, C. B. (1977). Influence of *Aestivum* wheat cytoplasm on hexaploid Triticale. *Can. J. Genet. Cytol.* **19**, 187–8.

GREGORY, W. C. (1966). Mutation breeding. In *Plant breeding* (ed. K. J. Frey) pp. 189–218. Iowa State University Press, Ames.

GRIFFING, B. (1956a). A generalized treatment of the use of diallel crosses in quantitative inheritance. *Heredity, Lond.* **10**, 31–50.

—— (1956b). Concept of general and specific combining ability in relation to diallel crossing systems. *Aust. J. biol. Sci.* **9**, 463–93.

—— (1960). Theoretical consequences of truncation selection based on the individual phenotype. *Aust. J. biol. Sci.* **13**, 307–43.

—— (1962a). Consequences of truncation selection based on combination of individual performance and general combining ability. *Aust. J. biol. Sci.* **15**, 333–51.

—— (1962b). Prediction formulae for general combining ability selection methods utilizing one or two random-mating populations. *Aust. J. biol. Sci.* **15**, 650–65.

—— (1963). Comparisons of potentials for general combining ability selection methods utilizing one or two random-mating populations. *Aust. J. biol. Sci.* **16**, 838–62.

—— (1967). Selection in reference to biological groups. I. Individual and group selection applied to populations of unordered groups. *Aust. J. biol. Sci.* **20**, 127–39.

—— (1968a). Selection in reference to biological groups. II. Consequences of selection in groups of one size when evaluated in groups of a different size. *Aust. J. biol. Sci.* **21**, 1163–70.

—— (1968b). Selection in reference to biological groups. III. Generalized results of individual and group selection in terms of parent offspring covariances. *Aust. J. biol. Sci.* **21**, 1171–8.

—— (1969). Selection in reference to biological groups. IV. Application of selection index theory. *Aust. J. biol. Sci.* **22**, 131–42.

—— (1975). Efficiency changes due to use of double-haploids in recurrent selection methods. *Theor. appl. Genet.* **46**, 367–86.

—— (1976a). Selection in reference to biological groups. V. Analysis of full-sib groups. *Genetics, Princeton* **82**, 703–22.

—— (1976b). Selection in reference to biological groups. VI. Use of extreme forms of non-random groups to increase selection efficiency. *Genetics, Princeton* **82**, 723–31.

—— and LANGRIDGE, J. (1963). Phenotypic stability of growth in the self-fertilized species *Arabidopsis thaliana*. In *Statistical genetics and plant breeding* (eds W. D. Hanson and H. F. Robinson) pp. 368–94. NAS-NRC Publ. No. 982.

GROTH, J. V. (1976). Multilines and 'super races': a simple model. *Phytopathology* **66**, 937–9.

GUPTA, P. K. and PRIYADARSHAN, P. M. (1982). Triticale: present status and future prospects. *Adv. Genet.* **21**, 255–345.

GUSTAFSSON, A. (1946). The effect of heterozygosity on variability and vigour. *Hereditas* **32**, 263–86.

—— (1947). Mutations in agricultural plants. *Hereditas* **33**, 1–100.

HALDANE, J. B. S. (1927). A mathematical theory of natural and artificial selection. VII. Selection intensity as a function of mortality rate. *Proc. Camb. phil. Soc. biol. Sci.* **27**, 131–6.

—— (1932). *The causes of evolution.* Longman Green, London.
—— (1954). *The biochemistry of genetics.* George Allen and Unwin, London.
—— (1956). The conflict between inbreeding and selection. 1. Self fertilisation. *J. Genet.* **54**, 56–63.
—— and JAYAKAR, S. D. (1963). Polymorphism due to selection of varying direction. *J. Genet.* **58**, 237–42.
HALLAUER, A. R. and SEARS, J. H. (1973). Changes in quantitative traits associated with inbreeding in a synthetic variety of maize. *Crop Sci.* **13**, 327–30.
HAMBLIN, J. and DONALD, C. M. (1974). The relationship between plant form, competitive ability and grain yield in a barley cross. *Euphytica* **23**, 535–42.
—— and ROSIELLE, A. A. (1978). Effect of intergenotypic competition on genetic parameter estimation. *Crop Sci.* **18**, 51–4.
HAMMER, R. E., PALMITER, R. D. and BRINSTER, R. L. (1984). Partial correction of murine hereditary growth disorder by germ-line incorporation of a new gene. *Nature, Lond.* **311**, 65–7.
HAMRICK, J. L. and ALLARD, R. W. (1972). Microgeographical variation in allozyme frequences in *Avena barbata. Proc. natn Acad. Sci. U.S.A.* **69**, 2100–4.
—— and —— (1975). Correlation between quantitative characters and enzyme genotypes in *Avena barbata. Evolution, Lancaster, Pa.* **29**, 438–42.
HANCOCK, T. W., JAMES, M. T., MAYO, O., CRADDOCK, M., PUCKRIDGE, R. J. and RATHJEN, A. J. (1983). Analysis of the Interstate Wheat Variety Trials. *Proc. Aust. Pl. Breeding Conf.*, Adelaide, Feb. 1983.
HANSON, A. (1982). *Burgundy.* London, Faber & Faber.
HANSON, H. (1977). The International Maize and Wheat Improvement Center (CIMMYT) outreach program. In *Resource allocation and productivity in national and international agricultural research* (eds T. W. Arndt, D. G. Dalrymple and V. W. Ruttan) pp. 306–22. University of Minnesota Press, Minneapolis.
HANSON, W. D. (1959). The break up of initial linkage blocks under selected mating systems. *Genetics, Princeton* **44**, 857–68.
HARLAN, J. R. and ZOHARY, D. (1976). Distribution of wild wheats and barley. *Science, N.Y.* **153**, 1074–80.
HARPER, J. L. (1977). *Population biology of plants.* Academic Press, London.
—— and WHITE, J. (1974). The demography of plants. *Ann. Rev. Ecol. Syst.* **5**, 419–63.
HARRINGTON, J. B. (1937). The mass-pedigree method in the hybridization improvement of cereals. *J. Am. Soc. Agron.* **29**, 379–84.
HARRIS, H. (1975). *The principles of human biochemical genetics* (2nd edn.) North-Holland, Amsterdam.
HARRIS, R. E., GARDNER, C. O. and COMPTON, W. A. (1972). Effects of mass selection and irradiation in corn measured by random S_1 lines and their test crosses. *Crop Sci.* **12**, 594–8.
HART, G. E. (1983). Hexaploid wheat (*Triticum aestivum* L. em Thell). In *Isozymes in plant genetics and breeding* Part B (eds S. D. Tanksley and T. J. Orton), pp. 35–56. Amsterdam, Elsevier.
—— and LANGSTON, P. J. (1977). Chromosomal location and evolution of isozyme structural genes in hexaploid wheat. *Heredity, Lond.* **39**, 263–77.
HARVILLE, D. A. (1975). Index selection with proportionality constraints. *Bio-*

metrics **31**, 223–5.

HATCHETT, J. G. and GALLUN, R. L. (1970). Genetics of the ability of the Hessian fly, *Mayetiola destructor*, to survive on wheats having different genes for resistance. *Ann. ent. Soc. Am.* **63**, 1400–7.

HAWKES, J. G. (1977). The importance of wild germplasm in plant breeding. *Euphytica* **26**, 615–21.

HAYAMI, Y. and AKINO, M. (1977). Organization and productivity of agricultural research systems in Japan. In *Resource allocation and productivity in national and international agricultural research* (eds T. M. Arndt, D. G. Dalrymple, and V. W. Ruttan). pp. 29–59. University of Minnesota Press, Minneapolis.

HAYASHI, S., SAKAI, K.-I. and MURAI, M. (1976). Genetic studies in natural populations of pinus. II. Geographical variation in relation to natural selection. *Mem. Fac. Agric. Kagoshima Univ.* **12**, 87–101.

HAYES, H. K. (1952). Development of the heterosis concept. In *Heterosis* (ed. J. W. Gowan) pp. 49–65. Iowa State University Press, Ames.

—— and GARBER, R. J. (1919). Synthetic production of high protein corn in relation to breeding. *J. Am. Soc. Agron.* **11**, 309–19.

HAYES, J. D. and FOSTER, C. A. (1976). Heterosis in self-pollinating crops, with particular reference to barley. In *Heterosis in plant breeding* (eds A. Jánossy and F. G. H. Lupton) pp. 239–56. (Proc. VII Congr. Eucarpia, Budapest, June, 1974). Elsevier, Amsterdam.

HAYMAN, B. I. (1953). Mixed selfing and random mating when homozygotes are at a disadvantage. *Heredity, Lond.* **7**, 185–92.

—— (1954). The theory and analysis of diallel crosses. *Genetics, Princeton* **39**, 789–809.

—— (1960). Heterosis and quantitative inheritance. *Hereditary, Lond.* **15**, 324–7.

—— and MATHER, K. (1953). The progress of inbreeding when homozygotes are at a disadvantage. *Heredity, Lond.* **7**, 165–83.

—— and —— (1956). Inbreeding when homozygotes are at a disadvantage: a reply. *Heredity, Lond.* **10**, 271–4.

HAYWARD, M. D. and VIVERO, J. L. (1984). Selection for yield in *Lolium perenne.* II. Performance of spaced plant selections under competitive conditions. *Euphytica* **33**, 787–800.

HAZEL, L. N. (1943). The genetic basis for constructing selection indexes. *Genetics, Princeton* **28**, 476–90.

HEIJBROEK, W. (1977). Partial resistance of sugarbeet to beet cyst eelworm (*Heterodera schachtii* schm.). *Euphytica* **26**, 257–62.

HEYLAND, K.-U. and FRÖHLING, J. (1977). Reaktionen vom Sommergersten-und Sommerweizensorten unterschiedlicher Krankheitsanfälligkeit auf den Einsatz systemischer Fungizide. *A. PflKrankh. PflPath. PflSchutz.* **84**, 451–67.

HILL, J. (1976). Genotype-environment interactions—a challenge for plant breeding. *J. agric. Sci., Camb.* **85**, 477–94.

—— and PERKINS, J. M (1969). The environmental induction of heritable changes in *Nicotiana rustica*: effects of genotype–environment interactions. *Genetics, Princeton* **61**, 661–75.

HILL, R. R. (1966). Designs to estimate effects of close substitution in alfalfa synthetics. *Crop Sci.* **6**, 471–3.

HILL, W. G. (1970). Theory of limits to selection with line crossing. In *Biomathematics, Vol. 1. Mathematical topics in quantitative genetics* (ed. K. Kojima) pp.

210–40. Springer, Berlin.

—— (1971). Investment appraisal for national breeding programmes. *Anim. Prod.* **13**, 37–50.

—— (1982). Predictions of response to artificial selection from new mutations. *Genet. Res.* **40**, 255–78.

—— and AVERY, P. J. (1978). On estimating the number of genes by genotype assay. *Heredity, Lond.* **40**, 397–403.

HINSON, K. and HANSON, W. D. (1962). Competition studies in soybeans. *Crop Sci.* **2**, 117–23.

HOCKING, R. R. (1976). The analysis and selection of variables in linear regression. *Biometrics* **32**, 1–49.

HOGARTH, D. M. (1977). Quantitative inheritance studies in sugarcane. III. The effect of competition and violation of genetic assumptions on estimation of genetic variance components. *Aust. J. agric. Res.* **28**, 257–68.

HOGG, R. V. (1977a). Statistical robustness: on its use in applications today. University of Iowa Department of Statistics Technical Report No. 60.

—— (1977b). An introduction to robust procedures. *Comm. Statist. Theory Meth. A.* **6**, 789–94.

HOGSETT, M. L. and NORDSKOG, A. W. (1958). Genetic-economic value in selection for egg production rate, body weight and egg weight. *Poult. Sci.* **37**, 1404–19.

HOHN, T., RICHARDS, K. and LEBEURIER, G. (1982). Cauliflower mosaic virus on its way to becoming a useful plant vector. *Curr. Topics Microbiol. Immunol.* **96**, 193–235.

HOLLAENDER, A. (1977). *Genetic engineering for nitrogen fixation.* Plenum Press, New York.

HOLLAND, D. A. (1969). Component analysis: an aid to the interpretation of data. *Exp. Agric.* **5**, 151–64.

HOLLIDAY, R. (1960). Plant population and crop yield. *Fld Crop Abstr.* **13**, 159–67, 247–54.

HOOKER, A. L. (1977). A plant pathologist's view of germplasm evaluation and utilization. *Crop Sci.* **17**, 689–94.

HOOYKAAS-VAN SLOGTEREN, G. M. S., HOOYKAAS, P. J. J., and SCHILPEROORT, R. A. (1984). Expression of Ti plasmid genes in monocotyledonous plants infected with *Agrobacterium tumefaciens. Nature, Lond.* **311**, 763–4.

HORA, F. B. (ed.) (1981). *Oxford encyclopaedia of trees of the world.* Oxford University Press.

HORNER, E. S., LUNDY, H. W., LUTRICK, M. C., and CHAPMAN, W. G. (1973). Comparisons of three methods of recurrent selection. *Crop Sci.* **13**, 485–9.

HOWARD, E. A. and DENNIS, E. S. (1984). Transposable elements in maize—the Activator–Dissociation (Ac–Ds) system. *Aust. J. biol. Sci.* **37**, 307–14.

HOWARD, H. W. (1968). The relation between resistance genes in potatoes and pathotypes of potato-root eelworm (*Heterodera rostochiensis*), wart disease (*Synchytrium endobioticum*) and potato virus X. *Abs. 1st Int. Congr. Plant Pathol.*, London, p. 92.

HUGHES, W. G. and BODDEN, J. J. (1977). Single gene restoration of cytoplasmic male sterility in wheat and its implications in the breeding of restorer lines. *Theor. appl. Genet.* **50**, 129–35.

HÜHN, M. (1975a). Estimation of broad sense heritability in plant populations: an improved method. *Theor. appl. Genet.* **46**, 87–99.

—— (1975b). Bemerkungen über die Bedeutung der Populationsgenetik und der ökologischen Genetik als Basis für forstgenetische und forstpflanzenzüchterische Arbeiten. *Silvae Genetica* **24**, 118–26.

Hull, E. G. (1945). Recurrent selection and specific combining ability in corn. *J. Am. Soc. Agron.* **37**, 134–45.

Hunter, H. and Leake, H. M. (1933). *Recent advances in agricultural plant breeding.* J. & A. Churchill, London.

Hurd, E. A. (1964). Root study of three wheat varieties and their resistance to drought. *Can. J. Plant Sci.* **44**, 240–8.

—— (1976). Plant breeding for drought resistance. In *Water deficits and plant growth* (ed. I. T. Kozlowski) Vol. IV. Academic Press, New York.

——, Townley-Smith, T. F., Mallough, D., and Patterson, L. A. (1973). Wakooma durum wheat. *Can. J. Plant Sci.* **53**, 261–2.

——, ——, Patterson, L. A., and Owen, C. H. (1972a). Wascana, a new durum wheat. *Can. J. Plant Sci.* **52**, 687–8.

——, ——, —— and —— (1972b). Techniques used in producing Wascana wheat. *Can. J. Plant Sci.* **52**, 689–91.

Hutchinson, J. B. (1965). *Essays on crop plant evolution.* Cambridge University Press.

—— (1974). *Evolutionary studies in world crops: diversity and change in the Indian subcontinent.* Cambridge University Press.

Innes, N. L., Brown, S. J., and Walker, J. T. (1974). Genetical and environmental variation for resistance to bacterial blight of Upland cotton. *Heredity, Lond.* **32**, 53–72.

—— and Hardwick, R. C. (1975). Possibilities of genetic improvement of *Phaseolus* beans in the U.K. *Outlook Agric.* **8**, 126–32.

International Atomic Energy Agency (1970). *Manual of mutation breeding.* Tech. Rep. Series No. 119. International Atomic Energy Agency, Vienna.

—— (1974). *Induced mutation for disease resistance in crop plants.* International Atomic Energy Agency, Vienna.

—— (1976). *Evaluation of seed protein alterations by mutation breeding.* International Atomic Energy Agency, Vienna.

Iruthayathas, E. E., Vlassak, K., and Laeremans, R. (1985). Inheritance of nodulation and N_2 fixation in winged beans. *J. Hered.* **76**, 237–42.

Izhar, S., Tabib, Y., and Swartzberg, D. (1984). Reciprocal transfer of male sterile and normal plasmons in Petunia. *Theor. Appl. Genet.* **68**, 455–7.

Jackson, J. E. (1967). The study of variation in crop yields from year to year. *Exp. Agric.* **3**, 175–82.

Jacquard, P. and Caputa, J. (1970). Comparison de trois modèles d'analyse des relations sociales entre espèces vegetales. *Ann. Amél. Pl.* **20**, 115–58.

Jain, S. K. (1975a). Population structure and the effects of breeding systems. In *Crop genetic resources for today and tomorrow* (eds O. H. Frankel and J. G. Hawkes) pp. 15–36. Cambridge University Press.

—— (1975b). Genetic reserves. In *Crop genetic resources for today and tomorrow* (eds O. H. Frankel and J. G. Hawkes) pp. 379–96. Cambridge University Press.

James, A. T. (1984). Plant tissue culture: achievements and prospects. *Proc. R. Soc. Lond. B* **222**, 135–45.

Jana, S. (1975). Genetic analysis by means of diallel graph. *Heredity, Lond.* **35**, 1–19.

—— (1976). Graphical analysis of tiller and ear production in a diallel cross of barley. *Can. J. Genet. Cytol.* **18**, 445–53.

—— and SEYFFERT, W. (1971). Simulation of quantitative characters by genes with biochemically definable action. III. The components of genetic effects in the inheritance of anthocyanins in *Matthiola incana* R. Br. *Theor. appl. Genet.* **42**, 329–37.

—— and —— (1972). Simulation of quantitative characters by genes with biochemically definable action. IV. The analysis of heritable variation by the diallel technique. *Theor. appl. Genet.* **42**, 16–24.

JAYAKAR, S. D. (1970). A mathematical model for interaction of gene frequencies in a parasite and its host. *Theor. pop. Biol.* **1**, 140–64.

JENKINS, M. T. (1940). The segregation of genes affecting yield of grain in maize. *J. Am. Soc. Agron.* **32**, 55–63.

JENKINSON, D. S. (1982). The nitrogen cycle in long-term field experiments. *Phil. Trans. R. Soc. Lond. B.* **296**, 563–71.

JENNINGS, D. L. (1970). Cassava in Africa. *Fld Crop Abs.* **23**, 271–8.

JENNINGS, P. R. and COCK, J. G. (1977). Centres of origin of crops and their productivity. *Econ. Bot.* **31**, 51–4.

JENNS, A. E. and LEONARD, K. J. (1985). Reliability of statistical analyses for estimating relative specificity in quantitative resistance in a model host–pathogen system. *Theor. Appl. Genet.* **69**, 503–13.

JENSEN, E. O., PALUDAN, K., HYLDIG-NIELSEN, J. J., JORGENSEN, P., and MARCKER, K. A. (1981). The structure of a chromosomal leghaemoglobin gene from soybean. *Nature, Lond.* **291**, 677–9.

JENSEN, N. F. (1970). A diallel selective mating system for cereal breeding. *Crop Sci.* **10**, 629–35.

—— (1978). Composite breeding methods and the DSM system in cereals. *Crop Sci.* **18**, 622–6.

JINKS, J. L. (1955). A survey of the genetical basis of heterosis in a variety of diallel crosses. *Heredity, Lond.* **9**, 223–38.

—— (1981). The genetic framework of plant breeding. *Phil. Trans. R. Soc. Lond. B* **292**, 407–19.

—— and POONI, H. S. (1976). Predicting the properties of recombinant inbred lines derived by single seed descent. *Heredity, Lond.* **36**, 253–66.

—— and —— (1981a). Comparative results of selection in the early and late stages of an inbreeding programme. *Heredity, Lond.* **46**, 1–7.

—— and —— (1981b). Properties of pure-breeding lines produced by dihaploidy, single seed descent and pedigree breeding. *Heredity, Lond.* **46**, 391–5.

—— and TOWEY, P. (1976). Estimating the number of genes in a polygenic system by genotype assay. *Heredity, Lond.* **37**, 69–81.

JOHANNSEN, W. (1903). *Über Erblichkeit in Populationen und in Reinen Linien.* Gustav Fischer, Jena.

—— (1909). *Elemente der Exakten Erblichkeitslehre.* Gustav Fischer, Jena.

JOHNSON, G. B. (1974). On the estimation of effective number of alleles from electrophoretic data. *Genetics, Princeton* **78**, 771–6.

—— (1977). Assessing electrophoretic similarity: the problem of hidden heterogeneity. *A. Rev. Ecol. Syst.* **8**, 309–28.

JOHNSON, H. W. (1972). Development of crop resistance to diseases and nematodes. *J. Environ. Qual.* **1**, 23–7.

JOHNSON, J. W. and PATTERSON, F. L. (1977). Interaction of genetic factors for fertility restoration in hybrid wheat. *Crop Sci.* **17**, 695–9.

Johnson, L. P. V. and Aksel, R. (1959). Inheritance of yielding capacity in a fifteen-parent diallel cross of barley. *Can. J. Genet. Cytol.* **1**, 208–65.

Johnson, R. (1984). A critical analysis of durable resistance. *A. Rev. Phytopathol.* **22**, 309–30.

Jones, F. G. and Parrott, D. M. (1965). The genetic relationship of pathotypes of *Heterodera rostochiensis* Woll. which reproduce on hybrid potatoes with genes for resistance. *Ann. appl. Biol.* **56**, 27–36.

Jørgensen, J. H. (1974). Mutational and independent origin of mutants induced by seed treatment of self-pollinated plants. In *Induced mutations for disease resistance in crop plants*, pp. 57–66. International Atomic Energy Agency, Vienna.

Kahler, A. L., Allard, R. W., and Miller, R. D. (1984). Mutation rates for enzyme and morphological loci in barley (*Hordeum vulgare* L.) *Genetics, Princeton* **106**, 729–34.

Kamiński, R. (1977). Variability and heritability of morphological and physiological characters of potato. *Genet. pol.* **18**, 115–23.

Kandra, G. and Maliga, P. (1977). Is bromodeoxyuridine resistance a consequence of cytokinin habituation in *Nicotiana tabacum*? *Planta* **133**, 131–3.

Kao, K. N., Constable, F., Michayluk, M. R., and Gamborg, O. L. (1974). Plant protoplast fusion and growth of intergeneric hybrid cells. *Planta* **120**, 215–27.

Kaplan, E. B. and Elston, R. C. (1972). A subroutine package for maximum likelihood estimation (MAXLIK). The University of North Carolina Institute of Statistics Mimeo Series, No. 823.

Karlin, S. and Feldman, M. W. (1970). Linkage and selection: two locus symmetric viability model. *Theor. pop. Biol.* **1**, 39–71.

Kasha, K. J. (1974). Haploids from somatic cells. In *Haploids in higher plants advances and potential* (ed. K. J. Kasha) pp. 67–87. University of Guelph.

Kassem, A. A., Tabl, M. M., Salem, A. E., and Khadr, F. H. (1976). Genetic variation in wheat following hybridization and treatment with gamma rays and ethylmethane sulfonate. I. Yield and yield components. *Egypt. J. Genet. Cytol.* **5**, 421–6.

Kaul, M. L. H. and Bhan, A. K. (1977). Mutagenic effectiveness and efficiency of EMS, DES and gamma-rays in rice. *Theor. appl. Genet.* **50**, 241–6.

Kays, S. and Harper, J. L. (1974). The regulation of plant and tiller density in a grass sward. *J. Ecol.* **62**, 97–105.

Keeble, F. and Pellew, C. (1910). The mode of inheritance of stature and of time of flowering in peas (*Pisum sativum*). *J. Genet.* **1**, 47–56.

Kehr, W. R. and Gardner, C. O. (1960). Genetic variability in Ranger alfalfa. *Agron. J.* **52**, 41–4.

Keim, D. L., Welsh, J. R., and McDonnell, R. L. (1973). Inheritance of photoperiodic heading response in winter and spring cultivars of bread wheat. *Can. J. Plant Sci.* **53**, 247–50.

Kemble, R. J., Mans, R. J., Gabay-Laughnan, S., and Laughnan, J. R. (1983). Sequences homologous to episomal mitochondrial DNAs in the maize nuclear genome. *Nature, Lond.* **304**, 744–7.

Kempthorne, O. (1954). The correlation between relatives in a random mating population. *Proc. R. Soc. B* **143**, 103–13.

—— (1955). The correlation between relatives in a simple autotetraploid population. *Genetics, Princeton* **40**, 168–74.

—— (1957). *An introduction to genetic statistics.* Iowa State University Press, Ames.

—— and CURNOW, R. N. (1961). The partial diallel cross. *Biometrics* **17**, 229–50.

KENNEDY, A. J. (1978). Analysis of sugar cane variety trials. I. Components of variance and genotype–environment interactions. *Euphytica* **27**, 767–75.

KERR, A. (1980). Biological control of crown gall through agrocin 84 production. *Plant Dis.* **64**, 25–30.

—— and HTAY, K. (1974). Biologial control of crown gall through bacteriocin production. *Physiol. Pl. Path.* **4**, 37–44.

KHALIFA, M. A. and QUALSET, C. O. (1974). Intergenotypic competition between tall and dwarf wheats. I. In mechanical mixtures. *Crop Sci.* **14**, 795–9.

KHAN, T. N. and ERSKINE, W. (1978). The adaptation of winged bean (*Psophocarpus tetragonolobus* (L.) DC.) in Papua New Guinea. *Aust. J. agric. Res.* **29**, 281–9.

KHUSH, G. S. (1977). Disease and insect resistance in rice. *Adv. Agron.* **29**, 265–341.

—— and COFFMAN, W. R. (1977). Genetic evaluation and utilization (G.E.U.) program. The rice improvement program of the International Rice Research Institute. *Theor. appl. Genet.* **51**, 97–110.

KIHARA, H. (1919). Über cytologische Studien bei einigen Getreidearten. I. Species-Bastarde des Weizens und Weizenroggen-Bastarde. *Bot. Mag., Tokyo* **33**, 17–38. [Cited by Morris, R. and Sears, E. R. (1967) The cytogenetics of wheat and its relatives. In *Wheat and wheat improvement* (eds K. S. Quisenberry and L. P. Reitz). American Society of Agronomy. No. 13 in the Series Agronomy.]

KILLICK, R. (1977). Genetic analysis of several traits in potatoes by means of a diallel cross. *Ann. appl. Biol.* **86**, 279–89.

KIMBER, G. and RILEY, R. (1963). Haploid angiosperms. *Bot. Rev.* **29**, 480–531.

KING, P. J. (1984). From single cells to mutant plants. *Oxford Surveys of Plant Molecular and Cell Biology* **1**, 7–32.

KINMAN, M. L. and SPRAGUE, J. F. (1945). Relation between number of parental lines and theoretical performance of synthetic varieties of corn. *J. Am. Soc. Agron.* **37**, 341–51.

KNIGHT, R. (1970). The measurement and interpretation of genotype–environment interaction. *Euphytica* **19**, 225–35.

—— (1971). A multiple regression analysis of hybrid vigour in single crosses of *Dactylis glomerata* L. *Theor. appl. Genet.* **41**, 306–11.

—— (1973). The relation between hybrid vigour and genotype–environment interaction. *Theor. appl. Genet.* **43**, 311–18.

—— (1978). Aspects of final selection. In *Australian wheat breeding today,* a treatise of the 2nd assembly of the Wheat Breeding Society of Australia (ed. G. J. Hollamby) p. 20. Roseworthy Agricultural College.

KNOTT, D. R. (1972a). Effects of selection for F_2 plant yield in subsequent generations in wheat. *Can. J. Plant Sci.* **52**, 721–6.

—— (1972b). Using race-specific resistance to manage the evolution of plant pathogens. *J. envir. Qual.* **1**, 227–31.

—— and DVOŘÁK, J. (1976). Alien germplasm as a source of resistance to disease. *A. Rev. Phytopathol.* **14**, 211–35.

—— and KUMAR, J. (1975). Comparison of early generation yield testing and a single seed descent procedure in wheat breeding. *Crop Sci.* **15**, 295–9.

—— and Srivastava, J. P. (1977). Inheritance of resistance to stem rust races 15B and 56 in eight cultivars of common wheat. *Can. J. Plant Sci.* **57**, 633–41.

Knox, R. B., Willing, R. R., and Pryor, L. D. (1972). Interspecific hybridization in poplars using recognition pollen. *Silvae Genetica* **21**, 65–9.

Kuehl, R. O., Rawlings, J. O., and Cockerham, C. C. (1968). Reference populations for diallel experiments. *Biometrics* **24**, 881–901.

Kupzow, A. J. (1976). Basic loci in cultivation of certain crops in the past and modern times. *Theor. appl. Genet.* **48**, 209–15.

Lacadena, J.-R. (1974). Spontaneous and induced parthenogenesis and androgenesis. In *Haploids in higher plants advances and potential* (ed. K. J. Kasha) pp. 13–32. University of Guelph.

Ladizinsky, G. (1975). Oats in Ethiopia. *Econ. Bot.* **29**, 238–41.

Lamacraft, R. R. (1974). Small plot variability in cereal experiments. M. Agric. Sci. Thesis, University of Adelaide.

Lande, R. (1976). The maintenance of genetic variability by mutation in a polygenic character with linked loci. *Genet. Res.* **26**, 221–35.

—— and Schemske, D. W. (1985). The evolution of self-fertilization and inbreeding depression in plants. I. Genetic models. *Evolution, Lancaster, Pa.* **39**, 24–40.

Larter, E. N. (1973). A look at yield trends in Triticale. In *Wheat, triticale and barley seminar* (ed. R. G. Anderson) pp. 215–20. CIMMYT, El Batán.

Latter, B. D. H. (1960). Natural selection for an intermediate optimum. *Aust. J. biol. Sci.* **13**, 30–5.

—— (1970). Selection in finite populations with multiple alleles. II. Centripetal selection, mutation and isoallelic variation. *Genetics, Princeton* **66**, 165–86.

Law, C. N. (1966). The location of genetic factors affecting a quantitative character in wheat. *Genetics, Princeton* **53**, 487–98.

—— (1967). The location of genetic factors controlling a number of quantitative characters in wheat. *Genetics, Princeton* **56**, 445–61.

—— (1972). The analysis of inter-varietal chromosomes in wheat and their first generation hybrids. *Heredity, Lond.* **28**, 169–79.

——, Snape, J. W., and Worland, A. J. (1983). Chromosome manipulation and its exploitation in the genetics and breeding of wheat. 15th Stadler Symposium Proc. University of Missouri, Columbia.

——, Worland, A. J., and Giorgi, B. (1976). The genetic control of ear-emergence time by chromosomes 5A and 5D of wheat. *Heredity, Lond.* **36**, 49–58.

Lazenby, A. and Rogers, H. H. (1965). Selection criteria in grass-breeding VI. Effects of defoliation on plants growing in small plots in field and controlled environment conditions. *J. agric. Sci., Camb.* **65**, 397–404.

Leather, G. R. (1983). Sunflowers (*Helianthus annuus*) are allelopathic to weeds. *Weed Sci.* **31**, 37–42.

Ledig, F. T. (1974). An analysis of methods for the selection of trees from wild stands. *Forest Sci.* **20**, 2–16.

Lefort, P. L. and Legisle, N. (1977). Quantitative stock-scion relationships in vine. Preliminary investigations by the analysis of reciprocal graftings. *Vitis* **16**, 149–61.

Leigh Brown, A. J. (1977). Physical correlates of an enzyme polymorphism. *Nature, Lond.* **269**, 803–4.

Leonard, K. J. (1977). Selection pressures and plant pathogens. *Ann. N.Y. Acad. Sci.* **287**, 207–22.

Lerner, I. M. (1954). *Genetic homeostasis.* Oliver and Boyd, Edinburgh.

Levin, D. A. (1975a). Interspecific hybridization, heterozygosity and gene exchange in *Phlox. Evolution, Lancaster, Pa.* **29**, 37–51.

—— (1975b). Pest pressure and recombination systems in plants. *Am. Nat.* **109**, 437–51.

Levings, C. S. and Dudley, J. W. (1963). Evaluation of certain mating designs for estimation of genetic variance in autotetraploid alfalfa. *Crop Sci.* **3**, 532–5.

Levy, L. W. (1981). A large-scale application of tissue culture: the mass propagation of Pyrethrum clones in Ecuador. *Env. Exp. Bot.* **21**, 389–95.

Lewis, D. (1951). Structure of the incompatibility gene. III. Types of spontaneous and induced mutation. *Heredity, Lond.* **5**, 399–414.

Lewis, E. B. (1978). A gene complex controlling segmentation in *Drosophila. Nature, Lond.* **276**, 565–70.

Li, C. C. (1967). Genetic equilibrium under selection. *Biometrics* **23**, 397–484.

—— (1975). *Path analysis—a primer.* Boxwood Press, Pacific Grove, California.

Libby, W. J. (1973). Domestication strategies for forest trees. *Can. J. For. Res.* **3**, 265–76.

Lim, S. T., Andersen, K., Shanmugam, K. T., O'Gara, F., Mielenz, J. R., Hershberger, C. L. and Valentine, R. C. (1979). Genetic engineering of symbiotic nitrogen fixation. *Biochem. Soc. Symp.* **44**, 81–7.

Liu, M.-C. and Chen, W.-H. (1978). Tissue and cell culture as aids to sugarcane breeding. II. Performance and yield potential of callus-derived lines. *Euphytica* **27**, 273–82.

Lloyd, D. G. (1975). The maintenance of gynodioecy and androdioecy in angiosperms. *Genetica* **45**, 325–39.

—— and Webb, C. J. (1977). Secondary sex characters in seed plants. *Bot. Rev.* **43**, 177–216.

Lonnquist, J. H. (1964). Modification of the ear-to-row procedure for the improvement of maize populations. *Crop Sci.* **4**, 227–8.

Luig, N. H. and Watson, I. A. (1961). A study of inheritance of pathogenicity in *Puccinia graminis* var. *tritici. Proc. Linn. Soc. N.S.W.* **86**, 217–29.

Lundqvist, A. (1958). Self incompatibility in rye. IV. Factors related to self-seeding. *Hereditas* **44**, 193–256.

—— (1966). Heterosis and inbreeding depression in autotetraploid rye. *Hereditas* **56**, 317–66.

—— (1969). Some effects of continued inbreeding in an autotetraploid highbred strain of rye. *Hereditas* **61**, 361–99.

Lupton, F. G. H. and Whitehouse, R. N. H. (1957). Studies on the breeding of self-pollinated cereals. I. Selection methods in breeding for yield. *Euphytica* **6**, 169–84.

Lush, J. L. (1937). *Animal breeding plans.* Iowa State University Press, Ames.

—— (1947). Family merit and individual merit as basis for selection. I, II. *Am. Nat.* **81**, 241–61; 362–79.

Lyon, M. F., Phillips, R. J. S., and Searle, A. G. (1954). The overall rates of dominant and recessive lethal and visible mutation induced by spermatogonial X-irradiation of mice. *Genet. Res.* **5**, 448–67.

Macindoe, S. L. and Walkden Brown, C. (1968). *Wheat breeding and varieties in Australia* (3rd edn). Science Bulletin No. 76. NSW Department of

Agriculture, Sydney.

MACKAY, G. R. (1972). On the genetic status of materials induced by pollination of *Brassica oleracea* with *Brassica campestris. Euphytica* **21**, 71–7.

MACKAY, T. F. C. (1984). Jumping genes meet abdominal bristles: hybrid dysgenesis-induced quantitative variation in *Drosophila melanogaster. Genet. Res.* **44**, 231–7.

MACKENZIE, G. S. (1832). On the culture of the potato. *Prize Essays and Trans. Highland Soc. of Scotland* 9.

MACKEY, J. (1974). Systematic approach to race-specific disease-resistance. In *Induced mutations for disease resistance in crop plants* pp. 9–22. International Atomic Energy Agency, Vienna.

MAHESWARAN, G. and WILLIAMS, E. G. (1984). Direct embryoid formation on immature embryoes of *Trifolium repens, T. pratense* and *Medicago sativa*, and rapid clonal propagation of *T. repens. Ann. Bot.* **54**, 201–11.

MANGELSDORF, P. C. (1965). The evolution of maize. In *Essays on crop plant evolution* (ed. J. B. Hutchinson) pp. 23–49. Cambridge University Press.

MANIATIS, T., FRITSCH, E. F., and SAMBROOK, J. (1982). *Molecular cloning: a laboratory manual.* Cold Spring Harbor Laboratory.

MANNING, H. L. (1956). Yield improvement from a selection index technique with cotton. *Heredity, Lond.* **10**, 303–22.

—— (1963). Realized yield improvement from twelve generations of progeny selection in a variety of Upland cotton. In *Statistical genetics and plant breeding* (eds W. D. Hanson and H. F. Robinson) pp. 329–49. NAS-NRC Publ. No. 982.

MARSHALL, D. R. and BROUÉ, P. (1973). Outcrossing rates in Australian populations of subterranean clover. *Aust. J. agric. Res.* **24**, 863–7.

—— and BROWN, A. H. D. (1975). Optimum sampling strategies in genetic conservation. In *Crop genetic resources for today and tomorrow* (eds O. H. Frankel and J. G. Hawkes) pp. 53–80. Cambridge University Press.

—— and BURDON, J. J. (1981). Multiline varieties and disease control. III. Continued use of overlapping and disjoint gene sets. *Aust. J. biol. Sci.* **34**, 81–95.

—— and PRYOR, A. J. (1978). Multiline varieties and disease control. I. The 'dirty crop' approach with each component carrying a unique single resistance gene. *Theor. appl. Genet.* **51**, 177–84.

—— and —— (1979). Multiline varieties and disease control. II. The 'dirty crop' approach with components carrying two or more genes for resistance. *Euphytica* **28**, 145–59.

MARTIN, F. W. (1976). Cytogentics and plant breeding of cassava: a review. *Pl. Breed. Abs.* **46**, 909–16.

MATERN, U., STROBEL, G. and SHEPHARD, J. (1978). Reactions to phytotoxins in a potato population derived from mesophyll protoplasts. *Proc. natn. Acad. Sci. U.S.A.* **75**, 4935–9.

MATHER, K. (1936). Types of linkage data and their value. *Ann. Eugen.* **7**, 251–64.

—— (1949). *Biometrical genetics.* Methuen, London.

—— (1951). *The measurement of linkage in heredity* (2nd edn). Methuen, London.

—— and JINKS, J. L. (1971). *Biometrical genetics* (2nd edn). Chapman and Hall, London.

—— and —— (1982). *Biometrical genetics* (3rd edn). Chapman and Hall,

London.

MATZINGER, D. F., COCKERHAM, C. C., and WERNSMAN, E. A. (1977). Single character and index mass selection with random mating in a naturally self-fertilizing species. In *Proc. Int. Conf. Quant. Genet. 1976* (eds O. Kempthorne, E. Pollack and T. B. Bailey, Jr.) pp. 503–18. Iowa State University Press, Ames.

MAXWELL, F. G. (1972). Morphological and chemical changes that evolve in the development of host plant resistance to insects. *J. environ. Qual.* **1**, 265–70.

MAYER, P. B. (1984). Is there urban bias in the Green Revolution? Report on a field trip to North Thanjavur. *Peasant Studies* **11**, 213–35.

MAYO, O. (1966). On the problem of self-incompatibility alleles. *Biometrics* **22**, 111–20.

—— (1971). Rates of change in gene frequency in tetrasomic organisms. *Genetica* **42**, 329–37.

—— (1975). Fundamental and population genetics. In *Textbook of human genetics* (eds G. R. Fraser and O. Mayo) pp. 3–63. Blackwell Scientific Publications, Oxford.

—— (1978). The existence and stability of a three-locus gametophytically-determined self-incompatibility system. *Adv. appl. Prob.* **10**, 14–15.

—— (1983). Problems of signficance tests in diallel analysis. *SABRAO J.* **15**, 147–9.

—— (1986). *The wines of Australia.* Faber & Faber, London.

—— and BROCK, D. J. H. (1978). Uses of polymorphism. In *The biochemical genetics of man*, 2nd edn (eds D. J. H. Brock and O. Mayo) pp. 421–66.

—— and HOPKINS, A. M. (1985). Problems of estimating the minimum number of genes contributing to quantitative variation. *Biom. J.* **27**, 181–7.

MAYR, E. (1963). *Animal species and evolution.* Belknap Press, Cambridge, MA.

MCCLINTOCK, B. (1951). Chromosome organisation and genic expression. *Cold Spring Harbor Symp. Quant. Biol.* **16**, 13–47.

—— (1956). Controlling elements and the gene. *Cold Spring Harbour Symp. Quant. Biol.* **21**, 197–216.

MCFADDEN, E. S. and SEARS, E. R. (1944). The artificial synthesis of *Triticum spelta.* (Abstr.) *Rec. Genet. Soc. Am.* **13**, 26–7.

MCGINNIS, W., GARBER, R. L., WIRZ, J., KUROIWA, A., and GEHRING, W. J. (1984*a*). A homologous protein-coding sequence in Drosophila homoeotic genes and its conservation in other metazoans. *Cell* **37**, 403–8.

——, LEVINE, M. S., HAFEN, E., KUROIWA, A., and GEHRING, W. J. (1984*b*). A conserved DNA sequence in homoeotic genes of the *Drosophila Antennapedia* and *Bithorax* complexes. *Nature, Lond.* **308**, 428–33.

MCKENZIE, R. I. H. and MARTENS, J. W. (1974). Breeding for stem rust resistance in oats. In *Induced mutation for disease resistance in crop plants*, pp. 45–8. International Atomic Energy Agency, Vienna.

MCMILLAN, I. and ROBERTSON, A. (1974). The power of methods for the detection of major genes affecting quantitative characters. *Heredity, Lond.* **32**, 349–56.

MCVITIE, J. A., GALE, M. D., MARSHALL, G. A., and WESTCOTT, B. (1978). The intra-chromosomal mapping of the Norin 10 and Tom Thumb dwarfing genes. *Heredity, Lond.* **40**, 67–70.

MICKE, A. (1976). Introduction. In *Induced mutation in crossbreeding*, pp. 1–4. International Atomic Energy Agency, Vienna.

MILLER, A. J. (1984). Selection of subsets of regression variables (with discussion). *J. R. Statist. Soc. A.* **147**, 389–425.

MITTON, J. B. (1978). Relationship between heterozygosity for enzyme loci and variation of morphological characters in natural populations. *Nature, Lond.* **273**, 661–2.

MODE, C. J. (1958). A mathematical model for the coevolution of obligate parasites and their hosts. *Evolution, Lancaster, Pa.* **12**, 158–65.

MOLISCH, H. (1937). *Der Einfluss einer Pflanze auf die Andere—Allelopathie.* Jena, Fischer.

MOLL, R. H., COCKERHAM, C. C., STUBER, C. W. and WILLIAMS, W. P. (1978). Selection responses, genetic–environmental interactions, and heterosis with recurrent selection for yield in maize. *Crop Sci.* **18**, 641–5.

——, ROBINSON, H. F., and COCKERHAM, C. C. (1960). Genetic variability in an advanced generation of a cross of two open-pollinated varieties of corn. *Agron. J.* **52**, 171–3.

MONTGOMERY, D. C. (1976). *Design and analysis of experiments.* John Wiley, New York.

MORENO-GONZALEZ, J. and GROSSMAN, M. (1976). Theoretical modification of reciprocal recurrent selection. *Genetics, Princeton* **84**, 95–111.

MORGAN, J. M. (1977). Differences in osmoregulation between wheat genotypes. *Nature, Lond.* **270**, 234–5.

MORIKAWA, T. (1978). Identification of monosomic and nullisomic oats using leaf peroxidase isozymes. *Jap. J. Genet.* **53**, 191–8.

MOSEMAN, J. G. (1959). Host-pathogen interaction of the genes for resistance in *Hordeum vulgare* and for pathogenicity in *Erysiphe graminis* f.sp. *hordei. Phytopathology* **49**, 469–72.

MOSER, J. W. JR (1972). Dynamics of an uneven-aged forest stand. *Forest Sci.* **18**, 184–91.

MU, S.-K., LIU, S.-Q., ZHOU, Y. K., QIAN, N. F., ZHANG, P., XIE, H.-X., ZHANG, F.-S. and YAN, Z. L. (1977). Induction of callus from apple endosperm and differentiation of the endosperm plantlet. *Scientia sin.* **20**, 370–6.

MUKHERJEE, C. (1976). Optimal strategy for replantation of coconut trees in Kerala. *Sankhyā C* **38**, 73–88.

MULITZE, D. K. and BAKER, R. J. (1985a). Evaluation of biometrical methods for estimating the number of genes. I. Effects of sample size. *Theor. Appl. Genet.* **69**, 553–8.

—— and —— (1985b). Evaluation of biometrical methods for estimating the number of genes. 2. Effects of type I and type II statistical errors. *Theor. Appl. Genet.* **69**, 559–66.

MÜLLER, G. (1977). Cross-fertilization in a conifer stand inferred from enzyme gene-markers in seeds. *Silvae Genetica* **26**, 223–6.

MULLER, H. J. (1927). The problem of genetic modification. Fifth International Congr. Genet., Berlin. *Z. indukt. Abstamm. u. VererbLehre* (suppl.) **1**, 234–60.

—— (1935). The origination of chromatin deficiencies as minute deletions subject to insertion elsewhere. *Genetica* **17**, 237–52.

MUNCK, L. (1976). Aspects of the selection, design and use of high lysine cereals. In *Evaluation of seed protein alterations by mutation breeding,* pp. 3–17. International Atomic Energy Agency, Vienna.

MUNGOMERY, V. E., SHORTER, R., and BYTH, D. E. (1974). Genotype x environment interactions and environmental adaptation. I. Pattern analysis—application to soya bean populations. *Aust. J. agric. Res.* **25**, 59–72.

MURDOCH, W. W., CHESSON, J., and CHESSON, P. L. (1985). Biological control in theory and practice. *Am. Nat.* **125**, 344–66.

NAMKOONG, G. (1966). Inbreeding effects on estimation of genetic additive variance. *For. Sci.* **12**, 8–13.

—— (1985). The influence of composite traits on genotype by environment relations. *Theor. Appl. Genet.* **70**, 315–7.

—— and SQUILLACE, A. E. (1970). Problems in estimating genetic variance by Shrikhande's method. *Silvae Genetica* **19**, 74–7.

NANCE, W. E. and GROVE, J. (1972). Genetic determination of phenotypic variation in sickle cell trait. *Science N.Y.* **177**, 716–18.

NARAIN, P. (1985). Homozygosity in a selfed population with an arbitrary number of linked loci. *J. Genet.* **59**, 254–66.

NATIONAL ACADEMY OF SCIENCES (1972). *Genetic vulnerability of major crops.* Washington.

—— (1975). *The winged bean—a high protein crop for the tropics.* Washington.

NELDER, J. A. (1962). New kinds of systematic designs for spacing experiments. *Biometrics* **18**, 283–307.

—— (1977). A reformulation of linear models. *J. R. statist. Soc. A.* **140**, 48–76.

NEWMAN, E. I. (1982). The possible relevance of allelopathy to agriculture. *Pestic. Sci.* **13**, 575–82.

NICHOLAS, F. W. (1980). Size of population required for artificial selection. *Genet. Res.* **35**, 85–105.

—— and ROBERTSON, A. (1980). The conflict between natural and artificial selection in finite populations. *Theor. Appl. Genet.* **56**, 57–64.

NITSCH, C. (1974a). La culture de pollen isolé sur milieu synthétique. *C.r. hebd. Séanc. Acad. Sci., Paris* **278D**,1031–4.

—— (1974b). Pollen culture—a new technique for mass production of haploid and homozygous lines. In *Haploids in higher plants: advances and potential* (ed. K. J. Kasha) pp. 123–35. University of Guelph.

NORDSKOG, A. W. (1978). Some statistical properties of an index of multiple traits. *Theor. appl. Genet.* **52**, 91–4.

NORONHA-WAGNER, M. and BETTENCOURT, A. J. (1967). Genetic study of the resistance of *Coffea* spp. to leaf rust. 1. Identification and behaviour of four factors conditioning disease reaction in *Coffea arabica* to twelve physiological races of *Hemileia vastatrix. Can. J. Bot.* **45**, 2021–31.

NUTMAN, P. S. (1969). Genetics of symbiosis and nitrogen fixation in legumes. *Proc. R. Soc. B.* **172**, 417–37.

NYBOM, N. (1970). Mutation breeding of vegetatively propagated plants. In *Manual of mutation breeding,* pp. 141–7. Tech. Rep. Series No. 119. International Atomic Energy Agency, Vienna.

OAKES, M. W. (1967). The analysis of a diallel cross of heterozygotes or multiple allelic lines. *Heredity, Lond.* **22**, 83–95.

OCHOA, C. (1975). Potato collecting expeditions in Chile, Bolivia and Peru, and the genetic erosion of indigenous cultivars. In *Crop genetic resources for today and tomorrow* (eds O. H. Frankel and J. G. Hawkes) pp. 167–73. Cambridge University Press.

OHTA, T. (1971). Associative overdominance caused by linked detrimental mutations. *Genet. Res.* **18**, 277–86.

—— (1975). Statistical analyses of *Drosophila* and human protein polymorphisms. *Proc. natn. Acad. Sci. U.S.A.* **72**, 3194–6.

—— and Kimura, M. (1970). Development of associative overdominance through linkage disequilibrium in finite populations. *Genet. Res.* **16**, 165–77.

Oka, H.-I. (1969). A note on the design of germplasm preservation work in grain crops. *SABRAO Newsl.* **1**, 127–34.

Okamoto, M. (1957). Asynaptic effect of chromosome V. *Wheat Inf. Serv.* **5**, 6.

—— (1962). Identification of the chromosomes of common wheat belonging to the A and B genomes. *Can. J. Genet. Cytol.* **4**, 31–7.

Oliver, A. J., King, D. R., and Mead, R. J. (1977). The evolution of resistance to fluoroacetate intoxication in mammals. *Search* **8**, 130–2.

Ollerenshaw, J. H. and Hodgson, D. R. (1977). The effects of constant and varying heights of cut on the yield of Italian ryegrass (*Lolium multiflorum* Lam.) and perennial ryegrass (*Lolium perenne* L.) *J. agric. Sci., Camb.* **89**, 425–35.

Oram, R. N. (1976). Increasing the quantity and quality of plant protein by genetic means. *Search* **7**, 136–9.

Owino, F. (1977). Genotype x environment interaction and genotypic stability in loblolly pine. IV Correlation studies. *Silvae Genetica* **26**, 176–9.

——, Kellison, R. C., and Zobel, B. J. (1977). Genotype x environment interaction and genotypic stability in loblolly pine. V. Effects of genotype x environment interaction on genetic variance component estimates and gain prediction. *Silvae Genetica* **26**, 131–4.

Page, A. R. and Hayman, B. I. (1960). Mixed sib and random mating when homozygotes are at a disadvantage. *Heredity, Lond.* **14**, 187–96.

Painter, R. H. (1951). *Insect resistance in crop plants.* Macmillan, New York.

Panse, V. G. (1940). Application of genetics to plant breeding II. The inheritance of quantitative characters and plant breeding. *J. Genet.* **40**, 283–302.

Papadakis, J. S. (1937a). Méthode statistique pour des expériences sur champ. *Bull Inst. Amél. Plant. à Salonique* No. 23.

—— (1937b). Est-ce seulement d'après le rendement en grain que se fait la sélection naturelle chez les plantes cultivées? *Bull. Inst. Amél. Plant. à Salonique* No. 26.

Park, Y. C. (1977a). Theory for the number of genes affecting quantitative characters. I. Estimation of and variance of the estimates of gene number for quantitative traits controlled by additive genes having equal effect. *Theor. appl. Genet.* **50**, 153–61.

—— (1977b). Theory for the number of genes affecting quantitative characters. II. Biases from drift, dominance, inequality of gene effects, linkage disequilibrium and epistasis. *Theor. appl. Genet.* **50**, 163–72.

Parry, M. S. and Rogers, W. S. (1968). Dwarfing interstocks: their effect on the field performance and anchorage of apple trees. *J. hort. Sci.* **43**, 133–46.

—— and —— (1972). Effects of interstock length and vigour on the field performance of Cox's Orange Pippin apples. *J. hort. Sci.* **47**, 97–105.

Parsons, P. A. (1959). Some problems in inbreeding and random mating in tetrasomics. *Agron. J.* **51**, 465–7.

Paterniani, E. and Vencovsky, R. (1977). Reciprocal recurrent selection in maize. (*Zea mays* L.) based on test crosses of half-sib families. *Maydica* **22**, 141–52.

Patterson, F. L. and Gallun, R. L. (1977). Linkage in wheat of the H_3 and H_6 genetic factors for resistance to Hessian fly. *J. Hered.* **68**, 293–6.

PATTERSON, H, D. and HUNTER, E. A. (1983). The efficiency of incomplete block designs in National List and Recommended List cereal variety trials. *J. agric. Sci. Camb.* **101**, 427–33.

——, Silvey, V., Talbot, M., and Weatherup, S. T. C. (1977). Variability of yields of cereal varieties in U.K. trials. *J. agric. Sci. Camb.* **89**, 239–45.

PEACOCK, W. J., DENNIS, E. S., GERLACH, W. L., LLEWELLYN, D., LÖRZ, H., PRYOR, A. J., SACHS, M. M., SCHWARTZ, D. and SUTTON, W. D. (1983). Gene transfer in maize; controlling elements and the *alcohol dehydrogenase* genes. In *Proceedings of Miami Winter Symposia* (eds K. Downey, R. W. Voellme, F. Ahmed, and J. Schultz). Vol. 20, pp. 311–25. New York, Academic Press.

PEARCE, S. C. (1969). Multivariate techniques of use in biological research. *Exp. Agric.* **5**, 67–77.

PEDERSON, D. G. (1969a). The prediction of selection response in a self-fertilizing species. I. Individual selection. *Aust. J. biol. Sci.* **22**, 117–29.

—— (1969b). The prediction of selection response in a self-fertilizing species. II. Family selection. *Aust. J. biol. Sci.* **22**, 1245–57.

—— (1974). Arguments against intermating before selection in self-fertilizing species. *Theor. appl. Genet.* **45**, 157–62.

PEIRCE, L. C. (1977). Impact of single seed descent in selecting for fruit size, earliness, and total yield in tomato. *J. Am. Soc. Hort. Sci.* **102**, 520–2.

PELHAM, J. (1966). Resistance in tomato to Tomato Mosaic Virus. *Euphytica* **15**, 258–67.

PERSON, C. and MAYO, G. M. E. (1974). Genetic limitations on models of specific interactions between a host and its parasites. *Can. J. Bot.* **52**, 1339–47.

PERUTZ, M. F. and LEHMANN, H. (1968). Molecular pathology of human haemoglobin. *Nature, Lond.* **219**, 902–9.

PFEIFFER, J. E. (1976). A note on the problem of basic causes. In *Origins of African plant domestication* (eds J. R. Harlan, J. M. J. de Wet, and A. B. L. Stemler) pp. 23–8. Mouton, The Hague.

PHIPPS, I. F., HOCKLEY, S. R., and PUGSLEY, A. T. (1943). Warigo—a disease-resistant wheat. *J. Aust. Inst. agric. Sci.* **9**, 17–20.

PHUNG, T. K. and RATHJEN, A. J. (1976). Frequency-dependent advantage in wheat. *Theor. appl. Genet.* **48**, 289–97.

—— and —— (1977). Mechanisms of frequency-dependent advantage in wheat. *Aust. J. agric. Res.* **28**, 187–202.

PICKERSGILL, B. (1971). Relationship between weedy and cultivated forms in some aspects of chilli peppers (genus *Capsicum*). *Evolution, Lancaster, Pa.* **25**, 683–91.

PIELOU, E. C. (1973). Geographic variation in host–parasite specificity. In *The mathematical theory of the dynamics of populations* (eds M. S. Bartlett and R. W. Hiorns) pp. 103–23. Academic Press, London.

POLACCO, J. C. and POLACCO, M. L. (1977). Inducing and selecting valuable mutations in plant cell cultures in a tobacco mutant resistant to carboxin. *Ann. N.Y. Acad. Sci.* **287**, 385–400.

PONTECORVO, G. (1975). 'Alternatives to sex': genetics by means of somatic cells. In *Modification of the information content of plant cells* (eds R. Markham, D. R. Davies, D. A. Hopwood and R. W. Horne) pp. 1–14. North-Holland, Amsterdam.

POWELL, J. R. (1975). Protein variation in natural populations of animals. *Evol. Biol.* **8**, 79–119.

POWERS, H. R. and SANDO, W. J. (1957). Genetics of host–parasite relationship in powdery mildew of wheat. (abs.) *Phytopathology* **47**, 453.

PRING, D. R. and LEVINGS, III, C. S. (1978). Heterogeneity of maize cytoplasmic genomes among male-sterile cytoplasms. *Genetics, Princeton* **89**, 121–36.

PRYOR, L. D. (1976). *Biology of Eucalypts.* Edward Arnold, London.

PUCK, T. T. (1981). Some new developments in genetic analysis of somatic mammalian cells. In *Control of cellular division and development B.* pp. 393–405. Alan R. Liss, New York.

PUGSLEY, A. T. (1966). The photoperiodic response of some spring wheats with special reference to the variety Thatcher. *Aust. J. agric. Res.* **17**, 591–9.

—— (1978). Genetic variability in wheat with respect to physiological traits. In *Australian wheat breeding today,* a treatise of the 2nd Assembly of the Wheat Breeding Society of Australia (ed. G. J. Hollamby) p. 40. Roseworthy Agricultural College.

PUTNAM, A. R. and DUKE, W. B. (1978). Allelopathy in agroecosystems. *A. Rev. Phytopathol.* **16**, 431–45.

RAJHATHY, T. (1977). Plant breeding evolving. *Can. J. Genet. Cytol.* **19**, 595–602.

RAMEY, T. B. and ROSIELLE, A. A. (1983). HASS cluster analysis: a new method of grouping genotypes in environments in plant breeding. *Theor. Appl. Genet.* **66**, 131–3.

RAMMAH, A. M. and BŐJTŐS, Z. (1976). Performance of some genotypes of lucerne under wide and narrow spaced planting. I. Heritability of forage yield and related traits and interrelationships among traits. *Acta agron. Hung.* **25**, 309–18.

RATHJEN, A. J. (1965). Genetic differentiation in *Phaseolus vulgaris* L. Ph.D. Thesis, University of Cambridge.

—— and LAMACRAFT, R. R. (1972). The use of computers for information management in plant breeding. *Euphytica* **21**, 502–6.

—— and PUGSLEY, A. T. (1978). Wheat breeding at the Waite Institute, 1925–1978. Biennial Report 1976–7 Waite Agricultural Research Institute, University of Adelaide, pp. 12–37.

RAWLINGS, J. O. and COCKERHAM, C. C. (1962). Analysis of double cross hybrid populations. *Biometrics* **18**, 229–44.

RENDEL, J. M. (1967). *Canalization and gene control.* Logos Press, London.

RHOADES, M. M. (1931). Cytoplasmic inheritance of male sterility in *Zea mays. Science N.Y.* **73**, 340–1.

RICE, E. L. (1974). *Allelopathy,* New York, Academic Press.

—— (1979). Allelopathy—an update. *Bot. Rev.* **45**, 15–190.

RIGGS, T. J. and KIRBY, E. J. M. (1978a). Developmental consequences of two-row and six-row ear type in spring barley. 1. Genetical analysis and comparisons of mature plant characters. *J. agric. Sci., Camb.* **91**, 199–205.

—— and —— (1978b). Developmental consequences of two-row and six-row ear type in spring barley. 2. Shoot apex, leaf and tiller development. *J. agric. Sci., Camb.* **91**, 207–16.

RILEY, R. (1960). The diploidisation of polyploid wheat. *Heredity, Lond.* **15**, 407–29.

—— (1965). Cytogenetics and the evolution of wheat. In *Essays on crop plant evolution* (ed J. B. Hutchinson) pp. 103–22. Cambridge University Press.

—— (1974). Developments in plant breeding. *Agric. Progr.* **49**, 1–16.

—— and CHAPMAN, V. (1958). Genetic control of the cytologically diploid behaviour of hexaploid wheat. *Nature, Lond.* **182**, 713–15.

——, —— and JOHNSON R. (1968). The incorporation of alien disease resistance in wheat by genetic interference with the regulation of meiotic chromosome synapsis. *Genet. Res.* **12**, 199–219.

ROBERTS, E. H. (1975). Problems of long-term storage of seed and pollen for genetic resources conservation. In *Crop genetic resources for today and tomorrow* (eds O. H. Frankel and J. G. Hawkes) pp. 269–96. Cambridge University Press.

—— and ELLIS, R. H. (1977). Prediction of seed longevity at sub-zero temperatures and genetic resources conservation. *Nature, Lond.* **268**, 431–3.

—— and —— (1982). Physiological, ultrastructural and metabolic aspects of seed viability. In *The physiology and biochemistry of seed development, dormancy and germination* (ed. A. A. Khan) pp. 465–85. Amsterdam, Elsevier.

ROBERTSON, A. (1956). The effect of selection against extreme deviants based on deviation or on homozygosis. *J. Genet.* **54**, 236–48.

—— (1959). The sampling variance of the genetic correlation coefficient. *Biometrics* **15**, 469–85.

—— (1960). A theory of limits in artificial selection. *Proc. R. Soc. B.* **153**, 234–49.

—— (1970). A theory of limits in artificial selection with many linked loci. In *Biomathematics, Vol. 1 Mathematical topics in population genetics* (ed. K. Kojima) pp. 246–88. Springer-Verlag, Berlin.

ROJAS, B. A. and SPRAGUE, G. F. (1952). A comparison of variance components in corn III. *Agron. J.* **44**, 462–6.

ROMERO LOPES, C., MUSCHE, R., and HEC, M. (1984). Identificação de pigmentos flavonõides e ãcidos fenõlicos nos cultivares Bourbon e Caturra de *Coffea arabica* L. *Rev. Brasil. Genet.* **7**, 657–69.

ROSIELLE, A. A., EAGLES, H. A. and FREY, K. J. (1977). Application of restricted selection indexes for improvement of economic value in oats. *Crop Sci.* **17**, 359–61.

ROTH, E. J. and LARK, K. G. (1984). Isopropyl-N(3-chlorophenyl) carbamate (CIPC) induced chromosomal loss in soybean: a new tool for plant somatic cell genetics. *Theor. Appl. Genet.* **68**, 421–31.

ROTHSTEIN, S. J., LAZARUS, C. M., SMITH, W. E., BAULCOMBE, D. C., and GATENBY, A. A., (1984). Secretion of a wheat α-amylase expressed in yeast. *Nature, Lond.* **308**, 662–5.

ROTILI, P., ZANNONE, L., and JACQUARD, P. (1976). Effects of association on the evaluation of lucerne populations. *Ann. Amél. Pl.* **26**, 139–55.

RUSSELL, W. A., SPRAGUE, G. F., and PENNY, L. H. (1963). Mutations affecting quantitative characters in long-time inbred lines of maize. *Crop Sci.* **3**, 175–8.

SAGE, G. C. M. and HOBSON, G. E. (1973). The possible use of mitochondrial complementation as an indicator of yield heterosis in breeding hybrid wheat. *Euphytica* **22**, 61–9.

SAKAI, K.-I. (1955). Competition in plants and its relation to selection. *Cold Spring Harb. Symp. quant. Biol.* **20**, 137–57.

—— and HATAKEYAMA, S. (1963). Estimation of genetic parameters in forest trees without raising progeny. *Silvae Genetica* **12**, 152–7.

SALEM, A. E., TABL, M. M. and KASSEM, A. A. (1976). Genetic variation in wheat following hybridization and treatment with gamma rays and ethylmethane sulfonate. II. Chemical and quality characters. *Egypt. J. Genet. Cytol.*

5, 427–32.

Samborski, D. J. and Dyck, P. L. (1968). Inheritance of virulence in wheat leaf rust on the standard differential wheat varieties. *Can. J. Genet. Cytol.* **10**, 24–32.

Sampson, D. R. and Tarumoto, I. (1976). Genetic variances in an eight parent half diallel of oats. *Can. J. Genet. Cytol.* **18**, 419–27.

Sanford, J. C., Weeden, N. F. and Chyi, Y. S. (1984). Regarding the novelty and breeding value of protoplast-derived variants of Russet Burbank (*Solanum tuberosum* L.) *Euphytica* **33**, 709–15.

Sarič, M. R., Zorzič, M., and Burič, D. (1977). Einfluss der Unterlage und des Reises auf die Ionenaufnahme und -verteilung. *Vitis* **16**, 174–83.

Sax, K. (1922). Sterility in wheat hybrids. II. Chromosome behaviour in partially sterile hybrids. *Genetics, Princeton* **7**, 513–52.

Scandalios, J. G. and Baum, J. A. (1982). Regulatory gene variation in higher plants. *Adv. Genet.* **21**, 347–70.

Scarisbrick, D. H., Wilkes, J. M. and Kempston, R. (1977). The effect of varying plant population density on the seed yield of Navy beans (*Phaseolus vulgaris*) in south-east England. *J. agric. Sci., Camb.* **88**, 567–77.

Schemske, D. W. and Lande, R. (1985). The evolution of self-fertilization and inbreeding depression in plants. II. Empirical observations *Evolution, Lancaster, Pa.* **39**, 41–52.

Schmidt, J. (1919). La valeur de l'individu a titre de générateur. Appréciée suivant la méthode du croisement diallèle. *C.r. Lab. Calsberg* **14**, 1–34.

Schnell, F. W. (1961). On some aspects of reciprocal recurrent selection. *Euphytica* **10**, 24–30.

Scholnick, S. B., Morgan, B. A. and Hirsh, J. (1983). The cloned dopa carboxylase gene is developmentally regulated when re-integrated into the *Drosophila* genome. *Cell* **34**, 37–45.

Schubert, K. R., Jennings, K. R. and Evans, H. J. (1978). Hydrogen reactions of nodulated leguminous plants. II. Effects on dry matter accumulation and nitrogen fixation. *Plant Physiol.* **61**, 398–401.

Schuler, J. F. and Sprague, G. F. (1956). Natural mutations in inbred lines of maize and their heterotic effect. II. Comparison of mother line vs mutant when outcrossed to unrelated inbreds. *Genetics, Princeton* **41**, 281–91.

Schwartz, D. and Laughner, W. (1969). A molecular basis for heterosis. *Science N.Y.* **166**, 626–7.

Scowcroft, W. R. (1977). Somatic cell genetics and plant improvement. *Adv. Agron.* **29**, 39–81.

Sears, E. R. (1952). Homoeologous chromosomes in *Triticum aestivum* (Abs.) *Genetics, Princeton* **37**, 624.

—— (1953). Nullisomic analysis in common wheat. *Am. Nat.* **87**, 245–52.

—— (1954). The aneuploids of common wheat. *Missouri Agric. Expt. Stn. Res. Bull.* 572.

—— (1956). The transfer of leaf-rust resistance from *Aegilops umbellulata* to wheat. *Proc. Brookhaven Symp. Biol.* **9**, 1–22.

—— (1958). The aneuploids of common wheat. *Proc. 1st Int. Wheat Genet. Symp.* 221–8.

—— (1962). The use of telocentric chromosomes in linkage mapping. (Abs.) *Genetics, Princeton* **47**, 983.

—— (1966). Chromosome mapping with the aid of telocentrics. *Proc. 2nd Int.*

Wheat Genet. Symp. Hereditas (Suppl.) **2**, 370–81.
—— (1977). An induced mutant with homoeologous pairing in common wheat. *Can. J. Genet. Cytol.* **19**, 585–93.
—— and OKAMOTO, M. (1958). Intergenomic chromosome relationships in hexaploid wheat. *Proc. Xth Int. Congr. Genet., Montreal,* **2**, 258–9.
—— and RODENHISER, H. A. (1948). Nullisomic analysis of stem-rust resistance in *Triticum vulgare* var. Timstein. *Genetics, Princeton* **33**, 123–4.
SEHGAL, S. M. (1977). Private sector international agricultural research: the genetic supply industry. In *Resource allocation and productivity in national and international agricultural research* (eds T. M. Arndt, D. G. Dalrymple and V. W. Ruttan) pp. 404–15. University of Minnesota Press, Minneapolis.
SEIF, E. and PEDERSON, D. G. (1978). Effect of rainfall on the grain yield of spring wheat, with an application to the analysis of adaptation. *Aust. J. agric. Res.* **29**, 1107–15.
SEN, D. (1981). An evaluation of mitochondrial heterosis and *in vitro* mitochondrial complementation in wheat, barley, and maize. *Theor. Appl. Genet.* **59**, 153–60.
SEYFFERT, W. (1966). Die Simulation quantitativer Merkmale durch Gene mit biochemisch definierbarer Wirkung. I. Ein einfaches Modell. *Züchter* **36**, 159–63.
—— (1971). Simulation of quantitative characters by genes with biochemically definable action. II. The material. *Theor. appl. Genet.* **41**, 285–91.
—— and FORKMANN, G. (1976). Simulation of quantitative characters by genes with biochemically definable action. VIII. Observation and discussion of non-linear relationships. In *Population genetics and ecology.* (eds S. Karlin and E. Nevo) pp. 431–40. Academic Press, New York.
SHAIKH, M. A. Q., AHMED, Z. U., MAJID, M. A. and WADUD, M. A. (1982). A high-yielding and high-protein mutant of chickpea (*Cicer arietum* L). derived through mutation breeding. *Env. Exp. Bot.* **22**, 483–9.
SHARMA, S. N. and PRASAD, R. (1978). Systematic mixed versus pure stands of wheat genotypes. *J. agric. Sci. Camb.* **90**, 441–4.
SHAW, C. H. (1984). Ti-plasmid-derived plant gene vectors. *Oxford Surveys in Plant Molecular and Cell Biology* **1**, 211–6.
SHEBESKI, L. H. (1967). Wheat breeding. In *Proc. Can. Centennial Wheat Symp.* (ed. K. F. Nelson) p. 253.
SHELBOURNE, C. J. A. (1969). *Tree breeding methods.* New Zealand Forest Service Technical Paper No. 55, Wellington, NZ Forest Service.
—— and LOW, C. B. (1980). Multi-trait index selection and associated genetic gains of *Pinus radiata* progenies at five sites. *N.Z. Forest Sci.* **10**, 307–24.
SHEPARD, J. F., BIDNEY, D., BARSBY, T. and KEMBLE, R. (1983). Genetic transfer in plants through interspecific protoplast fusion. *Science N.Y.* **219**, 683–8.
SHEPHERD, K. W. and MAYO, G. M. E. (1972). Genes conferring specific plant disease resistance. *Science N.Y.* **175**, 375–80.
SHORTER, R., BYTH, D. E., and MUNGOMERY, V. E. (1977). Genotype x environment interactions and environmental adaptation. II. Assessment of environmental contributions. *Aust. J. agric. Res.* **28**, 223–35.
SHRIKHANDE, V. J. (1957). Some considerations in designing experiments on coconut trees. *J. Ind. Soc. agric. Stat.* **9**, 82–99.
SIDHU, G. and PERSON, C. (1971). Genetic control of virulence in *Ustilago hordei* II. Segregations for higher levels of virulence. *Can. J. Genet. Cytol.* **13**, 173–8.

SIMMONDS, N. W. (1962). Variability in crop plants, its use and conservation. *Biol. Rev.* **37**, 442–65.

—— (1976a). *Evolution of crop plants.* Longman, London.

—— (1976b). Potatoes. In *Evolution of crop plants* (ed N. W. Simmonds) pp. 279–83. Longman, London.

—— (1977). Approximation for *i*, intensity of selection. *Heredity, Lond.* **38**, 413–14.

SIMON, P. W. and LINDSAY, R. C. (1983). Effects of processing upon objective and sensory variables of carrots. *J. Am. Soc. Hort. Sci.* **108**, 928–31.

SIMONS, M. D. (1972). Polygenic resistance to plant disease and its use in breeding resistant cultivars. *J. Environ. Qual.* **1**, 232–40.

SIMONSEN, Ø. (1976). Genetic variation in diploid and autotetraploid populations of *Lolium perenne* L. *Hereditas* **84**, 133–56.

—— (1977). Genetic variation in diploid and autotetraploid populations of *Festuca pratensis. Hereditas* **85**, 1–24.

SING, C. F., MOLL, R. H., and HANSON, W. D. (1967). Inbreeding in two populations of *Zea mays* L. *Crop Sci.* **7**, 631–6.

SINGH, R. B. and SHARMA, G. S. (1976). Induced polygenic variations in relation to gene action for yield and yield components in spring wheat. *Can. J. Genet. Cytol.* **18**, 217–23.

SINHA, S. K. and KHANNA, R. (1975). Physiological, biochemical, and genetic basis of heterosis. *Adv. Agron.* **27**, 123–74.

SLATKIN, M. (1978). Spatial patterns in the distribution of polygenic characters. *J. theor. Biol.* **70**, 213–28.

SLEPER, D. A., NELSON, C. J. and ASAY, K. H. (1977). Diallel and path coefficient analysis of tall fescue (*Festuca arundinacea*) regrowth under controlled conditions. *Can. J. Genet. Cytol.* **19**, 557–64.

SMITH, C. (1969). Optimum selection procedures in animal breeding. *Anim. Prod.* **11**, 433–42.

—— (1978). The effect of inflation and form of investment on the estimated value of genetic improvement in farm livestock. *Anim. Prod.* **26**, 101–10.

—— (1984). Genetic aspects of conservation in farm livestock. *Livestock Production Science* **11**, 37–48.

—— (1985). Scope for selecting among breeding stocks of possible economic value in the future. *Anim. Prod.* **41**, 403–12.

SMITH, H. F. (1936). A discriminant function for plant selection. *Ann. Eugen.* **7**, 240–50.

—— (1938). An empirical law describing heterogeneity in the yield of agricultural crops. *J. agric. Sci., Camb.* **28**, 1–23.

SMITH, P. F. (1975). Effect of scion and rootstock on mineral composition of mandarin-type citrus leaves. *J. Am. Soc. Hort. Sci.* **100**, 368–9.

SNAPE, J. W. (1982). Predicting the frequencies of introgressive segregants for yield and yield components in wheat. *Theor. appl. Genet.* **62**, 127–34.

——, LAW, C. N., and WORLAND, A. J. (1975). A method for the detection of epistasis in chromosome substitution lines of hexaploid wheat. *Heredity, Lond.* **34**, 297–303.

——, —— and —— (1977). Whole chromosome analysis of height in wheat. *Heredity, Lond.* **38**, 25–36.

—— and RIGGS, T. J. (1975). Genetic consequences of single seed descent in the breeding of self-pollinating crops. *Heredity, Lond.* **35**, 211–19.

Sneep, J. (1977). Selection for yield in early generations of self-fertilizing crops. *Euphytica* **26**, 27–30.

Soave, C., Suman, N., Viotti, A., and Salamini, F. (1978). Linkage relationships between regulatory and structural gene loci involved in zein synthesis in maize. *Theor. appl. Genet.* **52**, 263–7.

Somerville, C. R. (1984). The analysis of photosynthetic carbon dioxide fixation and photorespiration by mutant selection. *Oxford Surveys in Plant Molecular and Cell Biology* **1**, 103–31.

Southern, E. M. (1975). Detection of specific sequences among DNA fragments separated by gel electrophoresis. *J. Mol. Biol.* **98**, 503–17.

Sparrow, D. H. B. (1970). Some genetical aspects of malting quality. In *Proc. 2nd Int. Barley Genet. Symp.*, Pullman 1969, pp. 559–74.

Sprague, G. F. and Tatum, L. A. (1942). General versus specific combining ability in single crosses of corn. *J. Am. Soc. Agron.* **34**, 923–32.

Stam, P. (1977). Selection response under random mating and under selfing in the progeny of a cross of homozygous parents. *Euphytica* **26**, 169–84.

Stebbins, G. L. (1941). Apomixis in the angiosperms. *Bot. Rev.* **7**, 507–42.

Stevens, M. A., Kader, A. A., and Albright-Holton, M. (1977). Intercultivar variation in composition of locular and pericarp portions of fresh market tomatoes. *J. Am. Soc. Hort. Sci.* **102**, 689–92.

——, ——, —— and Algazi, M. (1977). Genotypic variation for flavour and composition in fresh market tomatoes. *J. Am. Soc. Hort. Sci.* **102**, 680–9.

Stigler, S. M. (1977). Do robust estimators work with *real* data? *Ann. Statist.* **5**, 1055–98.

Stuber, C. W., Moll, R. H., Goodman, M. M., Schaffer, H. E. and Weir, B. S. (1980). Allozyme frequency changes associated with selection for increased grain yield in maize (*Zea mays* L.). *Genetics, Princeton* **95**, 225–36.

Suge, H. (1978). The genetic contol of gibberellin production in rice. *Jap. J. Genet.* **53**, 199–207.

Summerfield, R. J., Most, B. H., and Boxall, M. (1977). Tropical plants with sweetening properties: physiological and agronomic problems of protected cropping. I. *Dioscoreophyllum cumminsii*. *Econ. Bot.* **31**, 331–9.

Suneson, C. A. (1956). An evolutionary plant breeding method. *Agron. J.* **48**, 188–90.

—— and Ramage, R. T. (1962). Competition between near isogenic genotypes. *Crop Sci.* **2**, 249–50.

—— and Stevens, H. (1953). Studies with bulked hybrid populations of barley. *USDA Tech. Bull.* No. 1067.

Sved, J. A. (1965). Genetical studies in tetraploids. Ph.D. Thesis, University of Adelaide.

—— (1971). Linkage disequilibrium and homozygosity of chromosome segments in finite populations. *Theor. pop. Biol.* **2**, 125–41.

—— (1972). Heterosis at the level of the chromosome and at the level of the gene. *Theor. pop. Biol.* **3**, 491–506.

—— (1976). Hybrid dysgenesis in *Drosophila melanogaster*: a possible explanation in terms of spatial organization of chromosomes. *Aust. J. biol. Sci.* **29**, 375–88.

——, Reed, T. E. and Bodmer, W. F. (1967). The number of balanced polymorphisms that can be maintained in a natural population. *Genetics, Princeton* **55**, 469–81.

Swift, J. (1974). A full and true Account of the Solemn Procession to the Gallows at the execution of William Wood, Esquire and Hard-ware-man. London.

Syme, J. R. (1970). A high-yielding Mexican semi-dwarf wheat and the relationship to yield of harvest index and other varietal characteristics. *Aust. J. exp. Agric. anim. Husb.* **10**, 350–3.

—— (1972). Features of high-yielding wheats grown at two seed rates and two nitrogen levels. *Aust. J. exp. Agric. anim. Husb.* **12**, 165–70.

Symon, D. E. (1954). Heterozygosity in subterranean clover (*Trifolium subterraneum* L.). *Aust. J. agric. Res.* **5**, 614–16.

Tai, G. C. C. (1976). Estimation of general and specific combining abilities in potato. *Can. J. Genet. Cytol.* **18**, 463–70.

Takebe, I., Labib, G. and Melchers, G. (1971). Regeneration of whole plants from isolated mesophyll protoplasts of tobacco. *Naturwissenschaften* **58**, 318–20.

Tallis, G. M. (1960). The sampling errors of estimated genetic regression coefficients and the errors of predicted genetic gains. *Aust. J. Statist.* **2**, 66–77.

—— (1962). A selection index for optimum genotype. *Biometrics* **18**, 120–2.

Tan, H. (1977). Estimates of general combining ability in *Hevea* breeding at the Rubber Research Institute of Malaysia. I. Phase II and IIIA. *Theor. appl. Genet.* **5**, 29–34.

——, Mukherjee, T. K. and Subramaniam, S. (1975). Estimates of genetic parameters of certain characters in *Hevea brasiliensis. Theor. appl. Genet.* **46**, 181–90.

—— and Subramaniam, S. (1976). Combining ability analysis of certain characters of young *Hevea* seedlings. *Proc. Int. Rubber. Conf., Kuala Lumpur* II, 13–16.

Taylor, I. B. and Evans, G. M. (1976). The effect of B chromosomes on homoeologous pairing in species hybrids. III Intraspecific variation. *Chromosoma* **57**, 25–32.

—— and —— (1977). The genotypic control of homoeologous chromosome association in *Lolium temulentum* × *Lolium perenne* interspecific hybrids. *Chromosoma* **62**, 57–67.

Thoday, J. M. (1961). Location of polygenes. *Nature, Lond.* **191**, 368–70.

—— (1971). Review of C. D. Darlington *Evolution of man and society. Heredity, Lond.* **27**, 304–6.

—— and Thompson, J. N. Jr (1976). The number of segregating genes implied by continuous variation. *Genetica* **46**, 335–44.

Thompson, M. (1971). Pollen incompatibility in filbert varieties. *Proc. Nut Growers' Soc., Oregon and Washington* **56**, 73–9.

—— (1974). Progress towards new filbert varieties. *Proc. Nut Growers' Soc., Oregon and Washington* **59**, 47–53.

Thornley, J. H. M. (1976). *Mathematical models in plant physiology.* Academic Press, London.

Timmis, J. N. and Steele-Scott, N. (1983). Sequence homology between spinach nuclear and chloroplast genes. *Nature, Lond.* **305**, 65–7.

Tollenaar, D. (1934). Untersuchungen über Mutation bei Tabak. I. Entstehungsweise und Wesen Kunstlich erzeugter Gen-Mutanten. *Genetica* **16**, 111–52.

—— (1938). Untersuchungen über Mutation bei Tabak. II. Einige Kunstlich erzeugter Chromosomen-Mutanten. *Genetica* **20**, 285–94.

TOWEY, P. and JINKS, J. L. (1977). Alternative ways of estimating the number of genes in a polygenic system by genotype assay. *Heredity, Lond.* **39**, 399–410.

TOWNLEY-SMITH, T. F. and HURD, E. A. (1973). Use of moving means in wheat yield trials. *Can. J. Plant Sci.* **53**, 447–50.

TRENBATH, B. R. (1977). Interaction among diverse hosts and diverse parasites. *Ann. N.Y. Acad. Sci.* **287**, 124–50.

TSAI, C.-K., CH'IEN, Y.-C., CHOU, Y.-L. and WU, S.-H. (1977). Regeneration of plants from tobacco protoplasts and some factors affecting the plant differentiation. *Scienta sin.* **20**, 458–68.

TUKEY, J. W. (1949). One degree of freedom for non-additivity. *Biometrics* **5**, 232–42.

TURELLI, M. (1984). Heritable genetic variation via mutation-selection balance: Lerch's zeta versus the abdominal bristle. *Theor. Pop. Biol.* **25**, 138–93.

TURNER, S. J., STONE, A. R., and PERRY, J. N. (1983). Selection of potato cyst-nematodes on resistant *Solanum vernei* hybrids. *Euphytica* **32**, 911–17.

TYSDAL, H. M. and CRANDALL, B. H. (1948). The polycross progeny performance as an index of the combining ability of alfalfa clones. *J. Am. Soc. Agron.* **40**, 293–306.

VALENTINE, J. (1984). Accelerated pedigree selection: an alternative to individual plant selection in the normal pedigree breeding method in the self-pollinated cereals. *Euphytica* **33**, 943–51.

VAN DER PLANK, J. E. (1963). *Plant diseases: epidemics and control.* Academic Press, New York.

—— (1968). *Disease resistance in plants.* Academic Press, New York.

VAN DER VEEN, J. H. and WIRTZ, P. (1968). EMS-induced genic male sterility in *Arabidopsis thaliana:* a model selection experiment. *Euphytica* **17**, 371–7.

VAVILOV, N. I. (1935). *Theoretical bases of plant breeding.* Moscow.

—— (1951). The origin, variation, immunity and breeding of cultivated plants. *Chronica Bot.* **13**, 1–364.

VELAZQUEZ, A. (1975). Somatic cell genetics. In *Textbook of human genetics* (eds G. R. Fraser and O. Mayo) pp. 269–325. Blackwell Scientific Publications, Oxford.

VERMA, J. P. and BORKAR, S. G. (1984). Reaction of mixed races of *Xanthomonas campestris* pv. *malvacearum* (E. F. Smith) Dye. *Current Science* **53**, 930–1.

VIRK, D. S., DHAHI, S. J., and BRUMPTON, R. J. (1977). Matromorphy in *Nicotiana rustica. Heredity, Lond.* **39**, 287–95.

VISSER, T. and MARCUCCI, M. C. (1984). The interaction between compatible and self-incompatible pollen of apple and pear as influenced by their ratio in the pollen cloud. *Euphytica* **33**, 699–704.

VOLIN, R. B. and BRYAN, H. H. (1976). Flora-Dade. A fresh market tomato with resistance to verticillium wilt. Circular S. 246, Agricultural Experiment Stations, University of Florida, Gainesville.

WADDINGTON, C. H. (1957). *The strategy of the genes.* Allen and Unwin, London.

—— and LEWONTIN, R. C. (1968). A note on evolution and changes in the quantity of genetic information. In *Towards a theoretical biology* 1. *Prolegomena* (ed. C. H. Waddington) pp. 109–10, Edinburgh University Press.

WADE, D. R. (1977). A physiological analysis of maximum yield potential in cereals. *A.D.A.S.Q. Rev.* **25**, 72–80.

WAGNER, R. (1975). Amélioration génétique de la vigne. *Ann. Amél. Pl.* **25**, 151–75.

Wall, A. M., Riley, R., and Chapman, V. (1971a). Wheat mutants permitting homoeologous meiotic chromosome pairing. *Genet. Res.* **18**, 311–28.

——, —— and Gale, M. D. (1971b). The position of a locus on chromosome 5B of *Triticum aestivum* affecting homoeologous meiotic pairing. *Genet. Res.* **18**, 329–39.

Walsh, E. J. (1974). Efficiency of the haploid method of breeding autogamous diploid species: a computer simulation study. In *Haploids in higher plants advances and potential* (ed. K. J. Kasha) pp. 195–209. University of Guelph.

Watson, I. A. (1970). Changes in virulence and population shifts in plant pathogens. *A. Rev. Phytopath.* **8**, 209–30.

—— and Singh, D. (1952). The future of rust resistant wheat in Australia. *J. Aust. Inst. agric. Sci.* **18**, 190–7.

Watson, J. D. and Tooze, J. (1981). *The DNA story.* A documentary history of gene cloning. San Francisco, W. H. Freeman.

Weatherup, S. T. C. (1980). Statistical procedures for distinctness, uniformity and stability variety trials. *J. agric. Sci., Camb.* **94**, 31–46.

Weber, W. E. (1984). Selection in early generations. In *Efficiency in plant breeding* (ed. W. Lange, A. C. Zeven and N. G. Hogenboom), pp. 72-81. Wageningen, Pudoc.

Wehrhahn, C. and Allard, R. W. (1965). The detection and measurement of the effects of individual genes involved in the inheritance of a quantitative character. *Genetics, Princeton* **51**, 109–19.

Weir, B. S. (ed.) (1983). *Statistical analysis of DNA sequence data.* New York, Marcel Dekker.

—— and Cockerham, C. C. (1973). Mixed selfing and random mating at two loci. *Genet. Res.* **21**, 247–62.

Westcott, B., Gale, M. D., and McVittie, J. A. (1978). A comparison of backcross and selfing methods for telocentric mapping in wheat. *Heredity, Lond.* **40**, 59–66.

Wetherill, G. B. and Ofusu, J. B. (1974). Selection of the best of *k* normal populations. *Appl. Statist.* **23**, 253–77.

White, E. M. (1982). The effects of mixing barley cultivars on incidence of powdery mildew (*Erysiphe graminis*) and on yield in Northern Ireland. *Ann. Appl. Biol.* **101**, 539–45.

Wilcox, M. D. (1983). Inbreeding depression and genetic variances estimated from self- and cross-pollinated families of *Pinus radiata. Silvae Genetica* **32**, 3–4.

Wilkes, H. G. (1977). Hybridization of maize and teosinte in Mexico and Guatemala and the improvement of maize. *Econ. Bot.* **31**, 254–93.

Wilkinson, G. N. (1984). Nearest neighbour methodology for design and analysis of field experiments. *Proc. XII Intern. Congr. Biometric Soc. Tokyo.* I. 69–79.

——, Eckert, S. R., Hancock, T. W. and Mayo, O. (1983). Nearest-neighbour (NN) analysis of field experiments (with discussion). *J. R. Statist. Soc. B.* **45**, 151–211.

—— and Mayo, O. (1982). Control of variability in field trials: an essay on the controversy between 'Student' and Fisher and a resolution of it. *Utilitas Math.* **21B**, 169–88.

Williams, E. R. (1986). A neighbour model for field experiments. *Biometrika* (in press).

WILLIAMS, G. D. J. (1973). Estimates of prairie provincial wheat yields based on precipitation and potential evapotranspiration. *Can. J. Plant. Sci.* **53**, 17–30.

WILLIAMS, R. J. (1977a). Identification of multiple disease resistance in cowpea. *Trop. Agric., Trinidad* **54**, 53–9.

—— (1977b). Identification of resistance to cowpea (yellow) mosaic virus. *Trop. Agric., Trinidad* **54**, 61–7.

WILLIAMS, W. (1959). Heterosis and the genetics of complex characters. *Nature, Lond.* **184**, 527–30.

WILLIAMS, W. A., TUCKER, C. L., and GUERRERO, F. P. (1978). Competition between two genotypes of lima bean with morphologically different leaf types. *Crop. Sci.* **18**, 62–4.

WILSON, P. and DRISCOLL, C. J. (1983). Hybrid wheat. In *Heterosis, reappraisal of theory and practice* (ed. R. Frankel) pp. 94–129. Berlin, Springer-Verlag.

WITTWER, S. H. (1974). Maximum production capacity of food crops. *BioSci* **24**, 216–24.

WOLFE, M. S. (1978). Some practical implications of the use of cereal variety mixtures. In *Plant disease epidemiology* (eds P. R. Scott and A. Rainbridge) pp. 201–7. Blackwell Scientific, Oxford.

WOOD, J. S. (1982). Mitotic chromosome loss induced by methylbenzimidazole-2-carbamate as a rapid mapping method in *Saccharomyces cerevisiae*. *Mol. Cell Biol.* **2**, 1080–7.

WOOD, J. T. (1977). 466 and all that. *Aust. J. Statist.* **19**, 185–93.

WRIGHT, A. J. (1971). The analysis and prediction of some two factor interactions in grass breeding. *J. agric. Sci., Camb.* **76**, 301–6.

—— (1973). The selection of parents for synthetic varieties of out-breeding diploid crops. *Theor. appl. Genet.* **43**, 79–82.

—— (1976a). The significance for breeding of linear regression analysis of genotype–environment interactions. *Heredity, Lond.* **37**, 83–93.

—— (1976b). Bias in the estimation of regression coefficients in the analysis of genotype–environment interaction. *Heredity, Lond.* **37**, 299–303.

—— (1977a). Predictions of non-linear responses to selection for forage yield under competition. *Theor. appl. Genet.* **49**, 201–7.

—— (1977b). Inbreeding in synthetic varieties of field beans (*Vicia faba* L.) *J. agric. Sci., Camb.* **89**, 495–501.

—— (1983). The effects of half-sib and mass selection on linkage disequilibrium in recurrent selection programmes. *Heredity, Lond.* **50**, 85–9.

—— and COCKERHAM, C. C. (1985). Selection with partial selfing. I. Mass selection. *Genetics, Princeton* **109**, 585–97.

WRIGHT, S. (1920). The relative importance of heredity and environment on determining the piebald pattern of guinea pigs. *Proc. natn. Acad. Sci., U.S.A.* **6**, 320–32.

—— (1921a). Systems of mating. *Genetics, Princeton* **6**, 111–78.

—— (1921b). Correlation and causation. *J. agric. Res.* **20**, 557–85.

—— (1922). The effects of inbreeding and crossbreeding on guinea pigs. *U.S. Dept. Agric. Bull.* 1121.

—— (1934a). The method of path coefficients. *Ann. Math. Stat.* **5**, 161–215.

—— (1934b). The results of crosses between inbred strains and guinea pigs, differing in number of digits. *Genetics, Princeton* **19**, 537–51.

—— (1952). The genetics of quantitative variability. In *Quantitative inheritance* (eds E. C. R. Reeve and C. H. Waddington). HMSO, London.

—— (1965). The interpretation of population structure by F-statistics with special regard to systems of mating. *Evolution, Lancaster, Pa.* **19**, 395–420.

—— (1968). *Evolution and the genetics of populations, Vol. 1. Genetic and biometric foundations.* University of Chicago Press.

—— (1969). *Evolution and the genetics of populations, Vol. 2. The theory of gene frequencies.* University of Chicago Press.

YAMAZAKI, T. (1984). The amount of polymorphism and genetic differentiation in natural populations of the haploid liverwort *Conocephalum conicum. Jap. J. Genet.* **59**, 133–9.

YAP, T. C., POH, C. T., and MAK, C. (1977). Comparative studies on selection methods in long bean. In *Animal Breeding Papers, Third Int. Congr. SABRAO,* Canberra, CSIRO. pp. 2c(i)-23-5.

YATES, F. (1936). A new method of arranging variety trials involving a large number of varieties. *J. agric. Sci., Camb.* **26**, 424–55.

—— (1947). Analysis of data from all possible reciprocal crosses between a set of parental lines. *Heredity, Lond.* **1**, 287–301.

—— (1967). A fresh look at the basic principles of the design and analysis of experiments. *Proc. 5th Berkeley Symp. Math. Statist. Pub.*, Vol. 4, pp. 777–90.

—— and COCHRAN, W. G. (1938). The analysis of groups of experiments. *J. agric. Sci., Camb.* **28**, 556–80.

YOUNG, C. C. (1983). Phytotoxic effects in a diploid and a tetraploid pasture. *Tech. Bull. Food &Fertilizer Centre Taiwan* No. 75.

YOUNG, S. S. Y. (1961). A further example of the relative efficiency of three methods of selection for genetic gains under less-restricted conditions. *Genet. Res.* **2**, 106–21.

ZALI, A. A. and ALLARD, R. W. (1976). The effect of level of heterozygosity on the performance of hybrids between isogenic lines of barley. *Genetics, Princeton* **34**, 765–75.

ZIRKLE, C. (1952). Early ideas on inbreeding and crossbreeding. In *Heterosis* (ed. J. W. Gowen) pp. 1–13. Iowa State University Press, Ames.

Subject index

Author index